B.Asis-oba
Duwankoj
Gadshikoj
Sjuren
Sjurtasch
Bijuk-Otarkoj
Orta Kesek
Salankoj
B.Tas-oba
Kamuschly
Kabarta
B.Jaila-basch
Bijuk-Karales
Kutschuk-Sjuren
B.Kaja-basch
Juchary-Karales
Albat
Mekensie
Tscherkes-Kermen
Chodsha-Ssala
B.El-Burun
B.Ssacharnaja Golowka
Adym-tschokrak
Schuli
B.Sybun-tepe
Aj-Todor
Now.Schuli
Uppa
Usenbasch
Werch.Tschorgun
Tschornaja Retschka
Kutschki
Nish.Tschorgun
B.Karakusch-Kaja
B.Betsch
Alsu
Kamary
Urkusta
Tschatal-Kaja
Tubaka
Baga
Saimnalych
Kitschik-Kol-burun
Bijuk-Muskomja
Sawatka

Funklenkpanzer

Funklenkpanzer: A History of German Army Remote- and Radio-Controlled Armor Units

Markus Jaugitz

Translated by David Johnston

J.J. Fedorowicz Publishing

Funklenkpanzer

Funklenkpanzer: A History of German Army
Remote- and Radio-Controlled Armor Units

Markus Jaugitz

Translated by David Johnston

Published by
J.J. Fedorowicz Publishing, Inc.
104 Browning Boulevard
Winnipeg, Manitoba
Canada R3K 0L7
Tel: (204) 837-6080
Fax: (204) 889-1960
e-mail: jjfpub@escape.ca
web: www.jjfpub.mb.ca

Printed in Canada
ISBN 0-921991-58-4

Signing Box by Matt Lukes

Printed by Friesens Printers, Altona , Manitoba, Canada

Other Titles by J.J. Fedorowicz Publishing

The Leibstandarte (1. SS-Panzer-Division):
Volumes I, II, III, IV/1 and IV/2
European Volunteers (5. SS-Panzer-Division)
Das Reich (2. SS-Panzer-Division):
Volumes I and II
The History of Panzer-Korps "G
The History of the 7. SS-Mountain-Division "Prinz Eugen"roßdeutschland":
Volumes 1, 2 and 3
Otto Weidinger
Otto Kumm
Manhay, The Ardennes: Christmas 1944
Armor Battles of the Waffen-SS, 1943-1945
Tiger: The History of a Legendary Weapon, 1942-1945
Hitler Moves East
Tigers in the Mud
Panzer Aces
Footsteps of the Hunter
History of the 12. SS-Panzer-Division "Hitlerjugend"
Grenadiers, the Autobiography of Kurt Meyer
Field Uniforms of German Army Panzer Forces in World War 2
Tigers in Combat, Volumes I and II
Infanterie Aces
Freineaux and Lamormenil—The Ardennes
The Caucasus and the Oil
East Front Drama—1944
The History of the Fallschirm-Panzer-Korps "Hermann Göring"
Michael Wittmann and the Tiger Commanders of the Leibstandarte
The Western Front 1944:
Memoirs of a Panzer Lehr Officer
Luftwaffe Aces
Quiet Flows the Rhine
Decision in the Ukraine: Summer 1943
Combat History of the Schwere Panzer-Jäger-Abteilung 653
The Brandenburgers — Global Mission
Field Uniforms of Germany's Panzer Elite
Soldiers of the Waffen-SS:
Many Nations, One Motto
In the Firestorm of the Last Years of the War
The Meuse First and Then Antwerp
Jochen Peiper: Commander, Panzer-Regiment "Leibstandarte"
Sturmgeschütze vor! Assault Guns to the Front!
Karl Baur: A Pilot's Pilot
Kharkov
Panzer Aces 2
Panzertaktik
The History of Panzerregiment "Großdeutschland"
Combat History of Sturmgeschütz-Brigade 276
Normandy 1944: German Military Organization Combat Power and Organizational Effectiveness
The Combat History of schwere Panzer-Abteilung 503

In Preparation:

The Combat History of schwere Panzer-Abteilung 508
Grenadiers (Reprint)
Tragedy of the Faithful: A History of the III. SS-Panzer-Corps

J.J. Fedorowicz Publishing

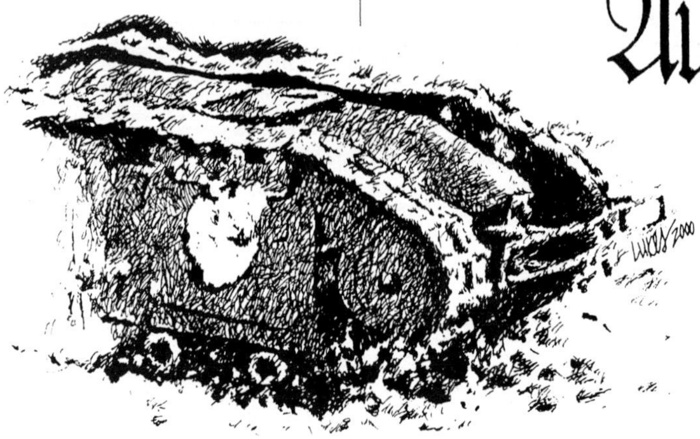

Autor Markus Jaugitz

(Vor- und Familienname)

(Unterschrift)

Signing Box by Matt Lukes

Publishers' Acknowledgements

Special thanks to David Johnston for his excellent work in translating the text from the original German.

Matt Lukes also deserves credit for another excellently rendered signing box. For those interested in owning original artwork by Matt, you can reach him at panzerknackr@infoserve.net.

Waldemar Trojca has also assisted immeasurably with his excellent organigrams and scanning of the photographs. Look for more of his work in several of our upcoming books.

We also wish to thank you, the reader, for purchasing this book and all of you who have written us with kind words of praise and encouragement. It gives us the impetus to continue translating the best available German-language books and produce original titles. Our catalog of books is listed on the preceding pages and can be viewed on our web site at *www.jjfpub.mb.ca*. We have also listed titles which are near production and can be expected in the near future. Many of these are due to your helpful proposals.

We look forward to your continued comments and constructive criticism.

John Fedorowicz — Mike Olive — Bob Edwards

Editors' Note

When translating German military terminology, modern American Army terminology is generally used wherever an equivalent term is applicable. In cases where there may be nuances where we think the reader might enjoy learning the German term, we have included it parenthetically.

In cases where the German term is commonly understood or there is no good, direct English equivalent, we have tended to retain the German term, e.g., *Schwerpunkt* (point of main effort), *Auftragstaktik* (mission-type orders) etc.

In an attempt to highlight the specific German terminology, we have italicized German-language terms and expressions. Since most of the terms are repeated several times, we have not included a glossary. Since we assume the reader will already have a basic understanding of German rank terms and the terminology used for vehicles, we have likewise not included any separate annexes to the book to explain them.

An additional note with regard to the German term *zustellen* is perhaps warranted. We have alternately translated this as either "to be placed under the operational control" of another unit or "to be attached". Technically, these are two different command and control relationships, neither of which actually exists in the German system. *Zustellen* actually implies "operational control," but since the phrase is rather unwieldy, we have usually translated it as "attached".

Table of Contents

About the Author

Markus Jaugitz was born in 1957 and for more than 20 years has studied the armored formations of the *Wehrmacht*. One of his principal areas of interest has been those units equipped with special equipment, the so-called "exotics". Markus Jaugitz is the regional director of the *Technische Hilfswerk*, a federal technical assistance organization which is involved in accidents and disasters in Germany and abroad. He is married and lives with his family in Mannheim.

Foreword

The idea of employing unmanned aircraft, missiles or ships is a very old one, however, it was not until the start of the Twentieth Century that rapid developments in the field of radio telephony made it possible to turn the concept into reality.

The militaries of the western nations showed a particular interest in these developments and before World War One efforts were concentrated in the navy. Motor vehicles and aircraft were in their infancy and therefore took second place. The Siemens Company began developing a wire-guided motorboat in Imperial Germany in 1911. The German navy carried out countless experiments with remote-control boats between 1914 and 1918, with control provided from the land but also from ships, aircraft and even airships. Numerous technical difficulties were encountered during the course of these trials, however. Toward the end of World War One the idea of an unmanned aircraft was also revived.

The collapse of the Imperial Empire after the cease-fire in November 1918 and the terms of the Versailles peace treaty brought a temporary halt to research and experiments in the field of remote-control.

In the 1930's the political situation in Europe changed from year to year. Germany's neighbors reacted to the Third Reich's expansionist policies by increasing defense spending and building extensive border fortifications. The various nations reworked operational principles and efforts were made to bring the equipment of the armed forces up to the latest technical standards. In this context it was especially important to also show interest in projects that had previously been rejected by many military authorities as unworkable.

In 1935 Great Britain began developing a remote-control tracked vehicle, project "Edward". Two years later in France, A career officer and inventor by the name of Pommelet produced plans for a tracked remote-control demolition-charge carrier.

Shortly before the outbreak of war, Germany also began development of remote-control vehicles. Then, in early summer 1940, the first army unit equipped with unmanned tracked vehicles was formed. This unit was the cornerstone of the development of a special force that, during the course of the war, grew to include several separate battalions and companies and saw employment on almost every front.

In the early years, when Germany was still victorious, the myth of a wonder weapon was born. The ranks of the special units included many young men who fascinated those in senior command positions with revolutionary proposals and technology such as image sensors, night-vision devices, mobile television cameras and much more. Technical problems and shortages of raw materials prevented many of these promising ideas from becoming reality, however.

In 1943, when the tide turned against Germany for good, the *Wehrmacht* went from the attack to the defense and withdrawal. From this point on the use of remote-control vehicles dwindled and increasingly lost importance.

Increasingly, the special units were employed as armored units with the remote-control demolition-charge carriers held in reserve for other purposes. Nevertheless, an entry in one's paybook as a member of a special unit was good "life insurance" and guaranteed a return to a replacement unit in Germany even in critical situations.

During the war members of the remote-control units were sworn to absolute secrecy and photography was forbidden. It is astonishing that, in spite of the threat of stiff penalties, some soldiers did manage to take photographs.

After many years of research I have succeeded in locating important documents and collecting many photographs. In this book I have attempted to describe the technical equipment and also fully describe the operational histories of the units which were equipped with it. Obtaining photographs of some units was extremely complicated because, towards the end of the war, it became increasingly difficult to obtain photographic materials.

I have also included other small combat units which, while not part of the remote-control force, were equipped with converted former remote-control vehicles.

I would like to express my gratitude to many former members of the remote-control force, who welcomed me to their reunions and who were always ready to answer my countless questions and volunteered their personal papers, documents and photographs. Without their efforts and support this book would not have been possible.

In spite of this book there are still many unclear areas and "secrets" which have yet to be explained. Readers are encouraged to contact me with any information which might help clear up these areas and I would like to thank them for their help in advance.

Markus Jaugitz
Mannheim, Winter 2000

Acknowledgements

I would like to thank the following persons and facilities:

Mr. John J. Fedorowicz for the opportunity to create a comprehensive history of this "secret force" and bring it to a large circle of readers.
Mr. Waldemar Trojca for producing the Organigrams
Mr. Jean Restayn for producing the color plates
Mr. Thomas L. Jentz for providing technical information and suggestions
Mr. Hilary L. Doyle for permission to use his scale drawings
Mr. Meinrad Lock organizer of the veterans association of former members of the radio-control and remote-control units
The staff of the Bundesarchiv Koblenz/Militärarchiv Freiburg
The staff of the Bundesarchiv Koblenz/Photo Archive
The staff of the Bundesarchiv Koblenz/Information Center
The staff of the Bundeswehr's Military Equipment Study Collection in Koblenz
The staff of the WASt in Berling
Mr. D. Fletcher and G. Holmes, Bovington Tank Museum, Great Britain

My most sincere thanks go to my wife Silvia
and my children for their understanding,
as I had to be separated from them so often.

The following persons assisted me with photos, important documents, maps, diary entries and much more.

Friedrich von Abendroth
Otto Achstätter (deceased) . . . . *Pz.Kp.(Fkl) 317; 3./Pz.Abt.(Fkl) 302*
Hermann Aichele (deceased) . . . . *2./Pz.Abt. 300; Pz.Kp. (Fkl) 313*
Rolf Aichele . . . . *Pz.Kp.(Fkl) 313; 3./s.Pz.Abt. 508*
Hans Bachmann . . . . *Pz.Kp.(Fkl) 311; 4./Pz.Abt.(Fkl) 302*
Klara BährLudwig Baisch . . . . *2./Pz.Abt. 300; Pz.Kp.(Fkl) 313*
Alfred Baldschun . . . . *Pz.Kp.(Fkl) 312; I./s.Pz.Abt.(Tiger/Fkl) 301*
Hans Baumann . . . . *Pz.Kp.(Fkl) 313; 3.1s.Pz.Abt. 508*
Artur Becker . . . . *Pz.Kp.(Fkl) 316; I./Pz.Abt.(Fkl) 302*
Hans-Axel Beer . . . . *4./Pz.Abt.(Fkl) 301; 1./Pz.Abt.(Fkl) 303*
Paul Belka . . . . *Vers.Kdo.(F.L.) Tropen; Pz.Kp.(Fkl) 317; 3./Pz.Abt.(Fkl) 302*
Werner Bischlepp . . . . *4./Pz.Abt.(Fkl) 301*
Franz Bitterwolf . . . . *2./MR.Abt. I*
Ernst Bold . . . . *I./MR.Abt. I*
Hans Bolz . . . . *Vers.Kdo.(F.L.) Tropen; Pz.Kp.(Fkl) 317; 3./Pz.Abt.(Fkl) 302*
Sophie Braam
Joachim Brandtner . . . . *I./Pz.Abt. 301; Pz.Kp. 315 (Fkl); 2./Pz.Abt.(Fkl) 302*
Eugen Bürck . . . . *Pz.Kp.(Fkl) 313; 3./s.Pz.Abt. 508*
Werner Busse . . . . *4./Pz.Abt.(Fkl) 301*
Heinz Cornelius . . . . *Pz.Kp.(Fkl) 319; Fkl-Pz.Zug/Pz.Abt.(Fkl) 303*
Günther Diehl . . . . *4./Pz.Abt.(Fkl) 301*
Werner Dubberstein . . . . *Pz.Kp.(Fkl) 317; 3./Pz.Abt.(Fkl) 302*
Alfred Dürr *3./s.Pz.Abt. 508*
Heinrich Dütsch . . . . *4./Pz.Abt.(Fkl) 301; l./Pz.Abt.(Fkl) 303*
Heiner F. Duske
Eckart Esser (deceased) . . . . *Pz.Kp.(Fkl) 311; 4./Pz.Abt.(Fkl) 302*
Fritz Fechtig *3./s.Pz.Abt. 508*
Siegfried Fellhauer . . . . *Stb./MR.Abt.1; Stb./Pz.Abt. 300*
Franz Fiedler (deceased) . . . . *Pz.Kp. 315 (Fkl); 2./Pz.Abt.(Fkl) 302*
Jus Fischer (deceased) . . . . *2./MR.Abt.1; 2./Pz.Abt. 300; Vers.Kdo.(F.L.) Tropen; Pz.Vers.und Ers.Abt. 300*
Wolfgang Fleischer
Dr. Justus Fleischhauer . . . . *Pz.Kp.(Fkl) 313; 3./s.Pz.Abt. 508*
Erich Franz . . . . *4./Pz.Abt.(Fkl) 301; I./Pz.Abt.(Fkl) 303*
Franz Franzen . . . . *Werkst.Zg./Pz.Abt.(Fkl) 301*

Dr. Günther Fritschken (deceased) . . . .*2./MR.Abt.1; 2./Pz.Abt. 300; Pz.Kp.(Fkl) 313*
Hans-Günther Fürbringer . . . .*4./Pz.Abt.(Fkl) 301; 1./Pz.Abt.(Fkl) 303*
Werner Galla . . . .*Vers.Kdo.(F.L.) Tropen; Pz.Kp.(Fkl) 317; 3./Pz.Abt.(Fkl) 302*
Artur Geuss . . . .*Pz. Kp. (Fkl) 31 1; 4./Pz.Abt. (Fkl) 302*
Hermann Gödde (deceased) . . . .*Pz.Kp.(Fkl) 312; I./s.Pz.Abt.(Tiger/Fkl) 301*
Werner Göltenbott . . . .*Pz.Kp.(Fkl) 316; 1./Pz.Abt.(Fkl) 302*
Hans Gradolph *Stb./MR.Abt.1*
Alfred Grams (deceased) . . . .*s.Pz.Abt.(Tiger/Fkl) 301*
Helmut Greiner . . . .*3./Pz.Abt. 300; Pz.Kp.(Fkl) 313*
Heinrich Großkopf . . . .*Pz.Kp.(Fkl) 314; 3./s.Pz.Abt. 504*
Walter Grumptmann . . . .*Pz.Kp. 315 (Fkl); 2./Pz.Abt.(Fkl) 302*
Franz Günther . . . .*Pz.Kp.(Fkl) 314*
Hans Haas . . . .*Pz.Vers.und Ers.Abt. 300*
Erwin Häling . . . .*Vers.Kdo.(F.L.) Tropen; Pz.Kp.(Fkl) 317; 3./Pz.Abt.(Fkl) 302*
Willi Hand *1./MR.Abt.1*
Hermann Hanke . . . .*Stb./Pz.Abt. 300; Pz.Vers.und Ers.Abt. 300*
Alois Happe . . . .*4./Pz.Abt.(Fkl) 301; l./Pz.Abt.(Fkl) 303*
Heinz Hauptmann . . . .*Werkst.Zg./Pz.Abt.(Fkl) 302; Werkst.Zg./Pz.Abt.(Fkl) 303*
Karl Hausch . . . .*Pz.Kp.(Fkl) 319; Fkl-Pz.Zug/Pz.Abt.(Fkl) 303*
Kurt Hecker . . . .*Vers.Kdo.(F.L.) Tropen; Pz.Kp.(Fkl) 314; 3./s.Pz.Abt. 504*
Hans Heim . . . .*1./MR.Abt.1; 1./Pz.Abt. 300;"Küstenjäger" - Div. Brandenburg*
Eberhard Held . . . .*Pz.Kp.(Fkl) 314; 3./s.Pz.Abt. 504*
Karl Hengstberger (deceased) *l./MR.Abt.1*
Johannes Hentschel . . . .*Pz.Kp.(Fkl) 314; 3./s.Pz.Abt. 504; Pz.Vern.Abt. I*
Herbert Herwig *3./s.Pz.Abt. 508*
Karl Herzog (deceased) . . . .*Pz.Kp.(Fkl) 312; 1./s.Pz.Abt.(Tiger/Fkl) 301*
Heinz Hinzmann . . . .*Vers.Kdo.(F.L.) Tropen; Pz.Kp.(Fkl) 317; 3./Pz.Abt.(Fkl) 302*
Gerhard Hochgreef . . . .*Pz.Kp.(Fkl) 312; 1./s.Pz.Abt.(Tiger/Fkl) 301*
Dieter Höland . . . .*Pz.Kp.(Fkl) 312; l./s.Pz.Abt.(Tiger/Fkl) 301*
Karl Hofmann (deceased) . . . .*2./MR.Abt.1; 2./Pz.Abt. 300; Pz.Kp.(Fkl) 313; 3./s.Pz.Abt. 508*
Gunter Hoffmann (deceased) . . . . . .*4./Pz.Abt.(Fkl) 301; 1./Pz.Abt.(Fkl) 303*
Günther Hopf . . . .*Pz.Kp.(Fkl) 314; 3./s.Pz.Abt. 504*
Hans-Karl Jakob . . . .*Pz.Kp.(Fkl) 319; 3./Pz.Abt.(Fkl) 303*
Walter Jasbec . . . .*Pz.Kp.(Fkl) 314; Fkl-Pz.Zug/Pz.Abt.(Fkl) 303*
Heinz Jenckel . . . .*l./Pz.Abt. 301; Pz.Kp. 315 (Fkl); 2./Pz.Abt.(Fkl) 302*
Karl-Heinrich Jessen . . . .*4./Pz.Abt.(Fkl) 301; I./Pz.Abt.(Fkl) 303*
Walter Jung . . . .*Pz.Kp.(Fkl) 319; 3./Pz.Abt.(Fkl) 303*
Otto Kapp . . . .*l./Pz.Abt. 301; Pz.Kp. 315 (Fkl); 2./Pz.Abt.(Fkl) 302*
Otto Kaiser . . . .*Pz.Kp.(Fkl) 319*
Wilhelm Keitel . . . .*Pz.Kp.(Fkl) 314; Pz.Kp.(Fkl) 319; 3./Pz.Abt.(Fkl) 303*
Alwin Ketter . . . .*Pz.Abt.(Fkl) 301*
Heinz Key (deceased) . . . .*Pz.Kp.(Fkl) 316; 1./Pz.Abt.(Fkl) 302*
Hermann Kienle . . . .*1./Pz.Abt. 301; Pz.Kp. 315 (Fkl); 2./Pz.Abt.(Fkl) 302*
Hans Kissling *Stb./MR.Abt.1*
Ernst Knaupp *Stb./MR.Abt.1; 1./Pz.Abt. 301*
Josef Knecht *2./MR.Abt.1*
Paul Knödler *3./s.Pz.Abt. 508*
Alfred Kornmeier . . . .*1./MR.Abt.1; Pz.Abt. 301; Pz.Kp. 315 (Fkl); Pz.Abt.(Fkl) 302*
Stefan König
Dr. Horst Kosche . . . .*Vers.Kdo.(F.L.) Tropen; Pz.Vers.und Ers.Abt. 300; Pz.Abt.(Fkl) 303*
Alfred Krauss . . . .*Pz.Kp.(Fkl) 313; 3./s.Pz.Abt. 508*
Franz Kunsleben (deceased) . . . .*Pz.Kp.(Fkl) 312; I./s.Pz.Abt.(Tiger/Fkl) 301*
Walter Kurnik . . . .*Pz.Kp.(Fkl) 31 1; 4./Pz.Abt.(Fkl) 302*
Otto Lehmann . . . .*Vers.Kdo.(FL) Tropen; Pz.Kp.(Fkl) 317; 3./Pz.Abt.(Fkl) 302*
Nicolas Leloup
Karl Ling . . . .*Pz.Kp.(Fkl) 313*
Emil Maier *l./MR.Abt.1*
Horst Martens . . . .*Pz.Kp.(Fkl) 316; l./Pz.Abt.(Fkl) 302*
Jeffrey D. McKaughan
Gerd Michaelis . . . .*Pz.Kp. 315 (Fkl); Pz.Abt.(Fkl) 302*
Harry Minge . . . .*Pz.Kp.(Fkl) 312; l./s.Pz.Abt.(Tiger/Fkl) 301*
Hugo Mohrhäuser . . . .*Pz.Kp.(Fkl) 313; 3./s.Pz.Abt. 508*

Wilhelm Müller .... *Pz.Kp.(Fkl) 313; 3./s.Pz.Abt. 508*
Karlheinz Münch
Bruno Müntz .... *1./Pz.Abt. 301; Pz.Kp. 315 (Fkl); 2./Pz.Abt.(Fkl) 302*
Heinz Nickel .... *Pz.Kp.(Fkl) 319; 3./Pz.Abt.(Fkl) 303*
Erika Niepenberg .... *Pz.Kp. (Fkl) 311; 4./Pz.Abt. (Fkl) 302*
Robert Off .... *1./Pz.Abt. 301; Pz.Kp. 315 (Fkl); Werkst.Zg./Pz.Abt.(Fkl) 302*
Kurt Offermann .... *Pz.Abt. 300*
Albert Osterried .... *Pz.Kp.(Fkl) 313; 3./s.Pz.Abt. 508*
Max Palitz (deceased) .... *3./Pz.Abt.(Fkl) 301; 3./s.Pz.Abt.(Tiger/Fkl) 301*
Gerhard Paust .... *1./Pz.Abt. 301; Pz.Kp. 315 (Fkl); 2./Pz.Abt.(Fkl) 302*
Johannes Pfleiderer .... *Stb./MR.Abt.1; 2./Pz.Abt. 300; Fkl-Pz.Zug/Pz.Abt.(Fkl) 303*
Hans Pieperhoff (deceased) .... *Pz.Abt.(Fkl) 301;Pz.Abt.(Fkl) 302*
Werner Pitschmann .... *Pz.Kp.(Fkl) 319; 3./Pz.Abt.(Fkl) 303*
Werner Poggensee .... *1./Pz.Abt. 301; Pz.Kp. 315 (Fkl); 2./Pz.Abt.(Fkl) 302*
Josef Prestel (deceased) .... *1./Pz.Abt. 301; Pz.Kp. 315 (Fkl); 2./Pz.Abt.(Fkl) 302*
Werner Regenberg
Dorothea Reinhard .... *NSU Werksarchiv*
Günther Rettig .... *3./Pz.Abt.(Fkl) 301; 3./s.Pz.Abt.(Tiger/Fkl) 301*
Eugen Sartory .... *MR.Abt.1; Pz.Abt. 300; Pz.Abt.(Fkl) 301*
Heinrich Schäfer .... *Pz.Kp.(Fkl) 316; I./Pz.Abt.(Fkl) 302*
Max Schick (deceased) .... *Stb./MR.Abt.1; Stb./Pz.Abt. 301, StbKp./Pz.Abt (Fkl) 303*
Willi Schiele *2./MR.Abt.1*
Josef Scherer .... *2./MR.Abt.1; 2./Pz.Abt. 300; Pz.Kp.(Fkl) 313*
Helmut Schleehuber *3./s.Pz.Abt. 508*
Wolfgang Schneider
Richard Schmidt (deceased) .... *Vers.Kdo.(F.L.) Tropen; Fkl-Pz.Zug/Pz.Abt.(Fkl) 303*
Franz Schott (deceased) .... *MR.Abt.1; Pz.Abt.(Fkl) 301; s.Pz.Abt.(Tiger/Fkl) 301*
Gottfried Schramm .... *Pz.Abt.(Fkl) 303*
Ludwig Schreiner .... *Pz.Abt.(Fkl) 301; Pz.Abt.(Tiger/Fkl) 301*
Hans Seeger .... *l./Pz.Abt. 301; Pz.Kp. 315 (Fkl); 2./Pz.Abt.(Fkl) 302*
Dr. Robert Sigmund .... *2./Pz.Abt. 301; Pz.Kp. 313 (Fkl); Pz.Abt.(Fkl) 302*
Theodor Specht .... *Pz.Kp.(Fkl) 312; Pz.Abt.(Fkl) 301; s.Pz.Abt.(Tiger/Fkl) 301*
Walter J. Spielberger
Willy Stolz .... *2./Pz.Abt. 301; Pz.Kp. 313 (Fkl); 3./s.Pz.Abt. 508*
Hans Hermann Stuke .... *Stb./s.Pz.Abt.(Tiger/Fkl) 301*
Gotthold Stumpf .... *Pz.Kp.(Fkl) 319; 3./Pz.Abt.(Fkl) 303*
Helmut Susenbeth .... *Vers.Kdo.(F.L.) Tropen; Pz.Vers.und Ers.Abt. 300*
Alfred Tamine .... *Pz.Abt. 300; Pz.Abt.(Fkl) 302*
Detev Terlisten
Fritz Trenkle (deceased)
Rudolf Tumma .... *Pz.Vers.und Ers.Abt. 300*
Johann Tzschöckel .... *Pz.Kp.(Fkl) 314; 3./s.Pz.Abt. 504*
Liselofte Vedder
Ruth Walther
Gerhard Weiß .... *4./Pz.Abt.(Fkl) 301*
Carl M. Wenzel .... *Vers.Kdo.(F.L.) Tropen; 4./Pz.Abt.(Fkl) 302; 3./Pz.Abt.(Fkl) 303*
Magdalene Wild .... *2./MR.Abt.1; 2./Pz.Abt. 300; Pz.Kp.(Fkl) 313; 3./s.Pz.Abt. 508*
Peter Winkler .... *Pz.Kp. 315 (Fkl); 2./Pz.Abt.(Fkl) 302*
Horst Werner .... *Vers.Kdo.(F.L.) Tropen; Pz.Kp.(Fkl) 317; 3./Pz.Abt.(Fkl) 302*
Georg Wimmer (deceased) .... *Pz.Vers.und Ers.Abt. 300*
Karl Wizgall .... *1./Pz.Abt. 300; I./Pz.Vers.und Ers.Abt. 300; Pz.Abt.(Fkl) 302*
Otto Wolfer .... *Pz.Kp.(Fkl) 313; 3.1s.Pz.Abt. 50*
Walter Zugwurst (deceased) .... *Pz.Kp. 319 (Fkl); 3./Pz.Abt.(Fkl) 303*

I would also like to thank all of those who assisted me in my research but whose names do not appear here because of publishing deadlines.

Chapter 1: The Equipment of the Remote- and Radio-Controlled Units

1.1 Description of Carriers

1.1.1 Borgward BI Mine Clearing Vehicle (Sd.Kfz. 300), Borgward BII, Borgward BIII, and Amphibious Borgward BII Ente (Duck)

On 19 October 1939 the Commander-in-Chief of the Army issued a requirement for a vehicle constructed out of concrete and with mine-clearing rollers for use by combat engineers in a mine-clearing role. The firm of Carl F. W. Borgward in Bremen was charged with its development and was instructed to complete a prototype by 15 January 1940.

Even before the prototype was completed, Borgward received a contract from the *Waffenamt* to build an initial batch of 50 vehicles with a three-piece mine roller trailer. These unmanned special vehicles were to be used as remotely-controlled mine clearing vehicles.

It was assumed that the mine-clearing vehicles would have only a short life and therefore it was desired to keep production costs as low as possible. Following this specification, Borgward designed a fully-tracked carriage with a concrete superstructure, and the vehicle received the official designation *Minenräumwagen Borgward BI* (*Sd.Kfz. 300*).

The chassis consisted of a rectangular hull with a four-cylinder Borgward Type 4 M 1.5 R II carburetor engine installed at the rear. Between the drive sprocket and idler wheel were three large roadwheels mounted on swing arms. Given the expected short life of the vehicle, the roadwheels were fitted with wooden wrappings instead of rubber ones.

Above the hull was the cast concrete superstructure which protected the engine, the drive train and the radio command receiver system. On the left side of the vehicle, between the second and third roadwheels, was a forward-facing lever which was used to regulate speed. In addition to a stop position, there were two settings for forward speeds of two and five kilometers per hour. A muffler was installed on the back of the superstructure in a horizontal position above the tracks.

The concrete superstructure was topped by a removable rectangular metal plate. At the rear of the plate there was a shielded opening through which cooling air was drawn into the engine compartment.

The production vehicles, which weighed 1.5 tons, were delivered between January and May 1940. However, because of production difficulties none were equipped with the radio command receiver system necessary for remote control.

On 3 April 1940 the Borgward company received a contract for a second series of mine-clearing vehicles. Production of these vehicles, which were designated the *Borgward BII*, began in July 1940.

The second series featured a number of improvements designed to eliminate shortcomings discovered in the first series. The number of roadwheels was increased to four per side and the position of the final drive was raised. In addition, a single return roller was installed on each side between the second and third roadwheels.

The total vehicle weight of the BII had risen to 2.3 tons, which made necessary the installation of a more powerful engine. The chosen replacement was the Borgward Type 6 M 2.3 RTBV carburetor engine which was also used in the Borgward 2300-series vehicle.

An armor plate protected the front of the vehicle against shellfire and shrapnel. The interior of the vehicle was capable of accommodating a demolition charge weighing up to 515 kilograms.

The speed-control lever remained on the left side of the vehicle. The muffler was mounted above the ventilation openings in the upper part of the rear superstructure. Several maintenance hatches were incorporated into the superstructure top plate, allowing access to the engine, transmission and the radio command receiver system.

The *BII* dispensed with the mine-roller trailer and the vehicles were employed as remotely-controlled demolition-charge carriers. Since the radio-controlled detonation system initially also proved unreliable, an interim measure was employed, whereby a movable metal rod was installed on the front of the vehicle. Through it, a pressure igniter activated the detonation mechanism.

Several prototypes of an amphibious version based on the *Borgward BII* were also built. This version was dubbed the *Ente* (Duck) by the troops. The concrete superstructure was replaced by a watertight compartment made of sheet metal. The return roller between the second and third roadwheels was deleted and a different type of roadwheel was used.

The top of the watertight compartment was designed to be removable. The exhaust system and muffler were integrated into the watertight compartment.

In the water the vehicle was propelled by a three-bladed propeller which was installed at the rear of the vehicle between two water rudders.

According to statements by former members of the mine-clearing battalion, at the end of June—beginning of July 1941 a prototype of the *Borgward BIII* was issued for troop trials. No further information concerning this type could be found. The *BIII* was said to have been a further development of the *BII* with the concrete superstructure replaced by one of steel plate construction.

The first remotely controlled demolitions carrier was the French "Véhicule Pommelet". By the end of May 1940 11 examples had supposedly been completed. According to a British report, several had been used against the Germans at Sedan. (Leon)

The *Borgward BI* mine-clearing vehicle (*Sd.Kfz. 300*) with attached mine rollers. (Spielberger)

The follow-on model of the *BI* was the mine-clearing vehicle, the *Borgward BII*. The *BII* was not outfitted with the mine-clearing rollers. (Spielberger)

An experimental version of the *Borgward BII* was the amphibious demolitions carrier, the *BII "Ente"*. (Spielberger)

1.1.2 Borgward BIV Explosives Carrier (Sd.Kfz. 301) Ausführungen A to C

After evaluating the use of mine-clearing vehicles at the beginning of the Russian campaign in June-July 1941, the *Waffenamt* issued a requirement for the development of a radio-controlled heavy demolition charge carrier.

In October 1941 the Carl F.W. Borgward company received a contract to design such a vehicle. The requested heavy demolition charge carrier was to have the capability of being guided by remote control to its target and — after it had dropped its demolition charge — quickly withdrawn from the blast zone. The vehicle was also to be equipped with smoke grenades and be capable of dropping decontamination agents should chemical warfare be employed.

In developing the new vehicle, the Borgward company was able to use components from the *VK 301/302* armored munitions carrier, and the first prototype was completed at the end of March 1942. The heavy demolition-charge carrier received the designation *Borgward BIV* (*Sd.Kfz. 301*).

Serial production of the *Ausführung A* began in April 1942 and 15 heavy demolition-charge carriers were delivered to *Panzer-Abteilung 300* that same month.

The vehicle had a gross weight of 3.6 tons and was powered by a Borgward Type 6 M 2.3 RTBV six-cylinder carburetor engine. This engine also powered the *Borgward BII* and the *VK 301/302* armored munitions carrier.

The vehicle's fuel capacity was 123 liters, which yielded a range of approximately 200 kilometers. Fuel consumption on roads was 58 liters per 100 kilometers. Maximum speed was 38 km/h on roads and 18 km/h off roads.

The BIV was equipped with a fluid drive, a gearshift transmission and a steering mechanism. The fluid drive, whose official designation was the *KSB-Trilok-Getriebe A II 260*, was an hydraulic drive which was installed directly in front of the engine.

The gearshift transmission consisted of two two-speed drives one behind the other installed in a common housing and flange-mounted to the steering mechanism. The forward-positioned drive sprockets were powered from the steering mechanism via the reduction-gear units. The drive sprockets had eleven movable rollers which engaged the caterpillar track. The tips of the drive sprockets were fitted with rubber segments. The drive sprockets moved the Type Kgw 52 200/140 tracks, each of which consisted of 50 links.

The driver's seat was situated in the right forward part of the hull and the instrument panel was mounted in front of the seat. Steering was by means of a steering lever, which was mounted on a cross-tube. Movement of the steering lever activated the steering brakes by means of connecting rods. A small adjustable windscreen was mounted on the superstructure roof for the driver.

Two detonation transformers were installed on the right and left sides of the hull floor and were connected to the ignition circuit. Should a radio-controlled *BIV* drive over a mine, the shock of the explosion caused an impulse to be transmitted via a spring system to the ignition circuit and the jettisonable charge was immediately detonated. The explosion of the 450-kilogram jettisonable charge was sufficient to destroy other mines in the area.

The delivery unit was mounted on the sloping front plate. The box was suspended on a lever on each side. Each lever was connected to a pull cable, which ran over a deflecting pulley and was severed by a demolition charge. As soon as the lever swung up the delivery unit was free to slide down. This process also set off a cord igniter. A burn length of up to 30 seconds could be selected by varying the length of the detonation cord. Once the fuse had burned down, the charge exploded.

The demolition charge consisted of 450 kilograms of Ekrasit explosive. This type of explosive was relatively safe to handle. If no detonator was fitted the charge could be struck by small-caliber fire without exploding. If the explosive caught fire, it slowly burned itself up.

A total of 628 *BIV Ausführung A* demolition charge carriers were produced in the period from April 1942 to June 1943, with 390 being built in a special program from the beginning of February to the end of June 1943.

During its production life, the *Ausführung A* was continually improved and modified:

1. The windscreen was deleted and replaced by a three-piece folding driver's shield of 8-mm plates.

2. Additional 8-mm plates were welded to the sides and rear of the hull to improve armor protection.

3. An emergency escape hatch was installed on the right side of the hull above the second roadwheel. The right-hand track covers were appropriately modified.

4. Change to dry (non-lubricated) tracks. The new track consisted of 77 Type 200/90 track links. New drive sprockets (three different types) were installed to move the dry track.

5. Rounded shields were fitted beneath the ends of the lever arms.

6. New system for actuating the lever arms. A threaded tube fitted with a detonating fuse was used in place of the pull cable. Detonation of the fuse destroyed the tube and allowed the lever arm to swing up.

7. The engine deck cover was changed. Two larger, adjustable ventilation flaps were installed, as well as a cutout on the left side of the cover for simplified passage of the exhaust pipe.

When units were withdrawn to their home bases, some *Borgward BIV's*, *Ausführung A*, built between April and October 1942 received some of the modifications listed above, especially the folding driver's shield and the 8-mm additional armor.

In July 1943 the Borgward company began production of the *Ausführung B* of the heavy demolition-charge carrier.

Externally, the only visible difference in the *Ausführung B* from late-model *Ausführung A BIV's* was the forward positioning of the antenna. The relocation of the antenna was the result of a revised interior arrangement.

The engine compartment had been redesigned. The air filter, which projected into the driver's compartment in the

Ausführung A, was installed in the engine compartment on the bulkhead in the *Ausführung B*, separating it from the driver's compartment. The reserve fuel tank, which had been located on the left side of the hull in the vicinity of the command receiver, was also deleted. A redesigned reserve fuel tank was installed in the right front corner of the engine compartment.

The radio command receiver and the igniter system were relocated from the engine compartment to the left side of the hull between the command receiver and the hull wall. The antenna base was mounted in the center of the forward superstructure roof next to the folding driver's visor.

During the course of production of the *Ausführung B*, the emergency escape hatch on the right side of the hull was deleted and a plate was welded into the opening. Eight-millimeter armor plates were welded onto the outside of the hull on both sides. The right fender again became a one-piece component resembling that of the initial *Ausführung A*. The final examples of the *Ausführung B* had a rectangular maintenance hatch installed in the sloping front plate.

The addition of the external armor plates raised the vehicle's gross weight to approximately four tons. In addition to this, the units requested that the vehicle have 20 mm of armor all around. This request could not be met by the *Ausführung B*, however, especially since the vehicle was already underpowered at a weight of four tons.

A total of 269 *Borgward BIV*, *Ausführung B*, demolition-charge carriers were produced in the period from July to November 1943.

In autumn 1943 prototypes of the *Ausführung C* began service trials with the *Panzer-Versuchs- und Ersatz-Abteilung 300 (Funklenk)*. After satisfactory conclusion of the trials, series production began in December 1943. The units' demand for heavier armor was answered and the vehicle's armored hull was constructed of 20-mm plates.

The distance between the roadwheels was increased, reducing the load on the running gear. The *Ausführung C* was powered by a Borgward Type 6 B 3.8 engine producing 78 horsepower.

The most obvious distinguishing features of the *Ausführung C* were its raised engine compartment and the relocation of the driver's position to the left side of the vehicle. This resulted in a complete rearrangement of the vehicle's interior.

The Borgward company received a contract from the *Waffenamt* to build more than 1,000 examples of the *Ausführung C*. Beginning in July 1944 difficulties arose with construction of the hull. After completion of 305 of the heavy demolition-charge carriers, production of the *Borgward BIV*, *Ausführung C*, was halted in October 1944.

A planned switch to the Sd.Kfz. 304 "Springer" was one reason for cancellation of the contract, as was the almost complete destruction of the Borgward factory in the 138th air raid on Bremen on 12 October 1944.

Factory prototype of the heavy demolitions carrier, the *BIV*. (Bundesarchiv)

The *Borgward BIV*, *Ausführung A.* (Spielberger)

Side view of the *BIV*, *Ausführung A* with removed track. (Spielberger)

A modified *Borgward BIV, Ausführung A.* This vehicle (chassis number 360376) was produced in 1942 and issued to *Panzer-Kompanie 315 (Funklenk).* The demolitions carrier was captured by the British at Normandy and then subjected to intensive evaluation. The results were published in Report No. 23 of the School of Tank Technology. DTD 3033 (Department of Tank Design) was affixed to this vehicle by the school to register it. (Jentz)

This and facing page : Rear and side views of the modified *BIV*. The number "224" and the tactical insignia of *Panzer-Kompanie 315 (Funklenk)* were applied to the vehicle's rear. (Jentz)

Before being employed, the ends of both of the swing arms which held the demolition charge were secured with bolts. (Jentz)

As soon as the swing arms rose, the demolition container slid down the angled front slope of the vehicle. (Jentz)

A view in the driver's compartment of the *BIV*, *Ausführung A*. (Jentz)

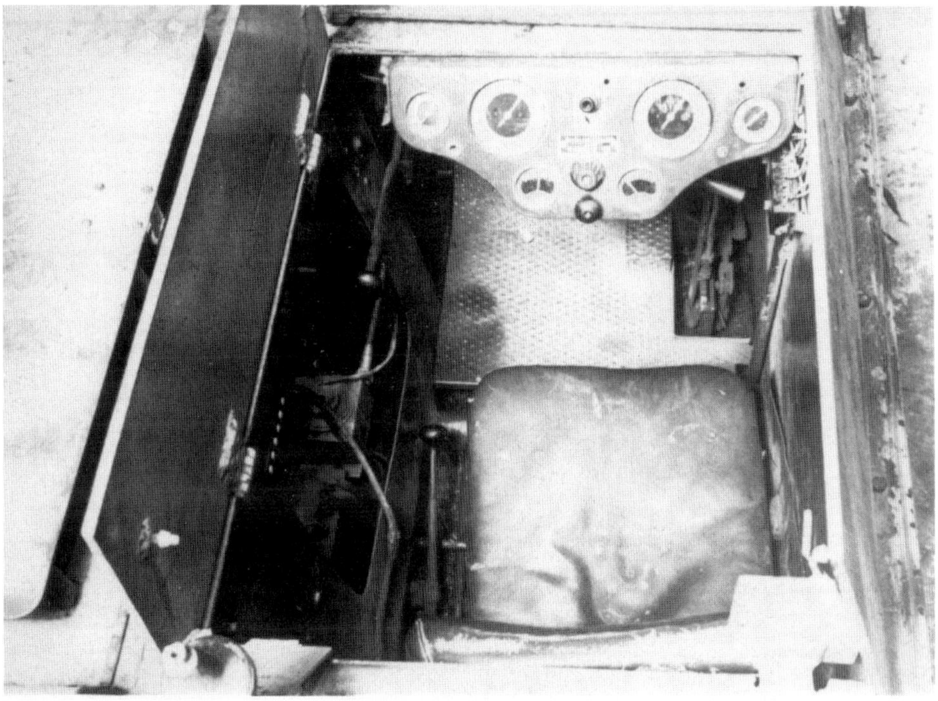

Drive sprocket of the *BIV*, *Ausführung A* had 11 rollers which engaged the rubber-cushioned track (*Kgw 53/200/140*). (Jentz)

The engine compartment of the BIV, Ausführung A. The engine-compartment cover has been removed. (Jentz)

All *BIV's* were test driven on Borgward's own property at Bremen-Uphusen. During the testing, the vehicles received a license plate (*HB = Hansestadt Bremen*). The pictures shown here are of a *Borgward BIV*, *Ausführung A*, however the rubber-cushioned tracks have been replaced with steel ones. The drive sprocket is also unusual and was only found on a few demolition carriers. (Franzen)

Representatives of the *Waffenamt* and Borgward inspect a vehicle on the test track. (Franzen)

A test driver in *Panzer* uniform stands in front of his *BIV*. (Franzen)

Left & above: On occasion, disabled vehicles had to be recovered. (Franzen)

Top left page 15: This *BIV*, *Ausführung A* has been outfitted with the long-range target-acquisition system, the *"Tonne P / Seedorf P"*. This photograph was taken during a vehicle demonstration for Hitler in 1943. (Trenkle)

Top right page 15: Demonstration of a *BIV*, *Ausführung A* for political and military leaders of the Third Reich. Propaganda Minister Goebbels and Armaments Minister Speer can be seen to the right. (Spielberger)

A *"Goliath"* next to a *Borgward BIV*, *Ausführung B*. The driver is mounting the antenna. (Regenberg)

It wasn't until after the fighting at Anzio in the early part of 1944 that the Allies were able to capture a *BIV* demolitions carrier and evaluate it. This is a *BIV*, *Ausführung B*. The demolitions carrier had been issued to the *2./Panzer-Abteilung (Funklenk) 301*. (Jentz)

Views of the driver's escape hatch on the right side and in the driver's compartment. (Jentz)

The *BIV* is evaluated intensively and its signals equipment reviewed. (Jentz)

This page and two left side photos page 18: The last version of the *Borgward BIV* was the *Ausführung C.* This demolitions carrier was captured by the Americans at the end of 1944 on the Ruhr Front. The driver's compartment was located on the left side of the vehicle. As a result, the engine compartment was also rearranged. (Jentz)

Another *BIV*, *Ausführung C* was captured in 1945 and evaluated extensively at the US Army Proving Grounds at Aberdeen, Maryland in the USA. (Jentz)

1.1.3 NSU Kettenkrad Demolition Charge Carrier and Explosives Carrier (Sd.Kfz. 304) "Springer"

In June 1941 the Borgward company completed an experimental model of a remote-control demolition-charge carrier based on the *NSU Kettenkrad*. The driver's seat and steering gear were moved to the rear of the vehicle. The side walls were raised and the resulting space was used to accommodate a demolition charge.

Like the mine-clearing vehicle, a driver was supposed to move the *Kettenkrad* to a line of departure, from where it would then be operated by remote control. The official type designation was *kleiner Kettenkraftwagen (ferngesteuert)* [small tracked vehicle (remote-control)]. This was abbreviated as *kl. Kettenkw (f.g.)*.

In spring 1942, at the urging of *Panzer-Abteilung 300*, the *NSU* company pushed ahead with development of the radio-controlled *Kettenkrad* demolition-charge carrier. In contrast to the experimental model, the following changes were made. A command receiver was installed in the *Kettenkrad*. A mechanism was installed on the front of the vehicle which allowed the steering lever to be locked. The one-piece engine cowling was raised and could be removed for maintenance. The rear of the *Kettenkrad* was completely redesigned and a box-shaped superstructure was installed in place of the rear seat. This was to house the demolition charge.

After a driver had moved the *NSU* demolition-charge carrier into position, the steering lever was locked down and the drive was switched to remote control. The vehicle was then guided to the target by remote control and was destroyed when the charge went off.

Several of these vehicles were committed in June 1942 during the fighting for Fortress Sevastopol. The results were not very encouraging. The *Kettenkrad* demolition-charge carrier often tipped over as a result of uneven weight distribution and narrow track width.

In spite of the problems that had been encountered, the *Heereswaffenamt* decided to continue development of a demolition-charge carrier based on the *Kettenkrad* after evaluating the operational use of this radio-controlled device.

At the beginning of 1943, the design bureau and test department of the *NSU* company were instructed to develop a medium demolition-charge carrier capable of transporting a 300-kilogram, non-jettisonable demolition charge.

The vehicle, now officially dubbed *"Springer"*, was assigned the *Sd.Kfz.* number 304. By 1 July 1943 three prototypes were ready for testing. These prototypes used the running gear of the *Kettenkrad*, however the number of roadwheels had been raised to five per side. The result was similar in appearance to the running gear of the prototype of the larger *HK 102 Kettenkrad*.

The control position, consisting of driver's seat and steering gear, was housed in the rear of the vehicle, while the exhaust was placed halfway along the right side of the vehicle. As with the *BIV*, the driver was provided with a folding three-piece armor shield for protection against small arms fire and shrapnel. At a conference held on 8 July 1943, Inspector-General of Armored Forces Guderian requested that a series of 500 vehicles be built. In addition, the production of the heavy demolition-charge carrier was to be scaled back in favor of the "*Springer*".

The *Sd.Kfz. 304* had a gross weight of 2.4 tons. The *Kettenkrad's* drive had not been designed for this, which led to overloading of the steering mechanism and inevitable problems. Despite lengthening the undercarriage to six roadwheels per side, test drives revealed excessive stress on the roadwheel arms. This shortcoming was overcome by strengthening the arms.

The first vehicles of the test series were delivered to the Army Test Station for Armored and Motorized Vehicles at Kümmersdorf at the end of April 1944. By July three vehicles, designated numbers 9 to 11, were under test. These trials included tests with a dry track produced by the Carl Ritscher company of Hamburg. At the same time *NSU* delivered a modified *Kettenkrad* track, the Type *Zpw 51/170/120* without rubber pads. Neither track proved satisfactory and trials were suspended.

Panzer-Versuchs- und Ersatz-Abteilung 300 (Funklenk) [Armored Testing and Replacement Battalion 300 (Radio Control)] also tested the "*Springer*" at Eisenach. In the autumn of 1944 *NSU* began building the production vehicle, and the *Heereswaffenamt* took delivery of the following quantities:

October 1944: 9
November 1944: 16
December 1944: 10
January 1945: 9
February 1945: 6

A small number of Sd.Kfz. 304's are said to have been employed in the final battles for central Germany.

Above and below: Experimental model of the first demolitions carrier based on the *NSU Kettenkrad*. The driver's seat and the steering mechanism are located in the rear.

Above and left: The driver's seat and the steering mechanism were left in their original positions on later models produced at the factory. What was new, however, was the elevated superstructure to the rear of the vehicle which accommodated the demolitions charge. (*NSU* Archives)

Below and left: Close-up views of the driver's compartment of the *NSU* demolitions carrier. The official designation of the vehicle was *kleines Kettenkrad — ferngelenkt*. (*NSU* Archives)

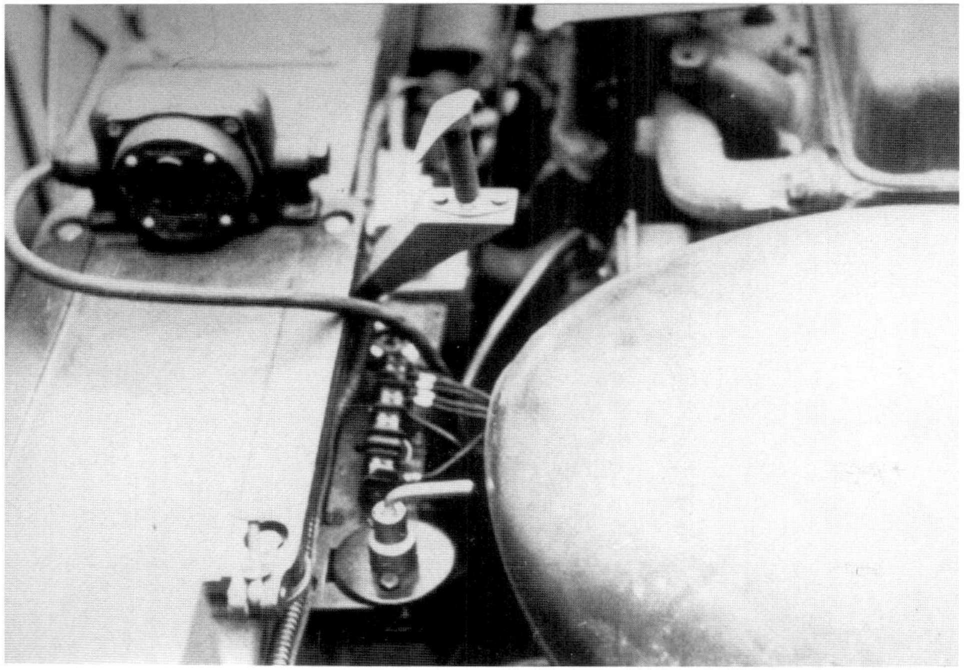

Pages 22-24: Another variant of the *NSU Kettenkrad HK 101* was an experimental model with a machine gun and fixtures for storing items, possibly ammunition. The seat for the machine gunner was forward facing. This experimental model was never used operationally and was later issued to *Panzer-Versuchs- und Ersatz-Abteilung 300 (Funklenk)*. (*NSU* Archives)

Top row: A further development of the *HK 101 Kettenkrad* was the *HK 102*. It was longer and wider than its predecessor. The *NSU "Springer"* demolitions-charge carrier (*Sd. Kfz. 304*) was developed using the *HK 102* as its basis. (NSU Archives)

Left: This early factory prototype of the *"Springer"* still had five roadwheels on each side. (Spielberger)

Factory prototype number 5 already had the six interleaved sets of roadwheels. The fender hasn't been mounted on this vehicle. (Spielberger)

Above and top of page 27: These pictures of prototype number 8 were taken at the German Army's testing grounds at Kummersdorf. (Spielberger)

Bottom page 27 and pages 28 and 29: Several *NSU "Springers"* fell into the hands of the Americans at the end of the war. These were captured at the garrison of *Panzer-Versuchs- und Ersatz-Abteilung 300 (Funklenk)* at Eisenach as well as the *NSU* plant at Neckarsulm. The vehicles were shipped to the United States and extensively evaluated. (Spielberger and McKaughan).

1.1.4 Borgward "Goliath" Light Explosive Charge Carrier (Sd.Kfz. 302)

At the beginning of 1940 the Frenchman Kégresse publicly revealed his latest invention, a light, tracked vehicle with rubber tracks. The vehicle was capable of being guided by remote control to obstacles and fortifications. After reaching the target, a small explosive charge carried in the vehicle was intended to destroy the object. Prior to the cease-fire in June 1940 a prototype of the vehicle was sunk in the Seine. It was later recovered by the Germans, who evaluated its technology.

In late 1940 the *Waffenamt* ordered the Borgward company to develop a small tracked vehicle capable of transporting an explosive charge weighing approximately 50 kilograms.

During the course of 1941, a factory model was developed which incorporated various components of the *Borgward BII.* In spring 1942, Borgward produced a prototype which subsequently went into production as the *Sd.Kfz. 302 "Goliath"* light demolition-charge carrier.

Two batteries powered two Bosch electric motors which moved the vehicle. The vehicle reached a speed of 8 to 10 kilometers per hour and had a range of 1.5 kilometers on roads and 0.8 kilometers off roads. Approximately 60 kilograms of explosives were positioned in the forward part of the vehicle. The rear of the vehicle contained a drum which held 750 meters of cable. A 5-mm armor plate was fitted on the front of the vehicle, the only part to be so protected.

The electrically-powered *Sd.Kfz. 302* possessed several shortcomings. One was its limited trench-crossing ability of 60 cm. The other was its restricted range which meant it had to be brought relatively close to its target, exposing the operators to enemy fire.

The first unit to receive the light demolition-charge carrier was the *3./Panzer-Abteilung 300*, which employed it successfully in the Battle of Sevastopol. Other *Sd.Kfz. 302's* were tested in Africa by the Versuchs-Kommando (Fernlenk) Tropen [Testing Command (Remote Control) — Tropical].

At the beginning of 1943, responsibility for the use of wire-controlled demolition-charge carriers was transferred from the radio-control units to the *Panzerpioniere. Panzer-Pionier-Kompanien (G) 811* to *814* were formed in the autumn of 1942. As per Table of Organization 727 (*Kriegsstärkennachweisung = KStN*), each was equipped with 162 *Sd.Kfz. 302's.*

In 1943 a prototype of a radio-controlled light demolition-charge carrier was completed. It was a *Sd.Kfz. 302* with a gasoline engine that had been developed by the *1./Panzer-Versuchs- und Ersatz-Abteilung 300.* The use of a radio receiver installation in a light demolition-charge carrier proved too costly, however, and the project was not pursued.

Demonstration of the tracked carrier designed by Kégresse in the spring of 1940. (Leon)

One of the factory test models of the light explosive charge carrier "Goliath" from the Borgward factory. (Spielberger)

A large number of wire-guided light demolitions-carrier *Sd.Kfz. 302* were issued to *Panzer-Abteilung 300* in June 1942 for employment at Sevastopol. (Seeger)

1.1.5 Other Developments

After the end of the French Campaign in the summer of 1940 the Germans gathered many vehicles, including damaged armored vehicles, in captured equipment collection points. The captured equipment was catalogued, repaired and modified, then assigned to German units for further use.

Demolition-Charge Carrier Based on the Bren Carrier

In 1936 the British Army introduced into service a fully-tracked vehicle for transporting personnel. The official designation for this vehicle, which was armed with a single Vickers machine gun, was Carrier, Machine Gun No. 1, Mark I.

An improved version, designated Carrier, Machine Gun No. 2, Mark I, appeared in 1937. Originally it was also supposed to be armed with a .303 Vickers machine gun, however following the introduction of the Bren light machine gun in 1938, either a Bren machine gun or a Boys antitank rifle was installed. From this point the machine was designated Carrier, Bren, No. 2, Mark I or Mark II. In 1939 a modified version of the Bren Carrier was introduced called the Scout Carrier.

Plans called for every infantry battalion to have 10 Bren Carriers. Weighing 3.75 tons, these tracked vehicles were powered by a Ford V8 motor. Maximum speed was 48 km/h.

Panzer-Abteilung 300 was issued a large number of captured Bren or Scout Carriers. The official German designation for the vehicles was *gepanzerter MG-Träger Br-Kenn-Nummer 731 (e)*. During repairs the drive and steering systems were modified to allow the vehicles to be wire controlled. Approximately 700 kilograms of explosives was poured into the free space in the rear of the superstructure. After conversion into a wire-controlled, demolition-charge carrier, the vehicle was designated *Ladungsträger auf Fahrgestell Bren (e)* while in service with *Panzer-Abteilung 300*.

Explosive Charge Carrier Based on the Utility Tractor

In 1935-36 the Belgian Army ordered an armored version of the Utility Tractor built by the British firm of Vickers-Carden-Lloyd. Two different types were delivered, the Model 1 or Type A, a cavalry version with a crew of three, and the Model 2 or Type B for the infantry with a crew of one.

Independent of this contract, the Belgian Army attempted to reach an agreement with Vickers to build the Utility Tractor under license. After negotiations were successfully concluded, the design and development company Familleureux was authorized to manufacture the Utility Tractor under license. The licensed version was also designated Model 2, however, it differed from its British predecessor in a number of respects. The most significant improvements introduced by the Familleureux model were a lengthened superstructure and revised arrangement of the drive sprockets, idler wheels and roadwheels which increased the track area in contact with the ground.

When placed into service by the Wehrmacht, the Utility Tractors were designated *Artillerie-Schlepper VA (Vickers-Armstrong) 601 b (belgisch)*.

The Utility Tractors issued to *Panzer-Abteilung 300* had explosives placed into the bins on the sides of the vehicle originally intended for transporting ammunition. Furthermore, the drive train was converted for remote control (wire). A drum with 1,200 meters of cable was installed on the rear of the vehicle.

The wire-controlled Utility Tractors were used up in remote-control operations in the summer of 1942.

A wire-guided Bren Carrier of the *3./Panzer-Abteilung 300* in the summer of 1942 on the Crimea. (Lock)

VW Schwimmwagen (Sondertyp 129)

On 8 April 1942 personnel of *Panzer-Abteilung 300* took delivery of two *VW Schwimmwagen* from the Porsche company in Stuttgart-Zuffenhausen. They were a variant of the Type 128 which was designated *Sondertyp 129* (Special Type 129).

On these vehicles the crew compartment was completely covered with sheet metal. An explosive charge was to be accommodated in the rear of the interior. The exhaust system was modified. Two vertically-arranged mufflers were shielded by rectangular metal casings. Takeoff-assist rockets could be attached on the left and right sides of the rear metal shield. These were supposed to be ignited shortly before the vehicle reached a riverbank in order to provide it with the necessary impetus when it touched ground.

The *Schwimmwagen Sondertyp 129* was tested on the River Spree, in the course of which one vehicle was written off. The idea of employing the *Schwimmwagen* as an amphibious demolition charge carrier was not pursued. The second vehicle was issued to *Panzer-Versuchs- und Ersatz-Abteilung 300*, where it was used for other purposes.

This page and page 34: Several views of the *VW-Schwimmwagen (Typ 129)* which was modified to be employed as an amphibious explosive-charge carrier. (*VW* Museum)

1.1.6 Command and Control Armored Vehicles

The *kleiner Panzerbefehlswagen* (small armored command vehicle) was developed on the chassis of the *Panzerkampfwagen I, Ausführung B.* The command vehicle was assigned the *Sd.Kfz.* number 265 and was originally intended as an armored command vehicle for company commanders as well as battalion and regiment commanders of armored units.

When *Minenräum-Kompanie 1* was formed in the summer of 1940 it received a number of *Sd.Kfz. 265's* as command and control vehicles. Installed in the interior of the tank was a command transmitter for the remote control of mine-clearing vehicles. The tank driver had an important role to play. Coordinating with the radio operator, he had to maintain the engine at a constant speed so as to ensure that sufficient battery voltage was available for the command transmitter.

Following the issuance of additional *kleine Panzerbefehlswagen* to *Minenräum-Abteilung 1*, the vehicles were modified by the unit. An additional antenna for the command transmitter was mounted on the superstructure roof. In addition, a panoramic telescope was installed in the top left hatch.

Eighteen of these vehicles were placed in service with *1.* and *2./Minenräum-Abteilung 1.* Another modified *kleiner Panzerbefehlswagen* was used by the battalion headquarters. A frame antenna was mounted on this vehicle. After *Minenräum-Abteilung 1* was redesignated *Panzer-Abteilung 300*, the surviving *Sd.Kfz. 265's* were turned in when the unit converted to the command *Panzer III.*

Panzerkampfwagen III, Sd.Kfz. 141, Ausführungen J – N

With the reorganization of the radio- and remote-control units in autumn 1941 and the associated formation of *Panzer-Abteilung 300*, it also became necessary to introduce a new type of command and control tank. Several requirements had to be met:

1. Installation of the radio command transmitter and accessories in the command and control tank.

2. Equipment similar to that of the armor unit to which the remote-control unit was supporting.

3. Self-defense capability during radio- or remote-control actions, by engaging targets such as enemy tanks or anti-tank guns.

All of these requirements were met by the *Panzerkampfwagen III, Ausführung J.* In addition, this vehicle was available in large numbers. *Panzer III's* issued to remote-control units were converted in the battalion maintenance facilities. The following modifications were made:

1. In place of the bustle rack, an armored equipment container designed to accommodate the radio command transmitter and accessories was attached to the rear of the turret. Since these containers were also manufactured by the units, several different types were seen.

2. Attachment of a shallow wooden personal-items container on the right track fender between the radio operator's vision slot and the rod antenna mount.

3. The tools previously mounted in this area on the track fender were moved. The engine crank, wire cutters and small tool box were remounted on the left track fender. The axe was moved to a position on the right fender next to the shovel. An additional replacement roadwheel was attached in front of the wooden container. The horn, the two side markers and the Notek headlight were moved forward on the track fenders to positions nearer the hinges.

In the period between spring 1942 and summer 1943, three units were issued the following versions of the *Leitpanzer III* (See box below):

In general, all *Panzerkampfwagen III's* were issued through normal supply channels, and thus they also included all production modifications and improvements introduced by the manufacturers.

Panzerkampfwagen III's which were no longer fit for combat duties were later employed by the training companies of the *Panzer-Versuchs- und Ersatz-Abteilung 300 (Funklenk).*

Unit	Version	Remarks
Panzer-Abteilung 300/301	*Panzerkampfwagen III, Ausführung J, Sd.Kfz. 141*	50-mm *KwK L/42*
	Panzerkampfwagen III, Ausführung J, Sd.Kfz. 141/1	50-mm *KwK 39 L/60*
	Panzerkampfwagen III, Ausführung L, Sd.Kfz. 141/1	50-mm *KwK 39 L/60* with frontal armor
Versuchs-Kommando (Fernlenk) Tropen	*Panzerkampfwagen III, Ausführung J, Sd.Kfz. 141*	50-mm *KwK L/42*
	Panzerkampfwagen III, Ausführung J, Sd.Kfz. 141	50-mm *KwK L/42* with frontal armor
Panzer-Kompanie (Fernlenk) 313	*Panzerkampfwagen III, Ausführung L, Sd.Kfz. 141/1*	50-mm *KwK L/42* with frontal armor and armored skirts
	Panzerkampfwagen III, Ausführung N, Sd.Kfz. 141/2	75-mm *KwK L/24* with armored skirts

Sturmgeschütz III, Sd.Kfz. 142/1, Ausführung G

The separate radio-control companies which were formed beginning in early 1943 and the reorganized *Panzer-Abteilung (Funklenk) 301* were equipped with the *Sturmgeschütz III, Ausführung G.*

The modified assault guns had to meet the same requirements as the *Leitpanzer III*, however, the *Sturmgeschütz III* offered some advantages which the *Panzer III* could not. The complete radio command transmitter system could be housed on the left side of the fighting compartment. Furthermore, the *Sd.Kfz. 142/1* was armed with a 75-mm *Sturmkanone 40 L/48*, which enabled it to successfully engage any contemporary enemy tank.

It has also been determined that most of the *Sturmgeschütz III's* used by these units were equipped with skirts.

Like the *Panzerkampfwagen III's*, all *Sturmgeschütz III's* were issued through normal channels and incorporated all improvements and modifications introduced by the manufacturers. Furthermore, in some cases the receiving units relocated some of the vehicles' tools.

The following table lists the units which were assigned Sturmgeschütz III, Ausführung G vehicles for use in the command and control role between spring 1943 and May 1945:

Unit	Version	Remarks
Panzer-Kompanie (Funklenk) 311 smoke	*Sturmgeschütz III, Ausführung G, Sd.Kfz. 142/1*	in some cases sections of fender were removed; grenade launchers
Panzer-Kompanie (Funklenk) 312 (*KFF*)	*Sturmgeschütz III, Ausführung G, Sd.Kfz. 142/1*	in some cases *8/ZW* hulls and driver's telescope plus deep wading muffler
Panzer-Kompanie (Funklenk) 314	*Sturmgeschütz III, Ausführung G, Sd.Kfz. 142/1*	some equipped with driver's telescope (*KFF*)
Panzer-Kompanie (Funklenk) 315	*Sturmgeschütz III, Ausführung G, Sd.Kfz. 142/1*	none
Panzer-Kompanie (Funklenk) 316	*Sturmgeschütz III, Ausführung G , Sd.Kfz. 142/1*	some with *8/ZW* hulls
Panzer-Kompanie (Funklenk) 317	*Sturmgeschütz III, Ausführung G, Sd.Kfz. 142/1*	bullet deflector in front of the commander's cupola
Panzer-Kompanie (Funklenk) 319	*Sturmgeschütz III, Ausführung G, Sd.Kfz. 142/1*	some with *8/ZW* hulls and deep wading muffler
Panzer-Abteilung (Funklenk) 301	*Sturmgeschütz III, Ausführung G, Sd.Kfz. 142/1*	some with *8/ZW* hulls and drivers telescope (*KFF*) plus smoke grenade launchers
Panzer-Abteilung (Funklenk) 302	*Sturmgeschütz III, Ausführung G, Sd.Kfz. 142/1*	assault guns from consolidated *Panzer-Kompanien 311, 315, 316* and *317*, and allocations from current production. Some with cast gun mantlet (*Saukopf*) and bullet deflector in front of commander's cupola
Funklenk-Panzer-Zug/ Panzer-Abteilung (Funklenk) 303	*Sturmgeschütz III, Ausführung G, Sd.Kfz. 142/1*	some with all-round-fire machine gun

Panzerkampfwagen VI, "Tiger" I, Sd.Kfz. 181

In an after-action report on the use of remote-control equipment during Operation *Zitadelle*, the commander of *Panzer-Abteilung (Funklenk) 301*, *Major* Reinel, proposed that the *Panzerkampfwagen VI "Tiger"* be used as a control tank. One of the reasons offered was the fact that as soon as the enemy realized demolition-charge carriers had been committed he tried with all means to engage or destroy the control tank. It was therefore argued that a tank should be used for the control mission which possessed sufficient armor protection and a long-range gun with which it could successfully engage enemy tanks at longer distances.

The radio-control company of the *Panzer-Lehr-Division* received its first ten *Panzerkampfwagen VI "Tigers"* in September 1943. Subsequently, the third companies of *schwere Panzer-Abteilungen 504* and *508*, which had been reorganized as radio-control companies, were also issued *"Tiger"* control tanks. In the late summer of 1944, *Panzer-Abteilung (Funklenk) 301* was completely re-equipped with *"Tigers"*.

The majority of the *"Tigers"* issued to radio-control units were mid- and late-production vehicles with steel roadwheels.

The radio command transmitter system was housed in the fighting compartment. On the *"Tigers"* of *schwere Panzer-Abteilungen 504* and *508* the antenna mount for the radio command transmitter was installed on the right side of the turret; those of *Panzer-Abteilung ("Tiger"/Funklenk) 301* were fitted on the right side of hull deck.

No further technical modifications were made to the tanks. The *"Tigers"* of *Panzer-Abteilung ("Tiger"/Funklenk) 301* were fitted with additional canister racks on the storage bin on the rear of the turret.

It is safe to assume that only 50 to 60 *Panzerkampfwagen VI "Tigers"* were converted as command vehicles for radio-controlled equipment. They were taken from the production line and incorporated all current factory modifications.

Other Command and Control Tanks

In addition to those already described, other tanks were evaluated by units for possible use in the control-tank role.

For instance, the radio-control company of the *Panzer-Lehr-Division* received the first five *Panzerkampfwagen "Tiger II's"* to be built.

Panzer-Abteilung 300 (Funklenk) tested a *Panzerkampfwagen V "Panther"* in the control-tank role. The results were less than satisfactory. The *"Panther"* was later used to test infrared devices.

The training companies of *Panzer-Versuchs- und Ersatz-Abteilung 300 (Funklenk)* were equipped for the most part with vehicles no longer fit for combat duty. Among its equipment were *Sturmgeschütz III*, *Ausführungen D* and *F/8*. It is possible that some of these vehicles took part in the fighting in central Germany in spring 1945.

The *kleiner Panzerbefehlswagen (Sd.Kfz. 265)* formed the basis for the first command and control tank used. There was a fixture for mounting the radio-control antenna on the hull; in addition, this vehicle has a 360-degree periscope at the commander's station.

A command and control *Panzer III, Ausführung J* with the 5 cm *Kwk 39 L/42* of the *1./Panzer-Abteilung 300* employed outside of Sevastopol in the summer of 1942. (Lock)

This command and control *Panzer III, Ausführung J* with the 5 cm *Kwk 39 L/42* of the *Versuchs-Kommando (Fernlenk) Tropen* has additional armor mounted on the turret and hull. (Lehmann)

After arriving in the northern sector of the Eastern front late in the summer of 1942, *Panzer-Abteilung 301* received six new command and control *Panzer III, Ausführung J* with the 5 cm *Kwk 39 L/60* and add-on armor.

The newly formed *Panzer-Kompanie (Funklenk) 311* received the first command and control armored vehicles based on the *Sturmgeschütz III, Ausführung G (Sd.Kfz. 142/1)* in the spring of 1943. The *Sturmgeschütz III* shown here still has the smoke grenades and uneven track-guard fenders. (Niepenberg)

The *Sturmgeschütz III's* of *Panzer-Kompanie 314 (Funklenk)* had specially formed mounting brackets on the hull sides for holding additional roadwheels. The additional armor mounted around the driver's position has been cut out in the vicinity of the driver's periscope port. Note the mounting bracket for the radio-control antenna on the forward portion of the fighting compartment roof. (Held)

After mid-1943, the majority of *Sturmgeschütz III's* of the radio-control companies were equipped with side-skirts of different styles. This assault gun was issued to *Panzer-Kompanie (Funklenk) 316*. (Lock)

The *"Tiger I's"* of the *3./schwere Panzer-Abteilung 504* and the *3./schwere Panzer-Abteilung 508* had a mounting bracket on the right-hand side of the turret for the radio-control antenna. (Herwig)

The radio-control antenna on the *"Tigers"* of *schwere Panzer-Abteilung (Tiger/Funklenk) 301* was affixed to the middle of the hull on the left-hand side of the vehicle. Each "Tiger" in the battalion also had mounting brackets for four "jerry" cans on the personal-gear stowage box on the turret rear. This *"Tiger"* has lost a track and the crew appears to be awaiting recovery assistance. (Höland)

1.1.7 Wire- and Radio-Control Equipment

Wire and radio control are components of remote control. The term remote control is understood to mean that a guided object carries out commands issued by a controller who is situated some distance away and is linked to the guided object by some form of communication.

Basically there are two methods of control. One involves the transmission of radio waves in the high frequency band, while the other transmits DC current and low-frequency signals by wire.

The following descriptions are intended to provide a brief overview of the radio-control methods used by remote- and radio-control units.

Wire Guidance

Wire control was used by the *"Goliath"* light demolition-charge carrier (*Sd.Kfz. 302*) and by the Bren Carrier and Utility Tractor demolition-charge carriers.

The electrical systems of these vehicles consisted of a command transmitter, the control cable and control set.

On the cover plate of the command transmitter were:

The drive switch with positions FORWARD (V), STOP (H) and REVERSE (R).
The steering switch with positions LEFT (L) and RIGHT (R).
The ignition switch with the ignition key and chain as well as three colored terminals for connecting the control cable.

The bottom part of the command transmitter contained nine removable flashlight batteries in series (standard version 4.5 V).

The control cable consisted of three braided insulated wires wound onto a cable drum. The interchangeable drums had the same dimensions as those for heavy field cable. Cable length was 750 meters for the *"Goliath"* and 1,200 meters for the Bren Carrier and Utility Tractor.

A command set was installed in a vibration-free mount. It contained a relay switch and one flashlight battery (4.5 V) as igniter.

Prior to operational use, the vehicles were moved to the start position. After a quick check of the electrical system, the control procedure was initiated. The controller took the command transmitter and, using the drive and steering switches, steered the vehicle toward the target. Only after the vehicle had reached a safe distance was the ignition key inserted into the ignition switch. When the target was reached, the ignition key was turned as far as it could go, which ignited the demolition charge. The demolition-charge carrier was destroyed in the explosion. The command cable was subsequently wound onto an empty cable drum for further use.

In the wire-guidance process, all commands for steering or initiation of the ignition process were achieved by different DC impulses. It was anticipated that these would be coded to allow for a change in impulse allocation.

The command transmitter and control equipment were provided by the firm of Buttermann & Schmitt of Berlin-Neuköln, the assembled and checked cable drums by the Osram company.

Radio Guidance

The technically more complicated radio-control system was used to operate the *Borgward BI* and *BII* mine-clearing vehicles, the *BIV*, *Ausführungen A* to *C* heavy demolition-charge carriers (*Sd.Kfz. 301*) and the *"Springer"* medium demolition-charge carrier (*Sd.Kfz. 304*).

Radio control was accomplished by the *FKL 8* radio control system. Developed by Blaupunkt, it transmitted multiple individual commands for the remote control of tracked vehicles.

The radio-control system consisted of two major equipment groups, the radio command-transmitter system and the radio command-receiver system, plus associated accessories.

The radio command-transmitter system, which was housed in the control tank, was comprised of:

The radio signal transmitter, consisting of command transmitter and radio transmitter (*Fu KS*)
The control box (*KoG*)
The *WR 81* power supply
The mast/rod antenna

In operational command and control armored vehicles both the mast antenna and the power supply were connected to the radio signal transmitter.

In the command-transmitter portion of the radio signal transmitter there were two tone groups, *"N"* and *"R"*, each containing five tone circuits. When employed, the controller had to connect the cable for the command transmitter to the upper (*"N"*) or lower (*"R"*) connector box in the rear bottom part of the transmitter, depending on the designator of the demolition-charge carrier. The switch installed on the device was then set to one of four settings (*O, IV, VI, IX*). If required, these numbers could be changed on the transmitter.

On the transmitter there was also a toggle switch which could be set to *"AUS"* (off), *"BEREIT"* (ready) or *"EIN"* (on). When the system was turned on the switch was moved from "OFF" to "READY". A red light came on after a warming-up phase. At that point, it was possible to switch the transmitter to "ON".

Radio control could then be accomplished by way of the control box.

The control box was made of cast aluminum and had two hooks on the back for suspending it from the belt. As well, there were lugs at the top of both sides for attaching a carrying strap.

The starter button was located on the right front beneath a cover plate. When the transmitter and receiver were ready, pressing the button started the motor of the demolition-charge carrier. On the top left there was another cover plate, beneath which was the push button for switching off the motor.

A toggle switch to the left of the starter button was used to select gears. Moving the switch forward transmitted the command FORWARD. If the switch was moved to the rear the vehicle changed to REVERSE. Shifting gears could only be accomplished when the vehicle was stationary.

A black ball lever was located in the center of the control box and when not under pressure it always returned immediately to the vertical start position. Pushing the lever forward caused the demolition-charge carrier to accelerate. Pulling the lever back caused it to slow down. Moving the lever to the right caused the vehicle to turn to the right, while moving the lever to the left caused the vehicle to turn left.

On the right side of the control box there was a headset jack. This enabled the operator to listen to the command tones. The toggle switch with the identifying letter *"Z"* was a safety switch which prevented inadvertent activation of the detonation system. The detonate command could only be issued if the *"Z"* switch was in the "up" position.

In addition, beneath a cover plate on the right side, there was a push button with the identification *"Z 1"*. Pushing this button immediately detonated the explosive charge and destroyed the entire vehicle.

Located beneath a cover plate on the left side of the control box was a push button with the identification *"Z 2"*. Pushing this button caused a radio signal to be sent which blew the explosive charges on the lever arm restraining cable or mounts, allowing the demolition charge box to slide down the sloping front of the vehicle. A contact-breaking device with delayed action lit the fuses located in the demolition charge. After the *"Z 2"* switch was activated, the vehicle automatically switched into reverse gear and moved out of the blast zone.

On the underside of the control box was the connecting slot for the cable to the command transmitter. The cable connection could be secured with a snap hook.

The radio command-receiver system, which was housed in the demolition-charge carrier, was comprised of:

The radio command receiver, consisting of command receiver and radio receiver (*FuKE*)
The switchbox (*SCHK*)
The detonating device (*ZGR*)
The mileage recorder (*WSSCH*)
The oil control system (*OSG*)
The smoke device (*NGR*)
The whip antenna

The switchbox was mounted at hip level of the *B IV's* driver and had seven cable connectors on the upper side. Behind this was the socket connector for the smoke device. Prior to leaving the vehicle before a mission, the driver used a special key to set a circular plate on the front of the switchbox to the *"HF"* position. This turned on the system for high-frequency operation.

The signals sent by the radio command transmitter were picked up by the radio or command receiver and processed in the switchbox. Signals for steering the demolition-charge carrier were transmitted to the oil control system as electrical impulses. Signals for jettisoning the demolition charge, blowing up the charge with the carrier or for generating smoke were transmitted to the detonation device or the smoke device.

In the oil-control system there was a system of five piston cylinders which were kept under hydraulic pressure by an oil pump. In each cylinder there was an electro-magnet through which the piston could be moved. The up and down movement of each piston produced pressure in a specific oil line, which acted on the various systems by hydraulic force. The oil-control system affected the speed, right-left steering and forward-reverse movement. The electro-magnets were activated by electrical impulses.

The detonating device served to set off the *Glühzünder 28* (electric igniter). A total of three igniters were attached. Two of these served to destroy the lever arm retaining cable or mounts; the third could immediately detonate the demolition charge and destroy the demolition-charge carrier in an emergency. The cable to the detonation transformers on the floor of the vehicle was also connected to the circuit of the igniter used to destroy the vehicle.

The mileage recorder was installed as a safety mechanism. It had a wind-up counter which could be used to set a certain driving distance in meters prior to the start of the remote-control procedure. Unless he received other instructions, the driver of the *BIV* set it for a safety radius of 100 meters. Only after the counter had run down could the detonation component of the jettison system be enabled. The mileage recorder could also be used to cancel this command. This was especially important if a *BIV* with an "enabled" jettison system had to be recalled to the command and control vehicle.

The entire process of checking the radio system and preparing it for operation is contained in a training manual (Source: L. Schreiner, *4./Panzer-Abteilung (Funklenk) 301*):

4./Panzer-Abteilung (Funklenk) 301 10 June 1943

"For Official Use Only"

BIV Checklist

Command: Conduct checks

Part I

Signals Mechanic: Transmitter Check

1. Connect the battery cable to the transmitter
2. Switch to "READY"
3. Screw on the antenna and
4. Attach the antenna cable
5. Check the voltage, at least 11.5 Volts
6. Switch to "ON" after the red lamp lights up
7. Check the anode potential by pressing the blue button. It must be at least 310 Volts
8. Set the transmitter to the highest amplitude while pressing the test button
9. Tighten the locking screw
10. Attach the control box
11. Note the tone group and ignition key
12. Once convinced that the transmitter is working properly, report to the squad leader: "Transmitter in order"

BIV Driver:

1. Ensure all cables coming from the switchbox are securely attached

2. Ensure automatic "STOP" is switched on

3. Ensure the radio receiver is seated firmly

4. Ensure antenna contact on the radio receiver

5. Ensure the plugs on the igniter device are seated firmly

6. Screw the antenna on

7. Switch to road gear (forward gear)

8. Turn the ignition key to "1"

9. Set the *HF* lever to "CHECK" position

10. During the checking process maintain foot on the brake and hand on the ignition key so that if *BIV* begins to move and brakes fail I can immediately switch to "0"

11. Only pull the hand brake if the *BIV* must be braked.

<u>Combat Engineer:</u> (all of the squad's *BIV's* in turn)

1. Check the voltage of the flashlight batteries in the igniter device

2. Check the ignition line together with the signals mechanic.

Part II

Gunner:

After receiving the signals mechanic's hand signals and using the control box, I issue the following commands, in order:

1. FO = a) Start if motor not running
   b) Slow if motor running

2. Ignite the entire system

3. Jettison (causing the carrier to reverse gear and full throttle)

4. Re = right

5. Li = left

6. FO = slower, and wait for automatic stop

7. FO = start

8. VO = switch to forward gear and full throttle

9. FO = slower

10. Ha = stop (forward gear engaged, brake on)

11. Listen to tones in control box; signals mechanic simultaneously confirms by raising a hand that the commands are being carried out correctly

12. After checking all BIV's, switch off the transmitter

<u>Signals Mechanic</u> (with the assistance of the *BIV* driver)

1. Check the battery voltage, at least 11.5 Volts

2. Check the anode voltage, at least 160 Volts with the field metering device touching the switchbox casing

3. Check the seating of the connectors

4. Check that transmitter and control box ignition keys work the same

5. Then insert test lights on igniter device

6. Issue commands in following order: *FO*, Total Ignite, Jettison, *Re*, *Li*, *FO* (wait for automatic stop and then) *FO*, *VO*, *FO*, *Ha*. Confirm the correct execution of each command by raising my hand and then giving the signal for the next command

7. At the same time I monitor the receipt of the tones

8. When completed, I report to the squad leader: "First *BIV* checked"

Part III

BIV Driver:

1. Press button *"B"*

2. *"FO"* to start

3. Brief *"F1"* to release the brakes

4. Switch ignition key to "0"

5. Switch *HF* to Manual position

6. Select cross-country gear

7. Remove whip antenna

Command: "Prepare to jettison"

Combat engineer:

1. Check the continuity of the electric caps (3).

2. Insert friction igniter and one electric cap into the wood

3. Pull friction igniter through the detonator hatch and secure it

4. Pull electric cap through the detonator hatch and close the hatch,

5. Attach electric caps to the clips of the terminal,

6. Screw on igniter and insert the wire through the tube without securing it and then close the access hatch.

7. Secure two electric caps on both sides of the restraining cable and connect them to the terminal.

8. Check to ensure that connectors are inserted in the igniter device.

B IV Driver:

1. Loosen the bolts on the jettison lever.

<u>Signals mechanic:</u> Check the continuity of Total Ignite and Jettison by inserting the test switch.

Command: "Prepare for radio operation":

Part I

Gunner at the Transmitter:

1. Switch to "Ready"

2. Check the voltage, at least 11.5 Volts

3. Switch to "ON" after the red light comes on

4. Check the anode voltage by pressing the blue button. It

must be at least 310 Volts.

5. Set the transmitter to maximum deflection while pressing the test button

6. Tighten the locking screw

7. Once convinced that the transmitter is working properly, I report to the squad leader: "Transmission ready"

BIV Driver:

1. Select road gear

2. Set ignition key to "1"

3. Set *HF* lever to test position and then immediately depress the foot brake with right foot

4. Press button "B" on switchbox, thus switching off automatic stop

5. Together with *BIV* signals mechanic, carry out mechanical checks by selecting: *FO*, *VO*, *FO*, *Rü*, *Re*, *Li*, *FO*, *VO*, *FO* and briefly *F1*. Switch the motor off

6. Switch the *HF* lever to Manual

7. Engage cross-country gear

8. Screw on the whip antenna

Signals mechanic: Monitor the mechanical check of the *BIV* by the driver.

Part II

1. *BIV* Driver:

   1. Screw the detonators into the charge
   2. Close and secure charge hatch
   3. Remove the screws from the jettison lever

2. Combat engineer:

   1. Check that the connector is properly seated on the igniter device
   2. Check to ensure that the screws have been removed from the jettison lever
   3. Check the seating of the detonators in the charge
   4. Satisfy myself that the friction igniter cable is not hung up somewhere.

Command: "Prepare for operations"

BIV Driver:

1. Position myself 30 meters behind the *Sturmgeschütz* in such a way that my movement will bypass the *Sturmgeschütz* and avoid all obstacles.

2. Leave the motor running

3. Remove the ignition key while in position "1" and put it in my pocket.

4. Attach the cable to the friction igniter

5. Place the *HF* lever in the "Operational" position and leave the *BIV* after having locked it.

(signed Busse)
Oberleutnant and Company Commander

Radio-command transmitter (frequency group N/R 075 with four settings: 0, II, IV and VII) as well as transformer and battery.

Radio-command transmitter (frequency group N/R 049 with four settings: 0, II, IV and VII).

The radio-control box.

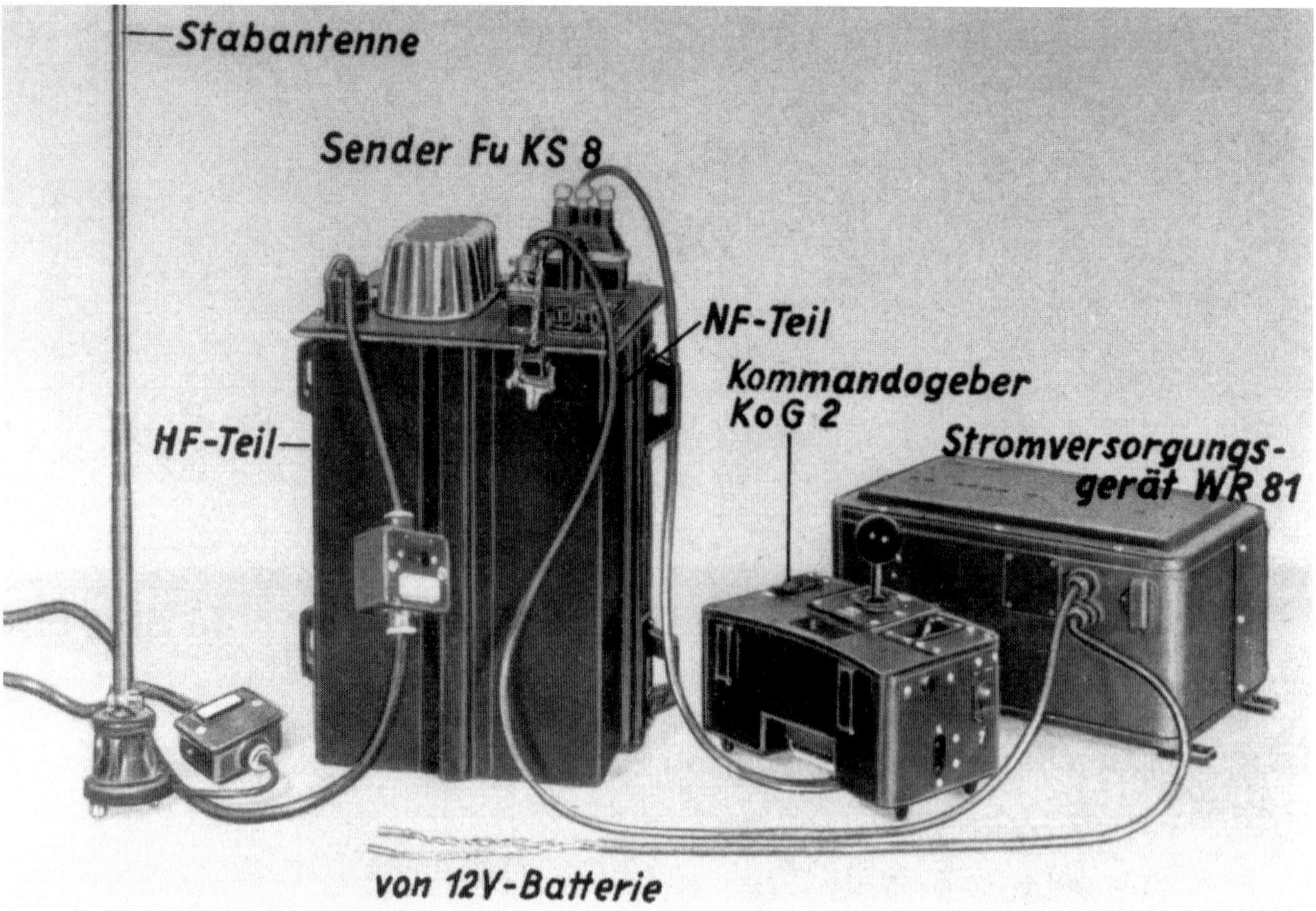

The radio-command receiver.

This transmitter was installed in the command and control tank.(Trenkle)

A crew member of a *Sturmgeschütz* of *Panzer-Abteilung (Funklenk) 302* guides a *BIV* to its target using the control box. (Jenckel)

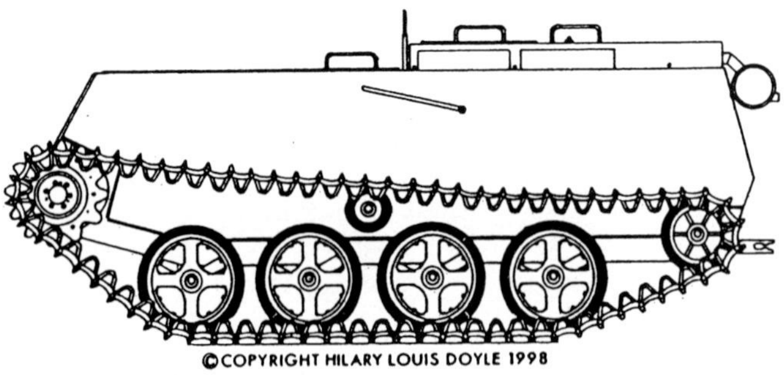

Borgward BI and *Borgward BII* Mine-Clearing Vehicles *(Sd.Kfz. 300)* and *kleiner Panzerbefehlswagen (Sd.Kfz. 265)* (from top to bottom) (All drawings copyright 1998 by Hilary Louis Doyle and used with permission.)

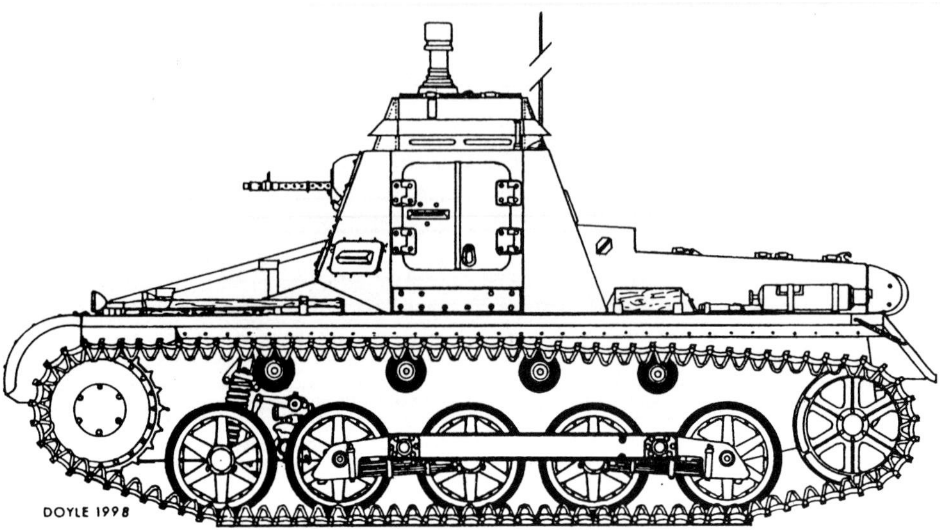

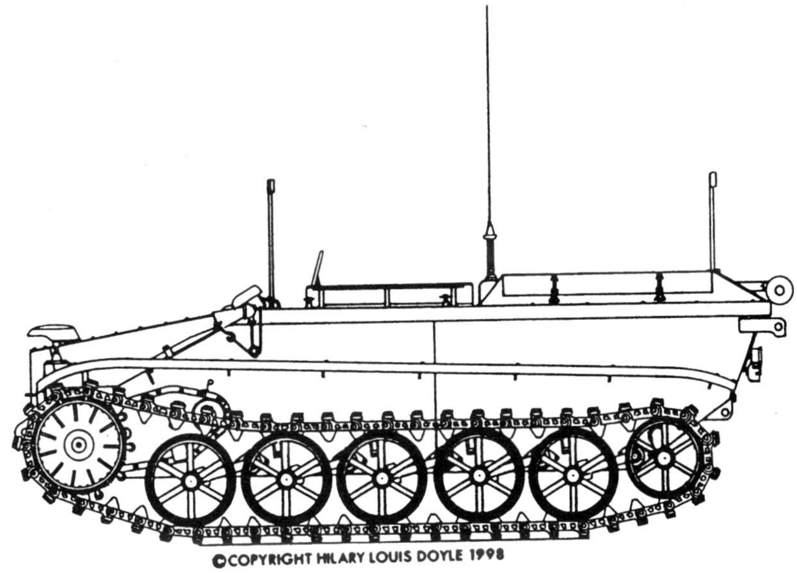

Borgward BIV a and *BIVc* demolition-charge carriers (above and below).(All scale drawings copyright 1998 by Hilary Louis Doyle and used with permission.)

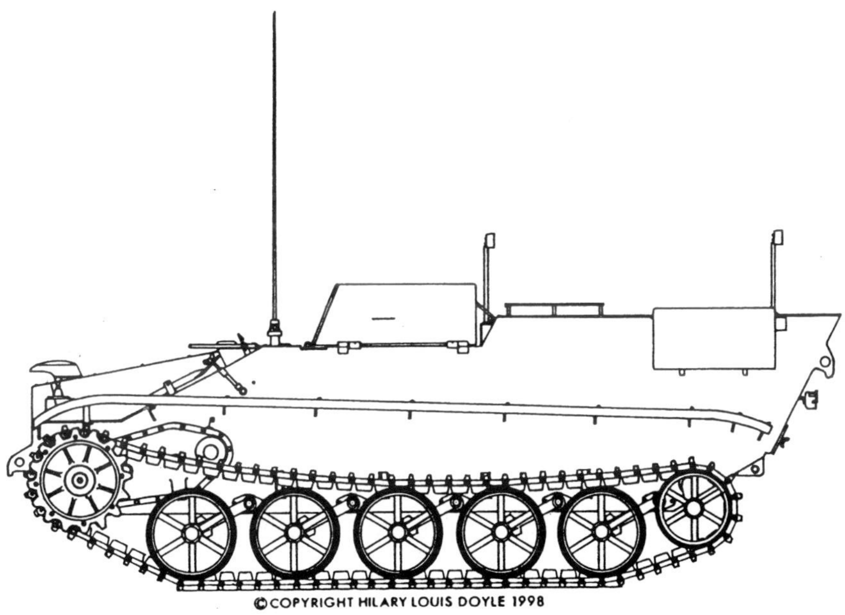

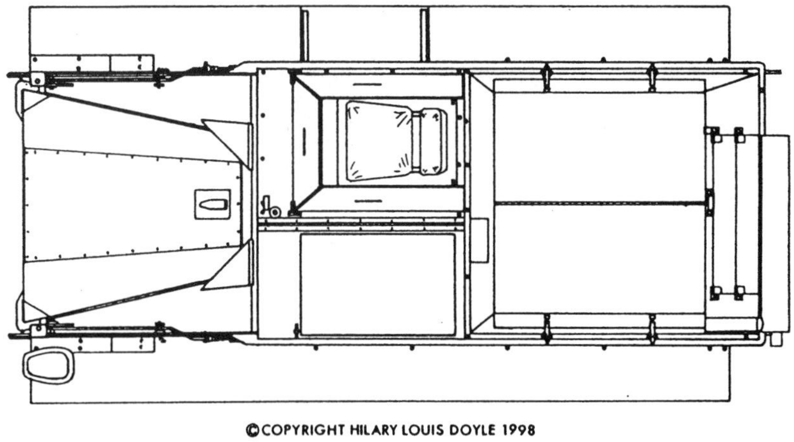

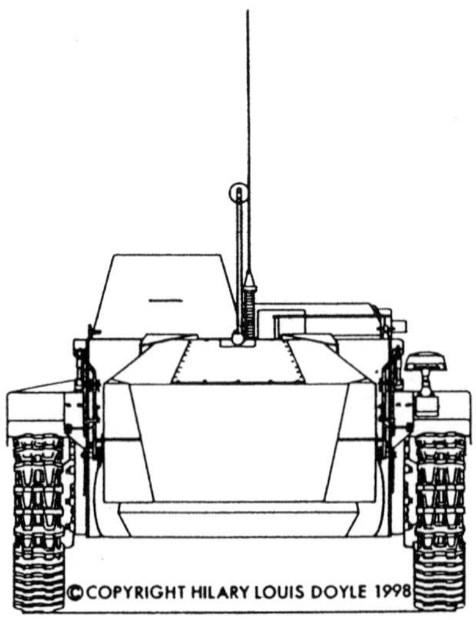

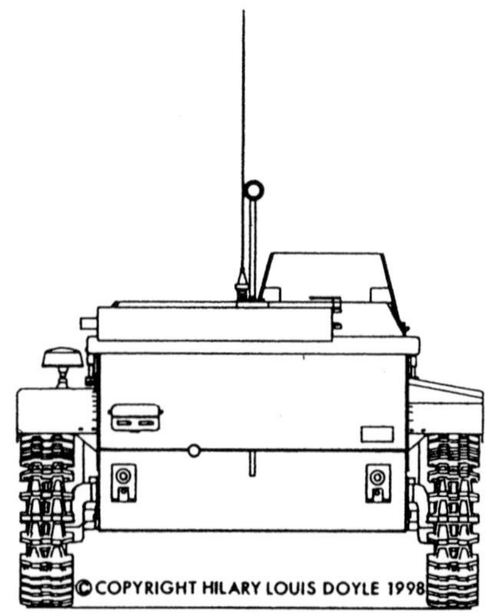

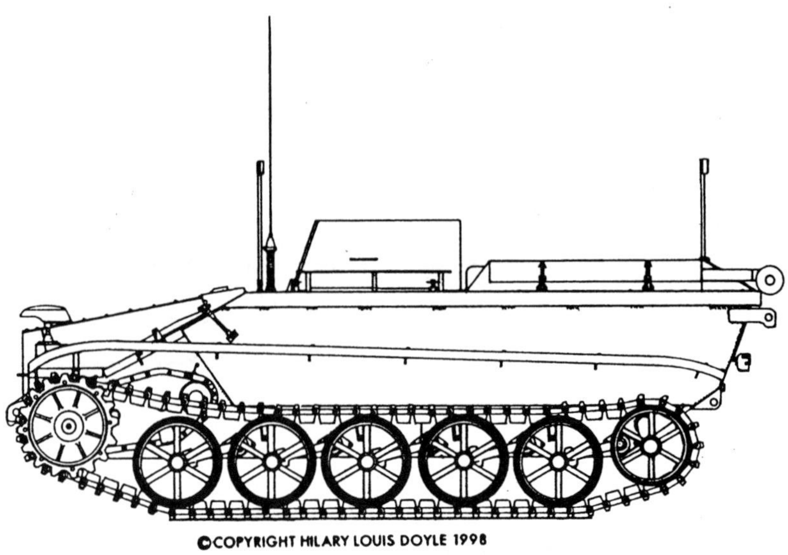

Borgward BIV Demolition-Charge Carrier *(Sd.Kfz. 301)*, *Ausführung B* (All scale drawings copyright 1998 by Hilary Louis Doyle and used with permission.)

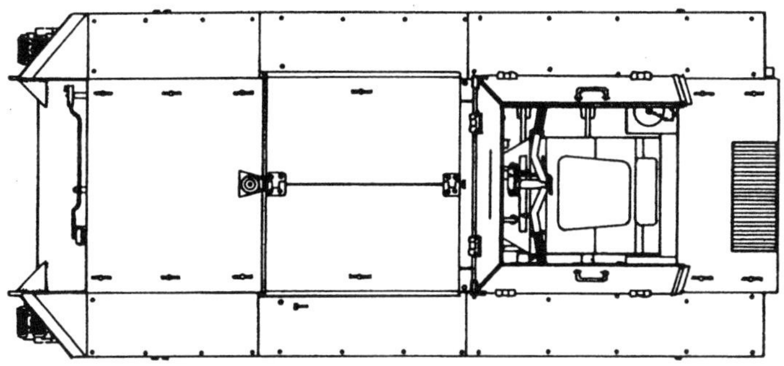

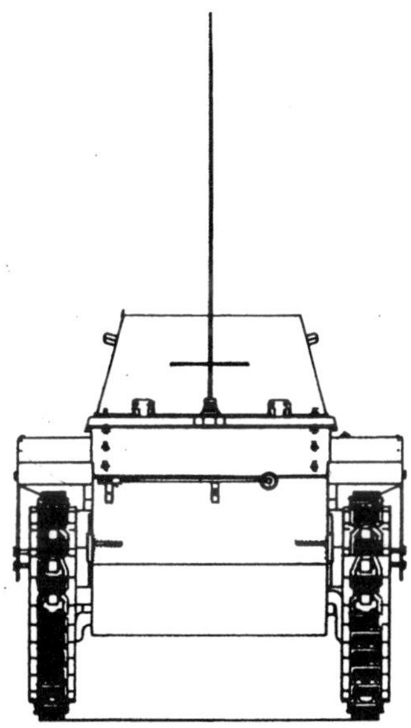

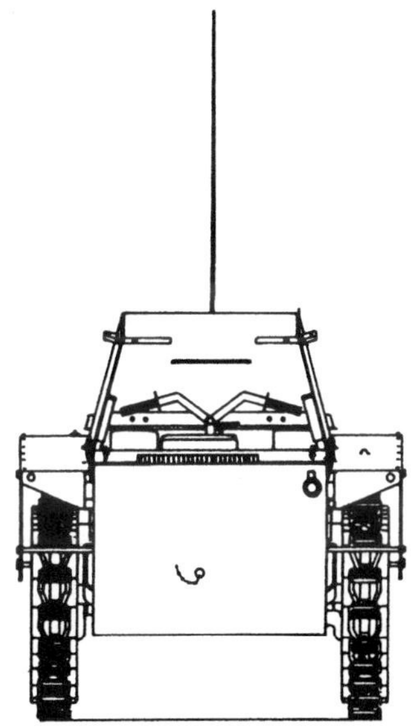

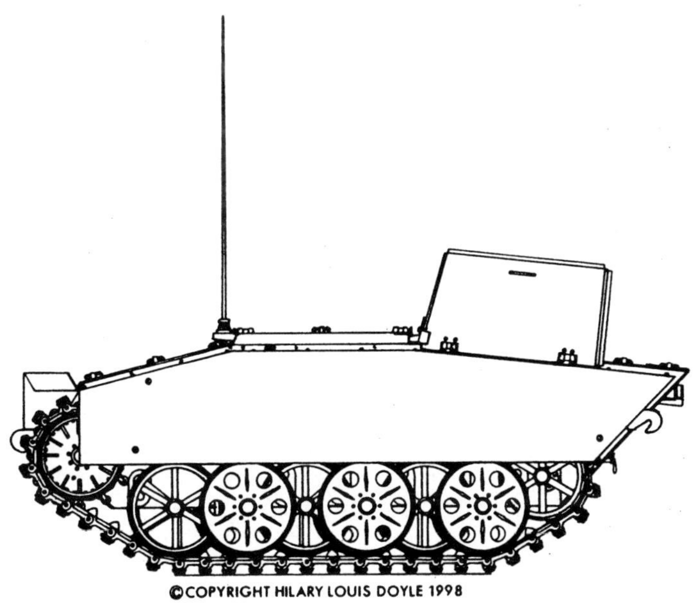

NSU "Springer" (*Sd.Kfz. 304*) DemolitionCharge Carrier (All scale drawings copyright 1998 by Hilary Louis Doyle and used with permission.)

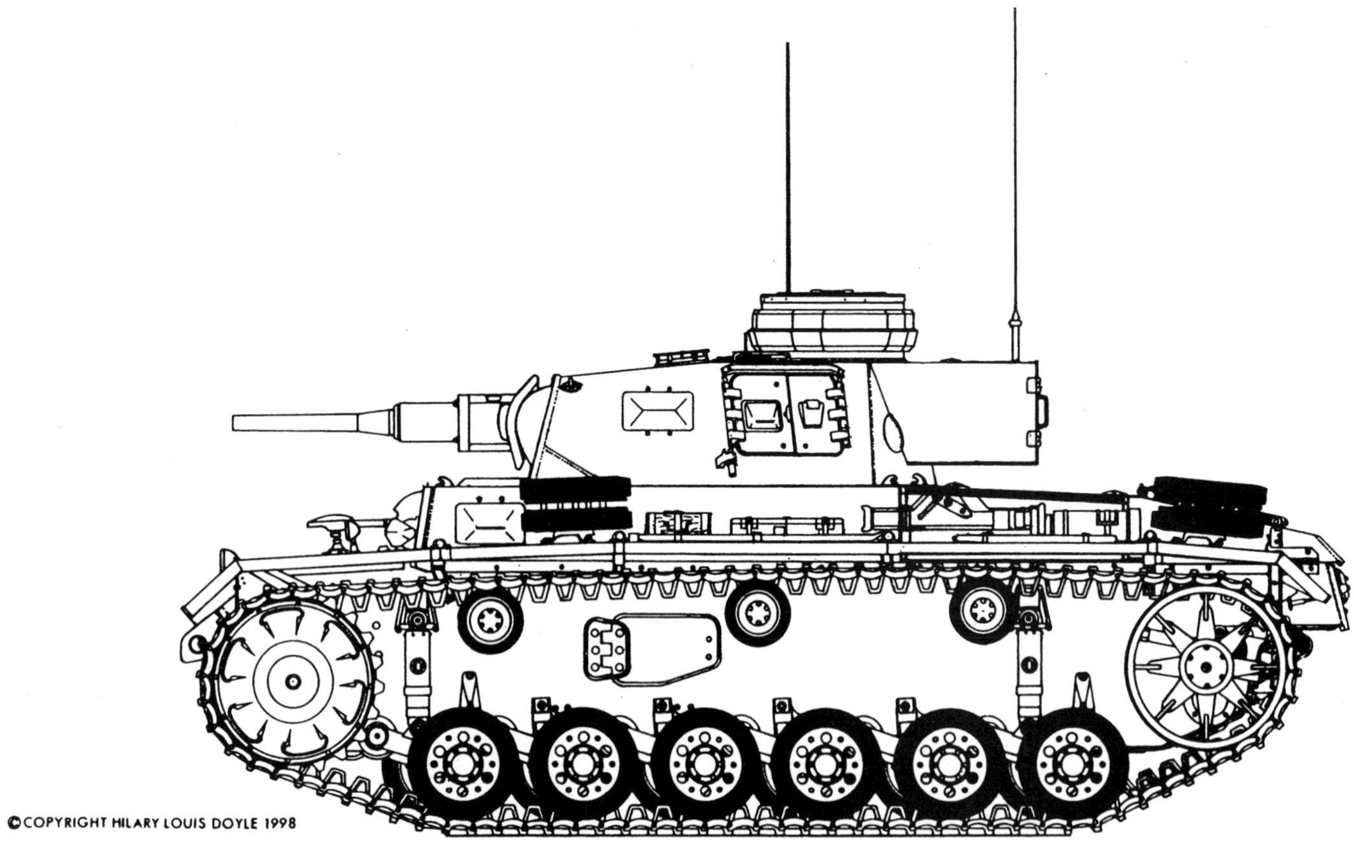

Panzerkampfwagen III, Ausführung J (top) and *Sturmgeschütz III*, Ausführung G (bottom) command and control vehicles. (All scale drawings copyright 1998 by Hilary Louis Doyle and used with permission.)

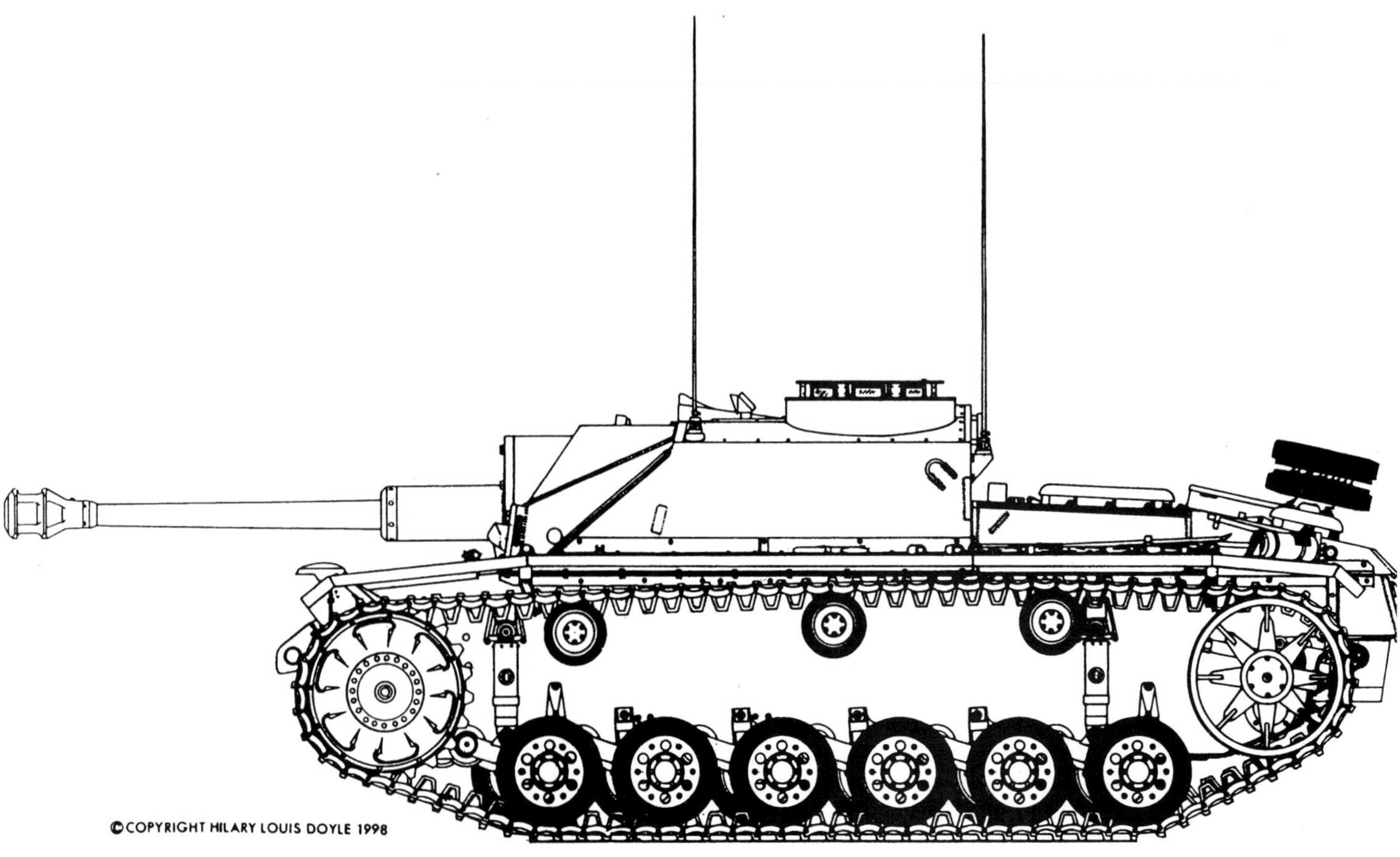

Chapter 2:
The Organization and Operational Use of the Radio-Controlled Units

2.1 Tables of Organization

The personnel strength plus the vehicle and equipment complement of a unit of the *Wehrmacht* (army, navy, air force) was contained and explained in a table of organization known as a *Kriegsstärkenachweisung* (*KStN*).

It was divided into columns with the headings Personnel (officers, civilian officials, noncommissioned officers, enlisted men), Weapons (rifles, pistols, machine guns, artillery pieces and mortars) and Vehicles (motor vehicles, prime movers, armored vehicles and motorcycles).

The unit was in turn broken down into its constituent units (for example company headquarters squad, platoon, squad, section, etc). Also listed were the ranks of personnel in the sub-units.

While the weapons column listed general classifications (rifle, pistol, etc), the vehicles column was more detailed, usually specifying the *Sd.Kfz.* or vehicle number.

Reference to a table of organization made it possible to accurately plan personnel and equipment requirements.

Every table of organization was assigned its own number. It also contained the date on which the table became effective. Over the years the tables of organization were constantly revised to meet existing requirements and changes.

The following table of organization was used for *Minenräum-Abteilung 1*, the first remote-control unit:

Stab/Minenräum-Abteilung 1 (*KStN 1107 a*): Headquarters of a mine-clearing battalion, dated 1 February 1941

1. and *2./Minenräum-Abteilung* 1 (*KStN 1159*): Mine-clearing company, dated 1 February 1941

When *Panzer-Abteilung 300* was formed, the tables of organization above were replaced as follows:

Stab/Panzer-Abteilung 300 (*KStN 1157*): Headquarters of a armor battalion (remote-control), dated 5 January 1942
1. and *2./Panzer-Abteilung 300* (*KStN 1159*): Light tank company (remote-control), dated 5 January 1942
3./Panzer-Abteilung 300 (*KStN 1160*): Medium tank company (remote-control), dated 5 January 1942

The reorganization of the remote-control/radio-control units in the winter of 1942-43 resulted in the following revisions. The following were used for an armor battalion (radio-control):

Stab/Panzer-Abteilung 301 (Funklenk) (*KStN 1107 f*): Headquarters of an armor battalion (radio-control), dated 1 February 1943
Stabs-Kompanie/Panzer-Abteilung 301 (Funklenk) (*KStN 1150 f*): Headquarters company of an armor battalion (radio-control), dated 1 February 1943
1. to *3./Panzer-Abteilung 301 (Funklenk)* (*KStN 1171 f*): Light tank company (radio control), dated 1 February 1943
Werkstattzug/Panzer-Abteilung 301 (Funklenk) (*KStN 1185 f*): Maintenance platoon (radio control), dated 1 April 1943

As a result of the formation of separate radio-control companies, *KStN 1171 f* was revised in early 1943, resulting in a *KStN 1171 f* Version A and a *KStN 1171 f* Version B. Version A was used by light radio-control tank companies equipped with *Sturmgeschütz III* command and control vehicles (*Sd.Kfz. 142/1*); Version B by radio-control tank companies equipped with *Panzer III* command and control tanks (Sd.Kfz. 141).

During 1943 an interim version of *KStN 1171 f* appeared which reduced authorizations. In addition to some personnel reductions, the number of one-tone prime movers (*Sd.Kfz. 10*), three-ton prime movers (*Sd.Kfz. 11*) and three-ton trucks was reduced.

A further revision of the radio-control table of organization took place on 1 June 1944. As of that date the following directives applied to *Panzer-Abteilungen (Funklenk) 301* and *302*:

Stab (KStN 1107 f): Headquarters of an armor battalion (radio control), dated 1 June 1944
Stabskompanie (KStN 1150 f): Headquarters company of an armor battalion (radio control), dated 1 June 1944
1. to *3./4. Kompanien (KStN 1171 f)*: Light tank company (radio control), dated 1 June 1944
Werkstattzug (KStN 1185 f): Maintenance platoon (radio control), dated 1 June 1944

KStN 1171 f — the light tank company (radio control), date 1/6/1944 — also applied to the separate radio-control companies. Comparing *KStN 1171 f* of 1 February 1943 with that of 1 June 1944 revealed a personnel reduction of about 25%, from 195 personnel to 137.

The final version of the table of organization for the radio-control battalion was issued on 1 October 1944. The most obvious change was another reduction in personnel and equipment strengths. On that date, *KStN 1171 f (f.G.)* (*f.G.* = *freie Gliederung* = free organization) for light tank companies (radio control) took effect. Personnel strength was 79 men.

The personnel and equipment of the maintenance sections and the combat and baggage trains of the three radio-control tank companies were combined into one supply company. This received the designation *KStN 1151 e (f.G.)* — supply company of a light armor battalion (radio control), dated

1 October 1944.

The third companies of *schwere Panzer-Abteilungen 504* and *508* were established based on *KStN 1176 f* — heavy tank company *"Tiger"* (radio control) of 1 February 1944. In contrast to *KStN 1171 f* (light radio-control tank company), *KStN 1176 f* had three platoons.

KStN 1176 f of 1 February 1944 applied both to units equipped with the *Tiger I (Sd.Kfz. 181)* and the *Tiger II (Sd.Kfz. 182)* and thus formed the basis for the *1. schwere Panzer-Kompanie (Funklenk)* of *Panzer-Lehr-Regiment 130* (*Panzer-Lehr-Division*).

The following directives applied to *schwere Panzer-Abteilung ("Tiger"/Funklenk) 301*, which was formed later:

Stab (*KStN 1107 f (f.G.)*): Headquarters and headquarters company of a heavy radio-control armor battalion (free organization), dated 1 August 1944

1. to *3. Kompanien* (*KStN 1176 f (f.G.)*): Heavy radio-control tank company (free organization), dated 1 August 1944.

Versorgungs-Kompanie (*KStN 1151 f (f.G.)*): Supply company of a heavy radio-control tank battalion (free organization), dated 1 August 1944.

Werkstatt-Kompanie (*KStN 1187 f*): Maintenance company of a heavy armor battalion, dated 1 November 1944.

In contrast to *KStN 1176 f* of 1 February 1944, *KStN 1176 f (f.G.)* — heavy radio-control tank company (free organization) of 1 August 1944 — again reduced the number of platoons to two.

In addition to the tables of organization listed above, there were three more which applied to remote-control or radio-control units:

KStN 1161: *Panzer-Feldersatz-Kompanie (Fernlenk)* (Armor replacement training battalion, remote control), dated 8 May 1942

KStN 101159: *Panzer-Versuchs- und Ausbildungs-Kompanie (Fernlenk)* (Armor experimental and training company, remote control), dated 8 May 1942

KStN: *Versuchs-Kommando (Fernlenk) Tropen* (Experimental detachment, remote control, tropical), dated 24 June 1942.

Kriegsstärkenachweisung (KStN) 1159 - Minenräumkompanie vom 01.02.1941

Kompanietrupp

1. Zug

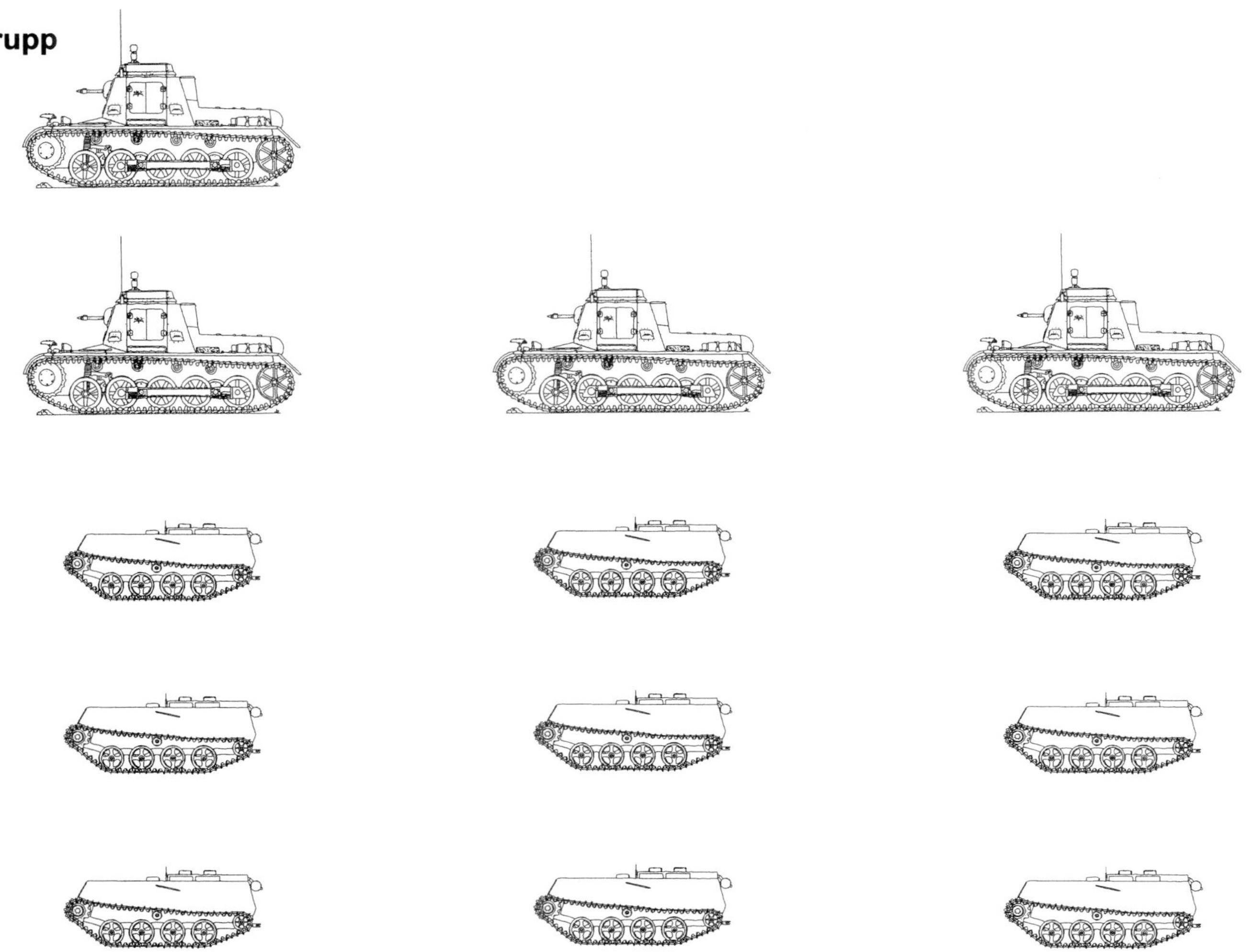

M.Jaugitz / Trojca 2000.

Kriegsstärkenachweisung (KStN) 1159 - Minenräumkompanie vom 01.02.1941

2. Zug

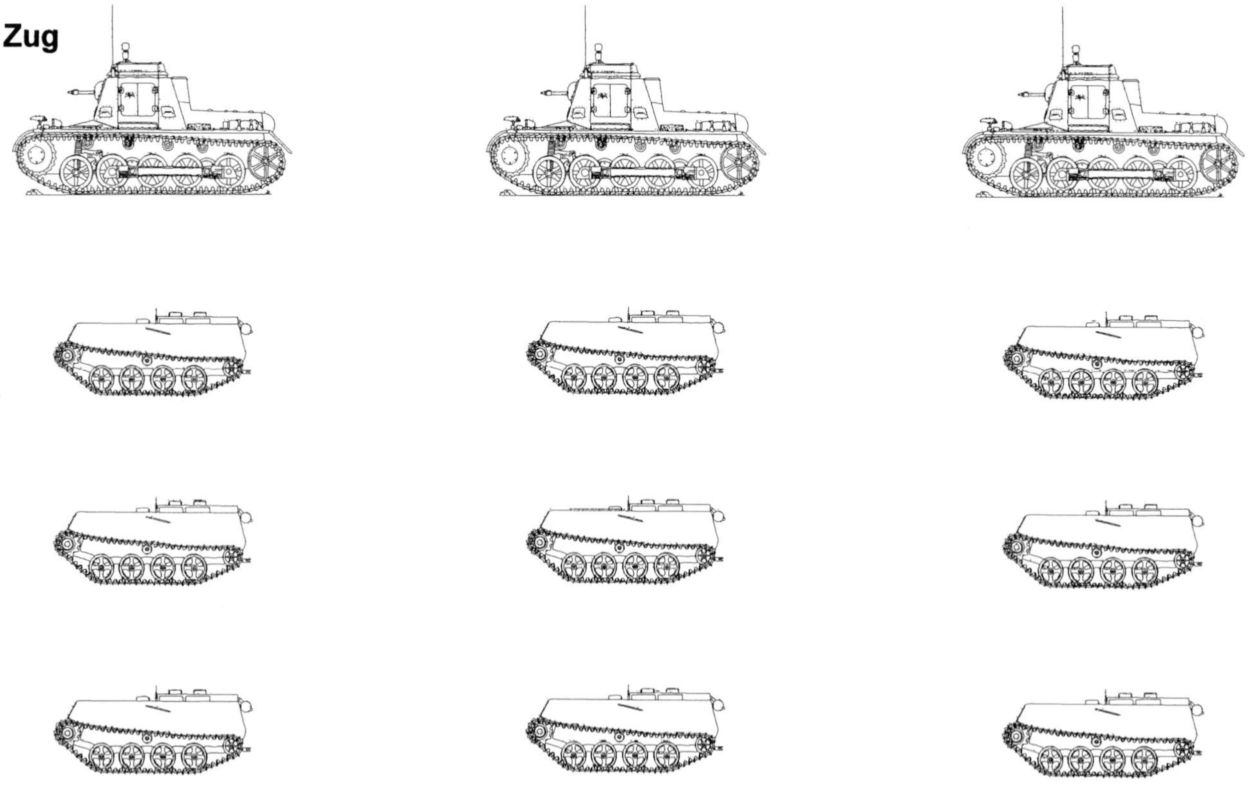

3. Zug

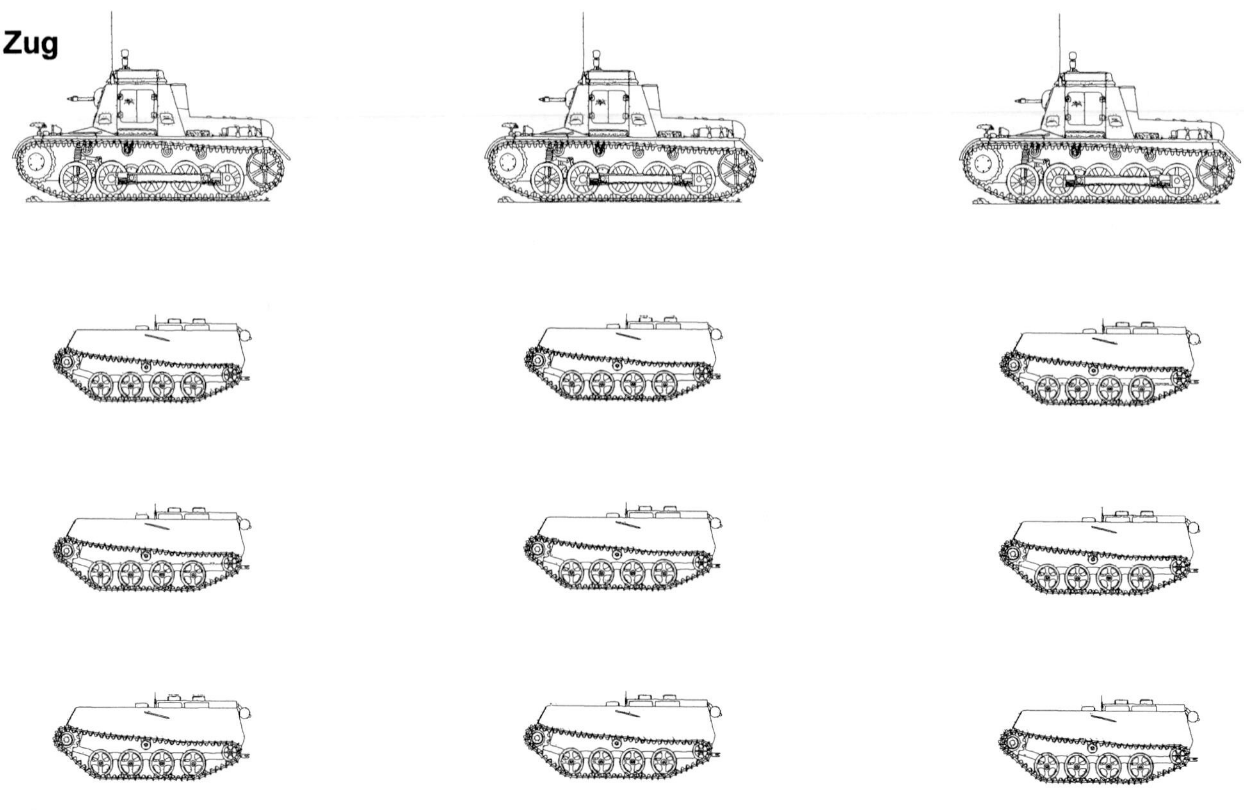

M.Jaugitz / Trojca 2000.

Kriegsstärkenachweisung (KStN) 1159 - leichte Panzer - Kompanie (FL) vom 05.01.1942

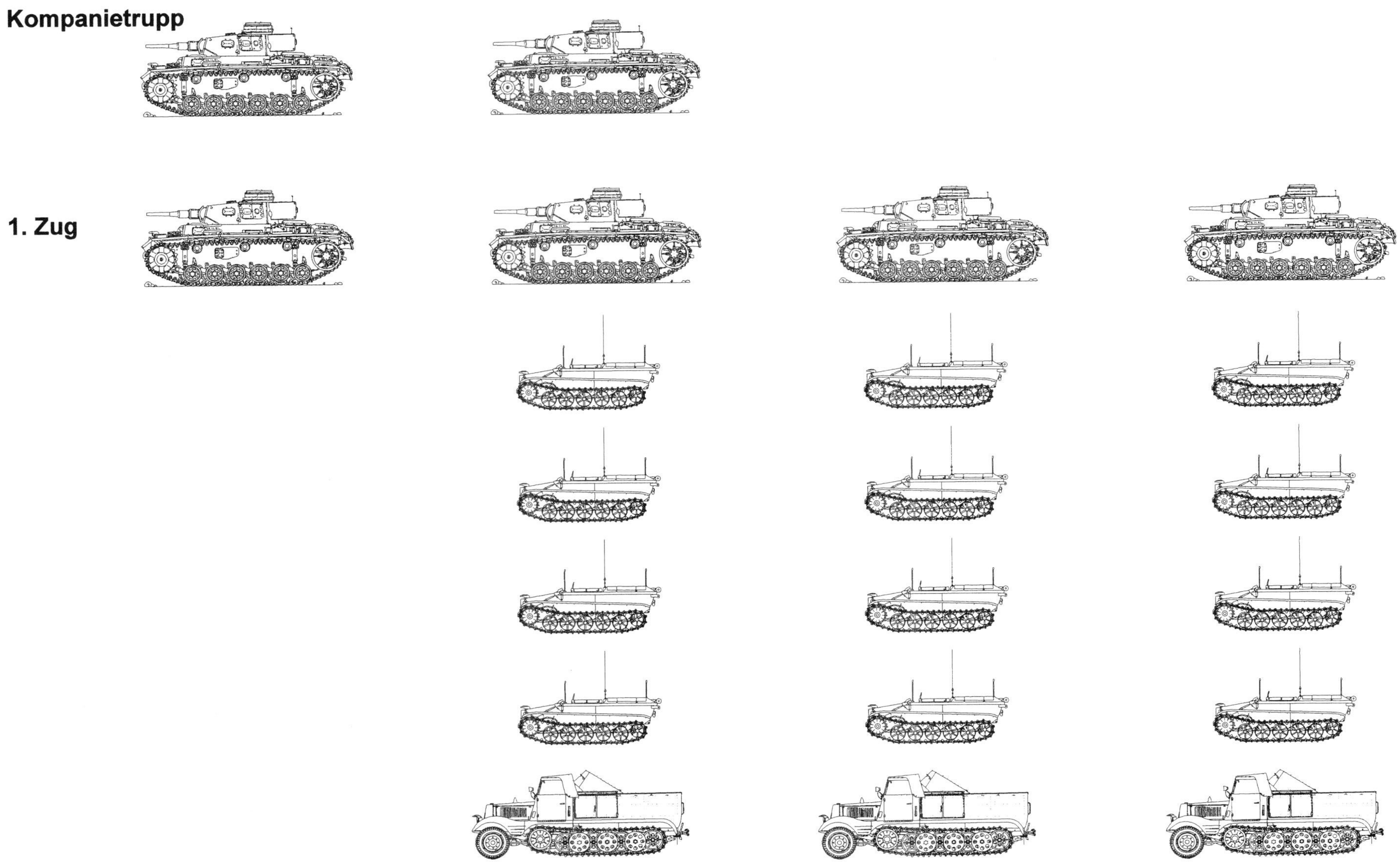

M.Jaugitz / 2000.

Kriegsstärkenachweisung (KStN) 1159 - leichte Panzer - Kompanie (FL) vom 05.01.1942

2. Zug

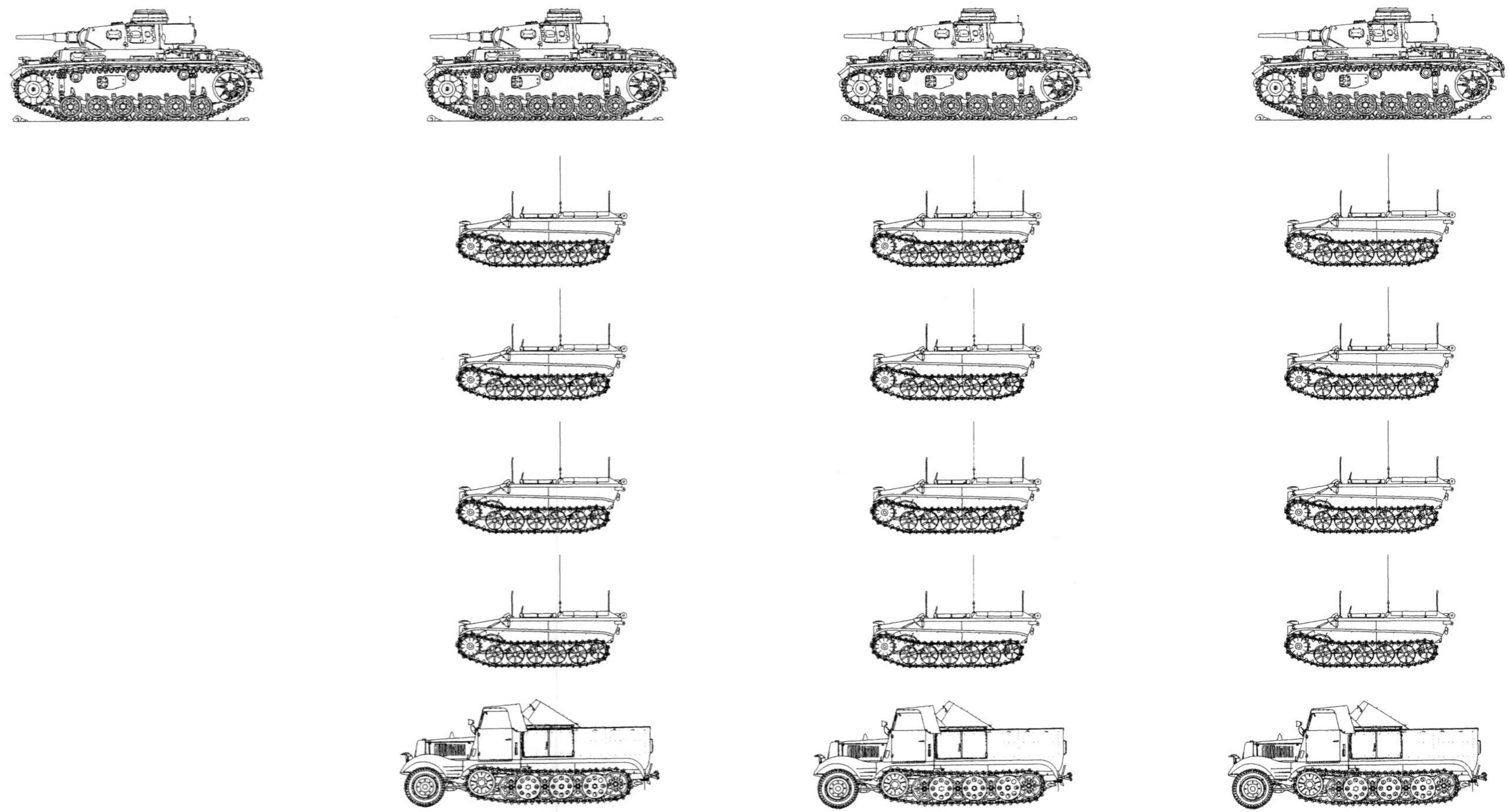

M.Jaugitz / Trojca 2000.

Kriegsstärkenachweisung (KStN) 1159 - leichte Panzer - Kompanie (FL) vom 05.01.1942

Sondergeräte - Reserve

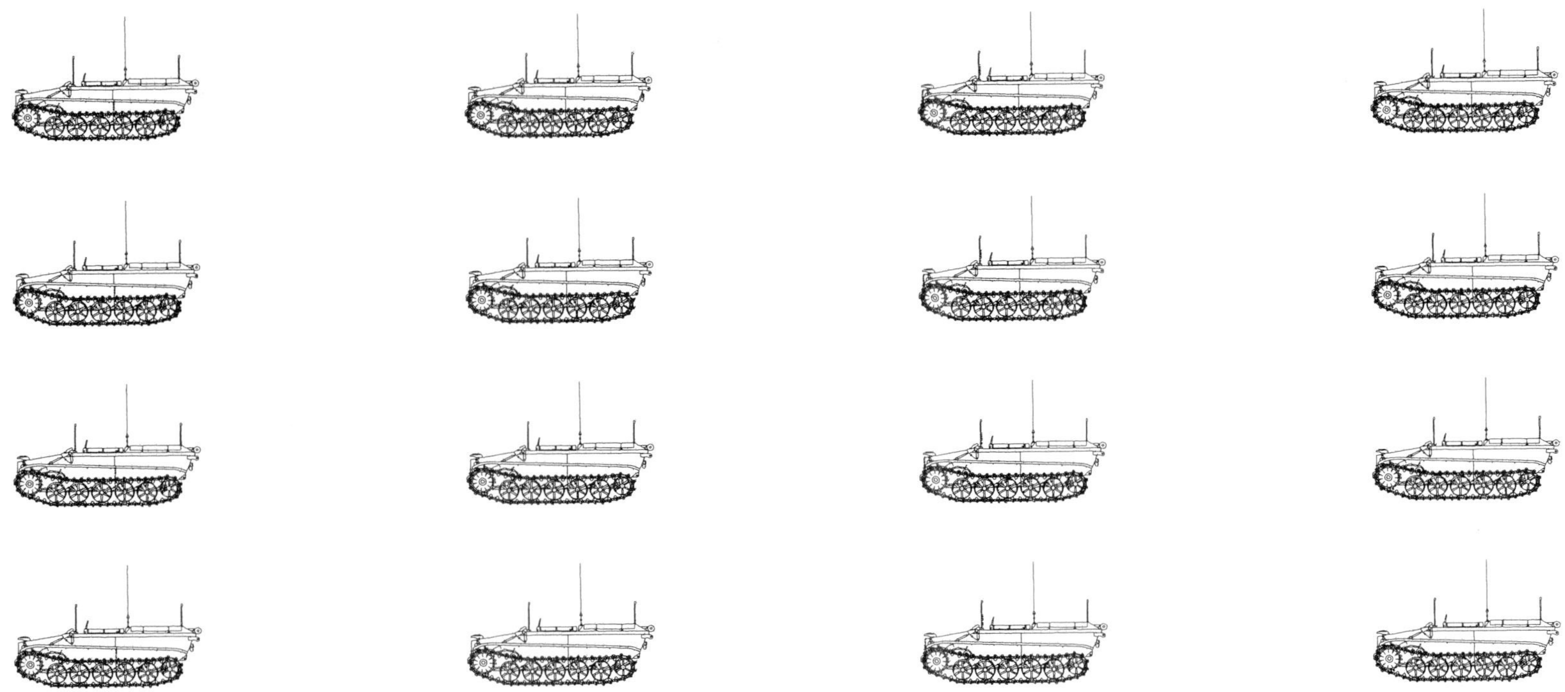

M.Jaugitz / Trojca 2000.

Kriegsstärkenachweisung (KStN) 1171 f (f.G.) - leichte Panzer - Kompanie f vom 01.10.1944

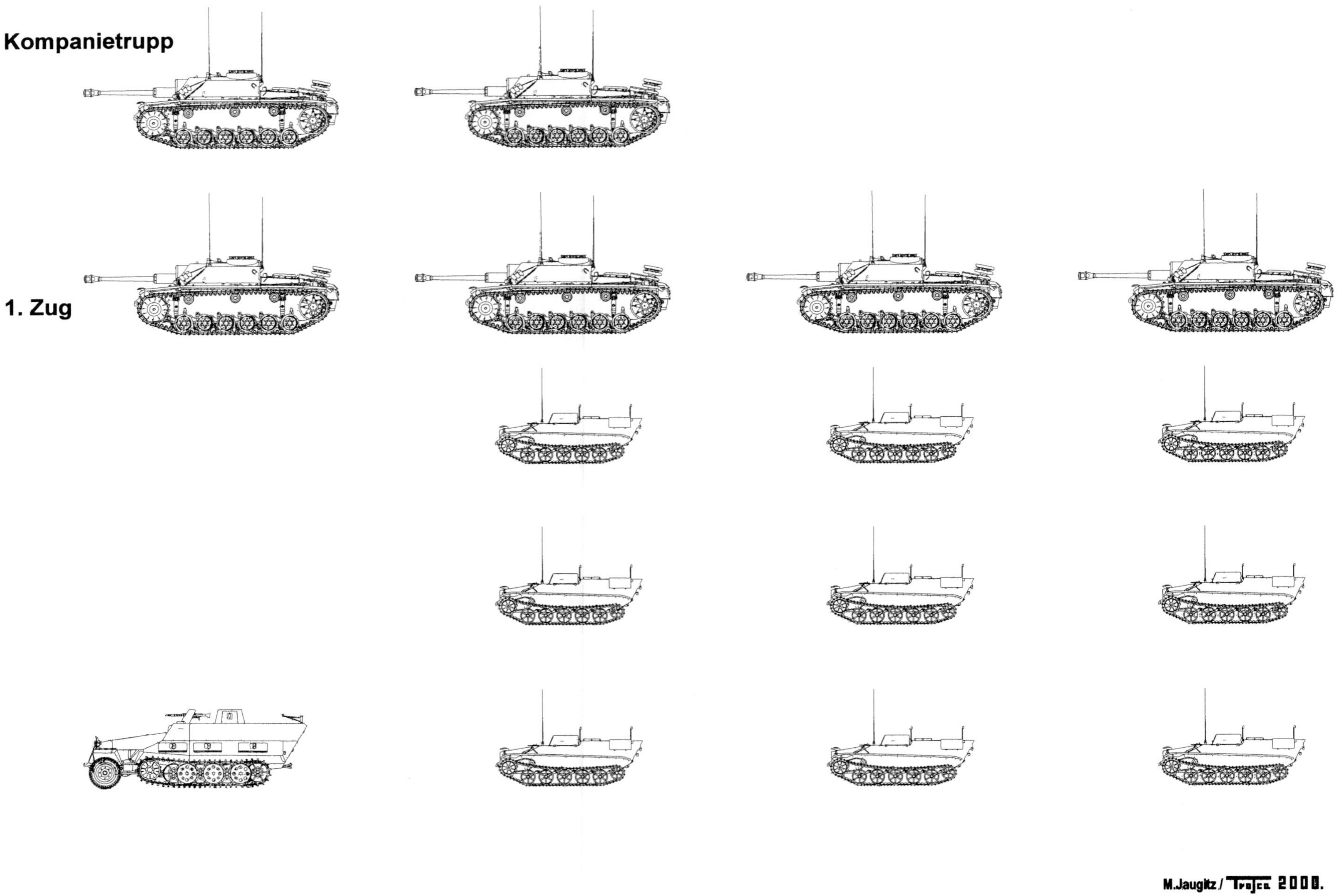

Kriegsstärkenachweisung (KStN) 1171 f (f.G.) - leichte Panzer - Kompanie f vom 01.10.1944

2. Zug

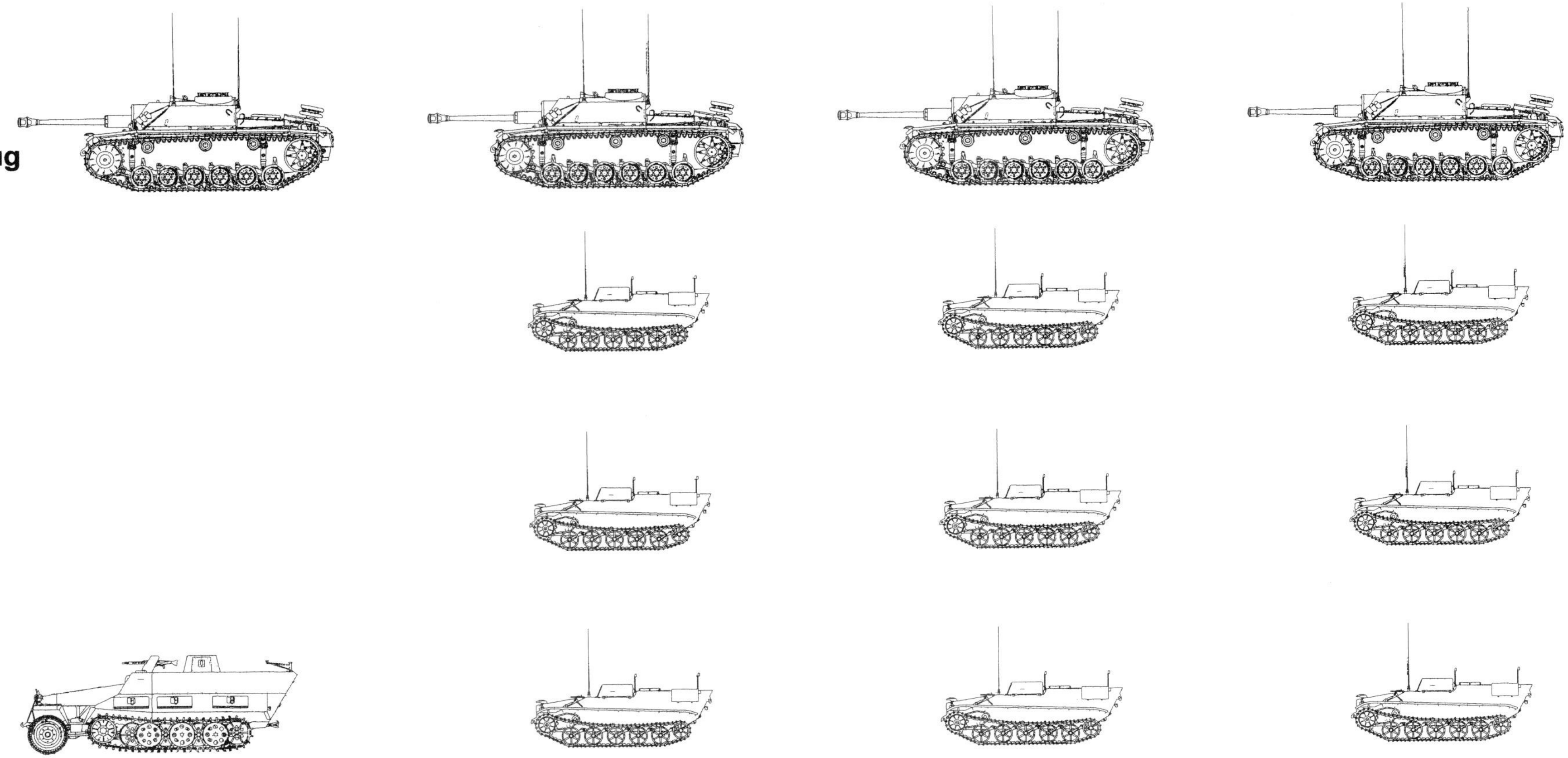

M.Jaugitz / Trojca 2000.

Kriegsstärkenachweisung (KStN) 1176 f - schwere Panzer - Kompanie "Tiger" (Fkl) vom 01.02.1944

Kompanietrupp

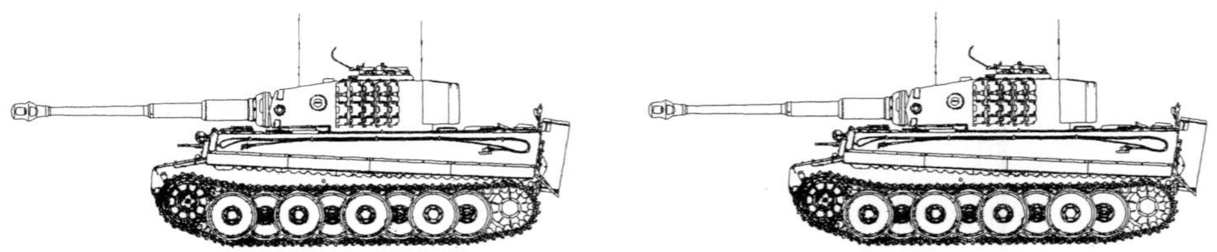

1. Zug

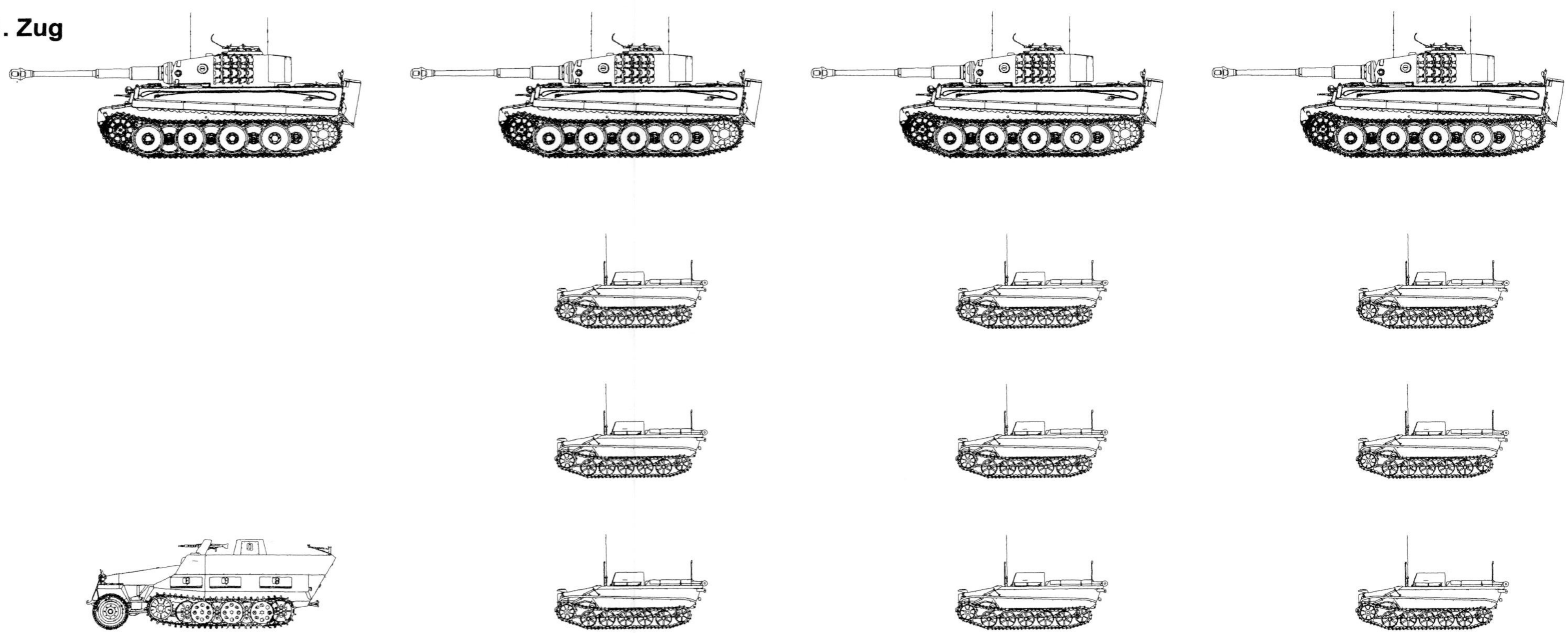

M.Jaugitz / 2000.

Kriegsstärkenachweisung (KStN) 1176 f - schwere Panzer - Kompanie "Tiger" (Fkl) vom 01.02.1944

2. Zug

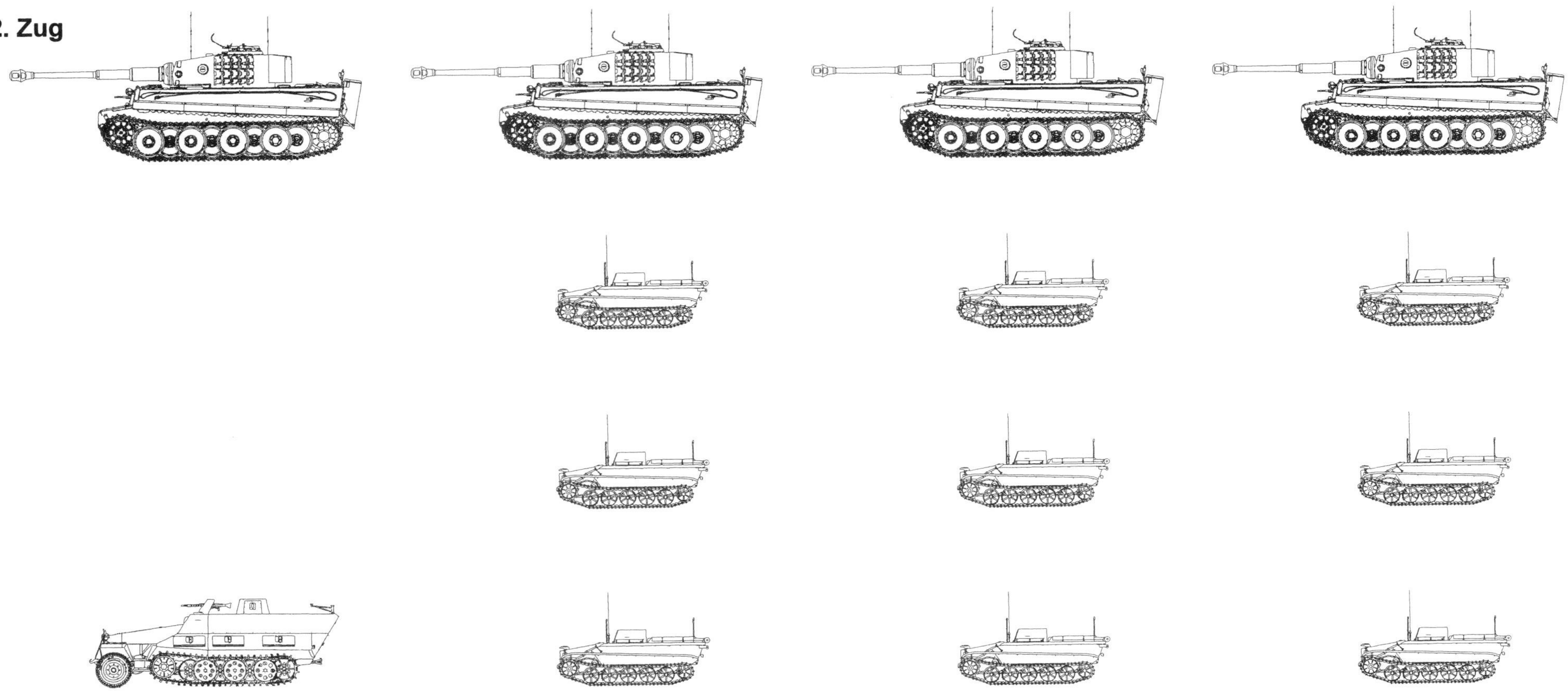

M.Jaugitz / TraJen 2000.

Kriegsstärkenachweisung (KStN) 1176 f - schwere Panzer - Kompanie "Tiger" (Fkl) vom 01.02.1944

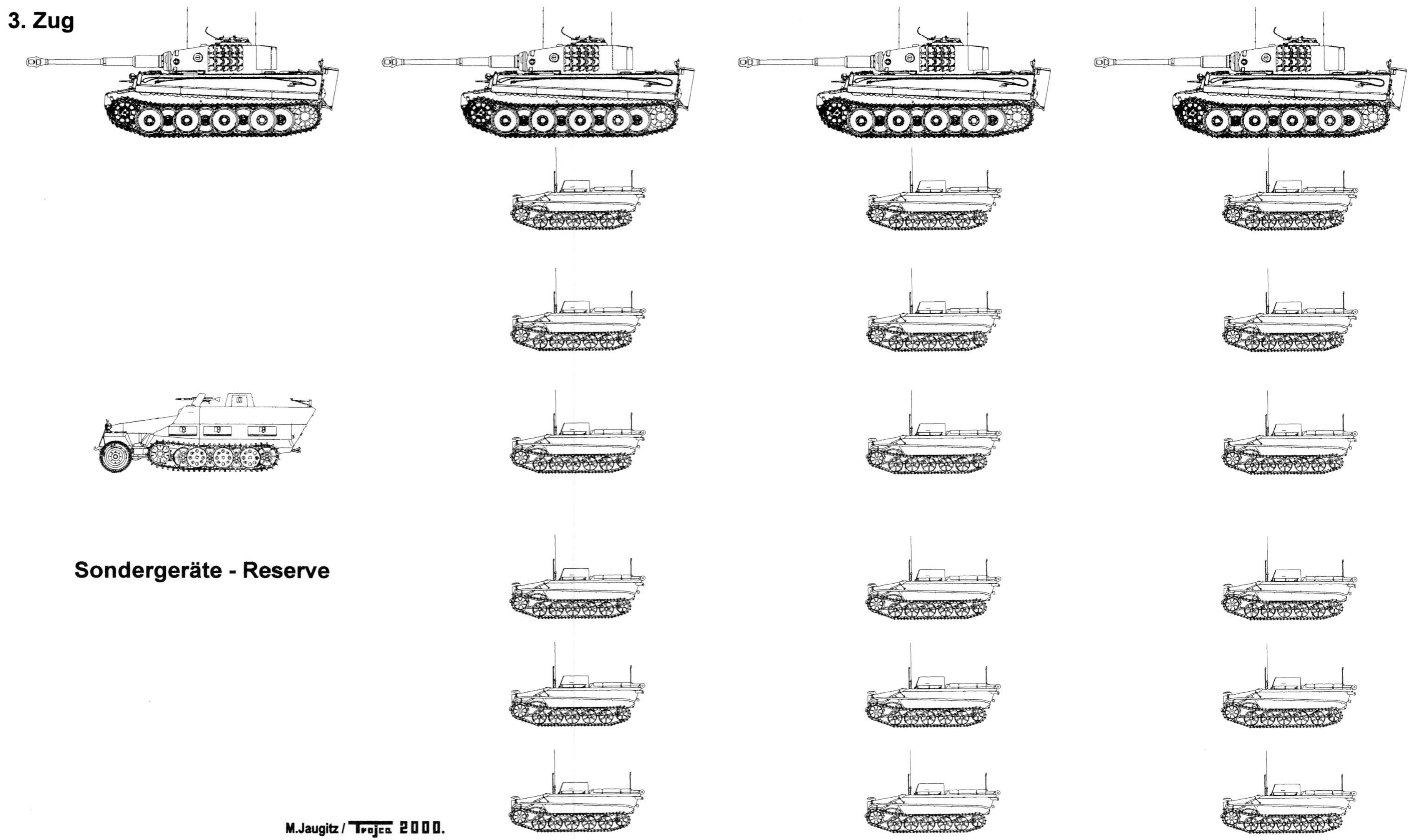

2.2 The Combat Units

2.2.1 Kompanie Glienecke

On 1 June 1940 a directive was issued instructing the *Panzertruppenschule Wünsdorf* to form a mine-clearing company. Assigned the Army Field Post Number 17031, it was a special formation of the *Wehrmacht* High Command and received the cover designation *"Kompanie Glienecke"*.

Interested applicants from various service branches were assembled at Groß-Glienecke under conditions of extreme secrecy. In July 1940, *Major* Klaus Müller assumed command of the company.

After being issued special vehicles and receiving brief training on them, the mine-clearing company was transferred from Berlin to Völklingen on the Saar at the end of July. The company was divided, with one part remaining in Völklingen and the other half being quartered in Blieskastel.

Beginning in late summer 1940 the unit carried out is first mine-clearing operations in the "Vosges Fortress Sector" in northern Alsace, using *Borgward BI* mine-clearing vehicles with towed mine rollers.

It was assumed that the enemy anti-personnel mines would be detonated either by the tracks of the mine-clearing vehicles running over them or by being "plowed up" by the towed mine rollers. It was expected that this method would result in only minimal damage to the vehicles. It was soon shown, however, that driving over a larger mine, such as an antitank mine, caused serious damage to the running gear and concrete superstructure, usually resulting in the vehicle being totally destroyed.

Since, as a rule, mines — no matter what the type — were usually laid in large numbers, it was hoped to exploit its detonation. Subsequently, large quantities of explosives were placed in the interior of the vehicle. When a mine was hit and detonated, it then set off the explosives in the mine-clearing vehicle. The size of the charge inevitably produced a shock wave which ignited and destroyed all other mines in the immediate vicinity of the vehicle.

The battalion's demolition experts were then requested to come up with a way to detonate an explosive charge in a mine-clearing vehicle by remote control. This idea could not be realized immediately, however, and the troops continued to improvise. On several of the newly-received Borgward BII mine-clearing vehicles a hole was knocked in the concrete superstructure on the left side of the vehicle above the drive sprocket and a metal rail was installed parallel to the vehicle front. If the vehicle drove into an obstacle, the rail was pushed back and set off the charge's detonating mechanism.

In September 1940 the sub-units were assembled at a new base at Weißenburg in Alsace. At that point in time the mine-clearing company's total strength was still about 400 men, which was far in excess of the authorized strength for a mine-clearing company.

After departing from Groß-Glienecke, *Kompanie Glienecke* arrived at Völklingen in the Saar at the end of July 1940. (Hand)

Büssing-NAG 500 S trucks were used to move the *Borgward BI* mine-clearing vehicles. (Hand)

Portable ramps continued to be part of the equipment of the company. (Hand)

The company was billeted in a former garrison of the *Reicharbeitsdienst* from August to September 1940. (Hand)

The vehicles remained in the open. The *kleine Befehlspanzer* in the foreground have not yet been equipped with the 360-degree optics. (Hand)

The mine-clearing vehicles remain on the trucks, covered with tarpaulins. (Hand)

Above: The mine-clearing vehicles were transported to the engagement area in the trucks. (Hand)

Top right: After arrival, the mine-clearing vehicles had to be off-loaded from the trucks using the portable ramps. (Hand)

Right: The numeral "30" has been painted on this mine-clearing vehicle. To help pin-point the vehicle in operation, a flag has been mounted on it. Note the relatively small size of the vehicle and its rollers. (Hand)

The *kleiner Befehlspanzer* and the *Minenräumwagen BI* are shown in a demonstration. (Hand)

Innumerable obstacles had been constructed as part of the Maginot Line. Here we see concrete barriers. (Hand)

"Dragon's teeth" were intended to impede the advance of armor. (Hand)

A mine-clearing vehicle loaded with explosives is guided to an obstacle and detonated. (Hand)

Top left: A gigantic cloud of smoke rises over the point of detonation. (Hand)

Above: After the smoke has cleared, the effects are visible. As a result of the successful detonation, a gap has been formed in the obstacle through which vehicles can move. (Hand)

Left: Within the framework of this exercise, a mine-clearing vehicle drove on to a mine. The track has been ripped apart and portions of the concrete superstructure have been destroyed. (Vehicle 360024) (Hand)

Several *Minenräumwagen* were destroyed in the course of the testing carried out in this sector of the Maginot Line. (Hand)

The remnants are collected and evacuated. The rest of the vehicle hull and running gear are pulled on to a truck. (Hand)

This *Sd.Kfz. 265 kleiner Befehlspanzer* has been used as a recovery vehicle for this operation. (Hand)

This *Borgward BI* has also been severely damaged. The forward-portion of the concrete superstructure has been ripped off. (Hand)

Top: The evacuation of other armored vehicles was also practiced. In this instance, a *Sd.Kfz. 265* tows another.

Opposite page top: The tow cable is attached. (Hand)

Opposite page bottom: Using a *Sd.Kfz. 265* as the "winch," the remainder of a mine-clearing vehicle is pulled on to the truck by means of a pulley. (Hand)

Right: This *kleiner Panzerbefehlswagen* is attempting to recover a truck which ran off the road in foggy weather.

Gliederung der "Kompanie Glienecke" (Stand: September 1940)

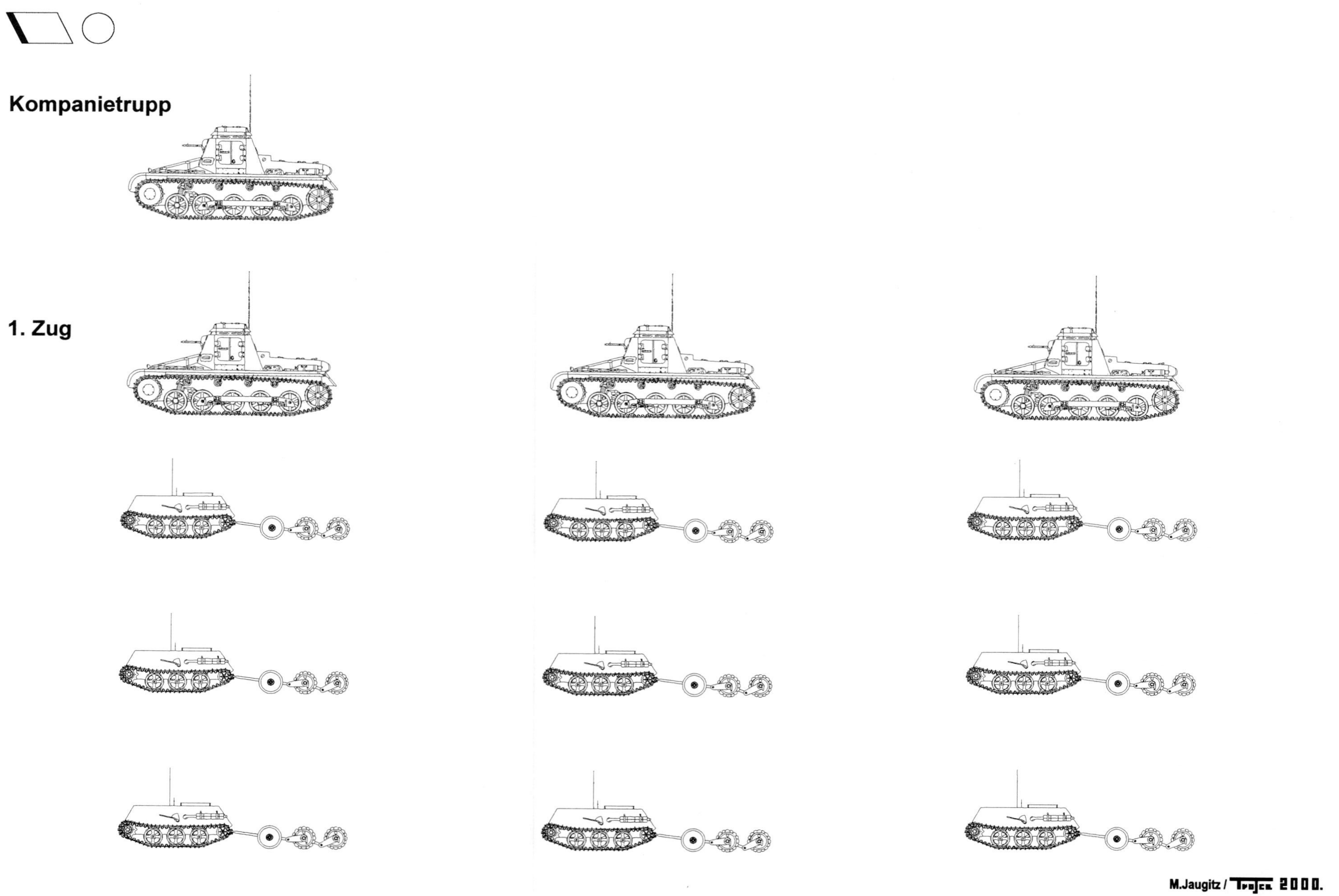

Gliederung der "Kompanie Glienecke" (Stand: September 1940)

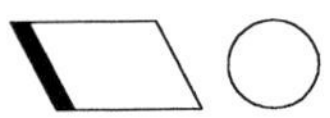

2. Zug

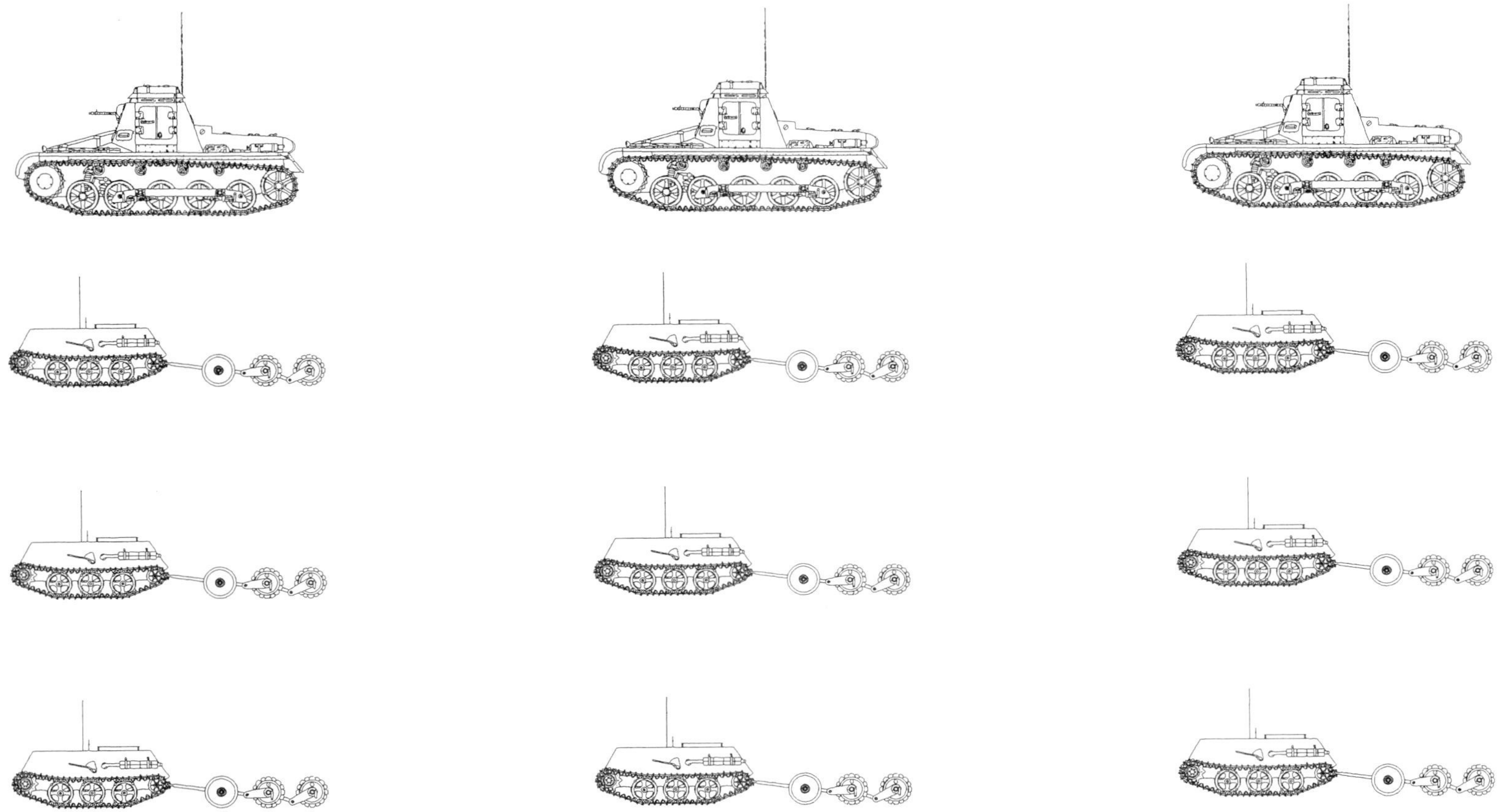

M.Jaugitz / Trojca 2000.

Gliederung der "Kompanie Glienecke" (Stand: September 1940)

3. Zug

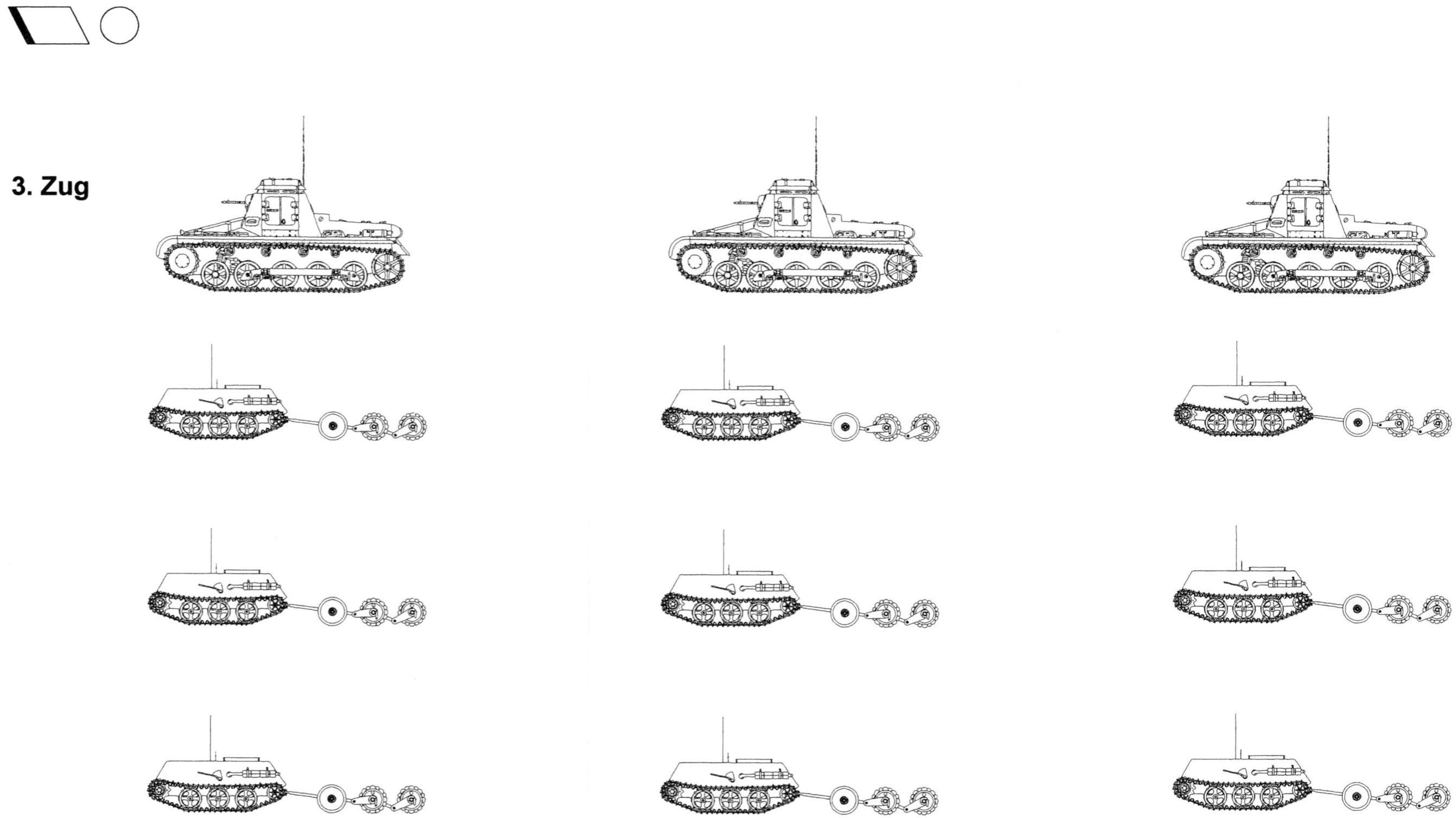

M.Jaugitz / Tractor 2000.

2.2.2 Minenräum-Abteilung 1

Beginning on 1 December 1940 at the Weißenburg base, *Minenräum-Kompanie 1* was reorganized as a battalion. The new unit was designated *Minenräum-Abteilung 1* and consisted of a headquarters with a signals and vehicle repair section (Army Field Post Number 17031) plus the *1. Minenräum-Kompanie* (Army Field Post Number 08914) and the *2. Minenräum-Kompanie* (Army Field Post Number 10089).

The battalion commander was *Major* Müller, who had been nicknamed *"Walzen-Müller"* on account of the mine-clearing vehicles with mine rollers (*Räumwalzen*) used by his unit.

The unit's personnel and equipment strengths were established when Tables of Organization 1107 (headquarters/mine-clearing battalion) and 1159 (mine-clearing company) dated 1 February 1941 took effect (see Chapter 2.1).

The following is the unit's list of officer positions in spring 1941:

Battalion Headquarters

Battalion commander: *Dipl.Ing. Major* Müller
Adjutant: *Leutnant* Senne
Headquarters company commander: *Oberleutnant* Strack
Battalion surgeon: *Dr.* Wirth

1./Minenräum-Abteilung 1: *Oberleutnant* Blau
Platoon leader: *Leutnant* Dettmann
Platoon leader: *Leutnant Graf* von der Recke
Platoon leader: *Leutnant* Peitsch

2./ Minenräum-Abteilung 1: *Hauptmann* Weicke
Platoon leader: *Leutnant* Fritschken
Platoon leader: *Leutnant* Fischer
Platoon leader: *Leutnant* Wispereit

The facility at Weißenburg was located in the former French "Hagenau fortress sector", the most heavily fortified part of the Maginot Line. Two large artillery groups, 54 heavy bunkers, 15 large underground positions, 3 heavy artillery observation bunkers plus hundreds of smaller fighting positions had been constructed in the years from 1930 to 1940 in a sector 31 kilometers wide. In front of and between the bunkers were six rows of antitank obstacles, barbed-wire obstacles and minefields, which were intended to prevent an enemy from breaking into or through the position.

The *Borgward BI* and *BII* mine-clearing vehicles were tested extensively in the area of these bunker systems from autumn 1940 until spring 1941. Furthermore, various trials were conducted with the amphibious version of the *Borgward BII*, the *"Ente"*, in a blocked-off tributary of the Rhine.

At the end of May, the unit began moving to the East-Prussian border area where the buildup to Operation *Barbarossa* was proceeding according to plan.

Details of the move are contained in the following order (source: *Bundesarchiv/Militärarchiv* Freiburg):

Secret!

Fortress Headquarters *"Blaurock"* — 11 June 1941
Abteilung I a N3. 3078/41 gKdos — 10 copies
3rd copy

Concerning: Fortress HQ *Blaurock* Operations Section *Nr. 0329/41 gKdos.Chefs.v.10.6.41*

SUBJECT: *Minenräum-Abteilung (mot.) 1*

On 15 and 16 June the wheeled elements of *Minenräum-Abteilung (mot.) 1* are to be moved from their former quartering areas southeast of Ortelsburg ahead of the army artillery of the *VIII. Armee-Korps* via Ortelsburg, Johannisburg, Arys, Lyck, Treuburg, Raczki and Suwalken to the assembly area of the *VIII. Armee-Korps*.

Minenräum-Abteilung will coordinate with the artillery of the *VIII. Armee-Korps* in the rest area east of Strasburg on 14 June.

It is intended that the tracked elements will be sent by rail to Suwalken on 18 June (order for rail movement to follow).

The battalion headquarters and the *2./Minenräum-Abteilung 1* will occupy an assembly area in the area west of Giby (at least 5 kilometers away from Giby) and remain under the control of the army. Depending on the situation, it will be committed with either the *VIII.* or *VII. Armee-Korps (mot.)* depending on how the situation develops.

Location to be assigned by *VIII. Armee-Korps*.

The *1./Minenräum-Abteilung 1* is under the operational control of the *VIII. Armee-Korps*.

An advance party from the battalion is to report to Headquarters, VIII. Armee-Korps in Lyck on 15 June for allocation of assembly areas and receipt of missions for the *1./Minenräum-Abteilung 1*.

For the Operations Staff
Chief of the General Staff
(signature illegible)

At a corps meeting held by the *9. Armee* on 13 June 1941 the enemy situation was again discussed. It was determined that no fortifications had been found except on the Njemen near Olita and that the Soviet forces near Grodno were stronger there than in the remainder of the planned area of operations.

At this meeting it was suggested that *Minenräum-Abteilung 1* be subordinated to *Panzergruppe 3*, commanded by *General* Hoth. *Panzergruppe 3* consisted of the *LVII. Armee-Korps* and *XXXIX. Armee-Korps* as well as group-level units and attached army-level units.

The *LVII. Armee-Korps* consisted of the *12. Panzer-Division*, *19. Panzer-Division* and *18. Infanterie-Division (mot.)*, while the *XXXIX. Armee-Korps* was comprised of the *7. Panzer-Division*, *20. Panzer-Division*, the *14. Infanterie-Division (mot.)* and the *20. Infanterie-Division (mot.)*.

After arriving in the area of *Panzergruppe 3*, the unit was split up before operational commitment. The headquarters and the *2./Minenräum-Abteilung 1* were placed under the operational control of the *XXXIX. Armee-Korps*, while the *1./Minenräum-Abteilung 1* was attached to the *28. leichte*

Division (VIII. Armee-Korps).

Prior to the commencement of operations, the superior headquarters were made aware of the doctrinal principles for the operations of mine-clearing vehicles. (See Chapter 4.2: *"Merkblatt über den Einsatz von Minenräum-Wagen."*) The attack on the Soviet Union began at 0315 hours on 22 June 1941.

From the line of departure at Suvalki, the headquarters and the *2./Minenräum-Abteilung 1* advanced with other units of the *XXXIX. Armee-Korps* northeast in the direction of Kalvaria. The 3rd Platoon of the company took the lead, followed by a grenadier platoon of *Schützen-Brigade 7* and a combat engineer platoon of *Panzer-Pionier-Bataillon 58* of the *7. Panzer-Division*. The mine-clearing vehicles ground through a plowed-up outpost section, however, it was not mined.

As part of the *Kampfgruppe* of *Infanterie-Regiment 83* (*28. leichte Division*), the *1./Minenräum-Abteilung 1* advanced southeast via Sopockinie toward Grodno, which was reached on 24 June. After that operation, the *1./Minenräum-Abteilung 1* left the *VIII. Armee-Korps* and remained for the time being in the Grodno area.

Meanwhile, units of the *XXXIX. Armee-Korps* had reached Vilna on 25 June. Based on the situation, platoons of the *2./Minenräum-Abteilung 1* were placed under the control of the *7. Panzer-Division* and the *20. Panzer-Division*. Krevo was reached on 26 June and Minsk on 29 June. On 2 July the *XXXIX. Armee-Korps* advanced northeast in the direction of Vitebsk, with the headquarters and 2nd Company of *Minenräum-Abteilung 1* initially remaining at the corps' disposal in the Minsk area. This order was soon rescinded, however.

The morning report issued by the *20. Panzer-Division* at 1045 hour on 1 July reveals that the *2./Minenräum-Abteilung 1* had already twice repulsed enemy reconnaissance advances into the sector of *Schützen-Regiment 59* by detonating remotely-controlled mine-clearing vehicles.

On 3 July the *1./Minenräum-Abteilung 1* received orders to the effect that it had been placed under the operational control of the *LVII. Armee-Korps* and was to proceed to Michaliszki.

On 5 July the *2./Minenräum-Abteilung 1* reached Lepel. After it was suspected that the Soviet Army was bolstering its forces to defend the east bank of the Düna, a platoon of the *2./Minenräum-Abteilung 1* was attached to the *7. Panzer-Division* and two platoons to the *20. Panzer-Division*. After a successful crossing of the Düna at Ulla and Beschenkowitschi, Vitebsk was taken on 8 July.

On 8 July the *Minenräum-Abteilung 1* was released from its attachments, whereupon the mine-clearing battalion left the operational area of *Panzergruppe 3*.

Panzergruppe 3 Group Command Post, 8 July 1941

Operations Section

1.) *Minenräum-Abteilung 1* to assemble in Vilna for return to the rear.

2.) The *1./Minenräum-Abteilung 1*, formerly attached to the *LVII. Armee-Korps*, will leave the *LVII. Armee-Korps* and proceed to Vilna.

3.) Advance party to proceed to Vilna ahead of the main body and report to the Vilna local-area command.

4.) Five armored vehicles under the command of an officer are attached to the headquarters of *Panzergruppe 3*. Report to corps headquarters in Lepel.

Warning order of above by telephone

Distribution: Normal

(signed Spatz) (Source: *Bundesarchiv/Militärarchiv* Freiburg)

The unit proceeded via Vilna to Treuburg, where it was loaded aboard trains for movement to Wünsdorf.

In the weeks that followed, overdue maintenance work was carried out on vehicles and equipment and the experiences gained in the unit's brief tour of duty in the campaign against the Soviet Union were evaluated.

On 10 September 1941, *Dipl.Ing. Major* Müller was transferred. The battalion's new commander was *Hauptmann* Weicke, former commander of the *2./Minenräum-Abteilung 1*.

On 15 September 1941, *Minenräum-Abteilung 1* was redesignated *Panzer-Abteilung 300*.

Gliederung der Minenräum - Abteilung 1 (Stand: Juni 1941)

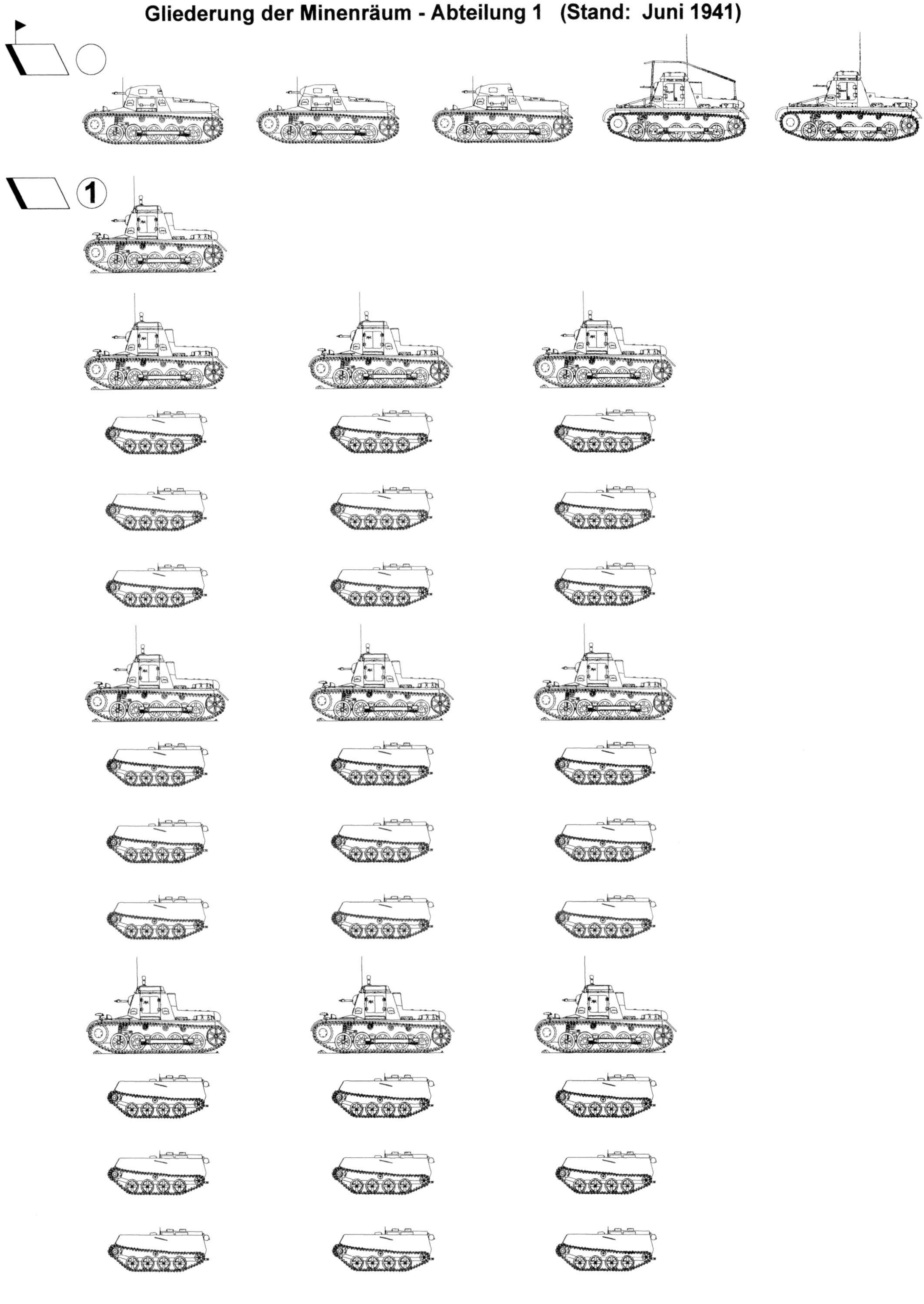

M.Jaugitz / 2000.

Gliederung der Minenräum - Abteilung 1 (Stand: Juni 1941)

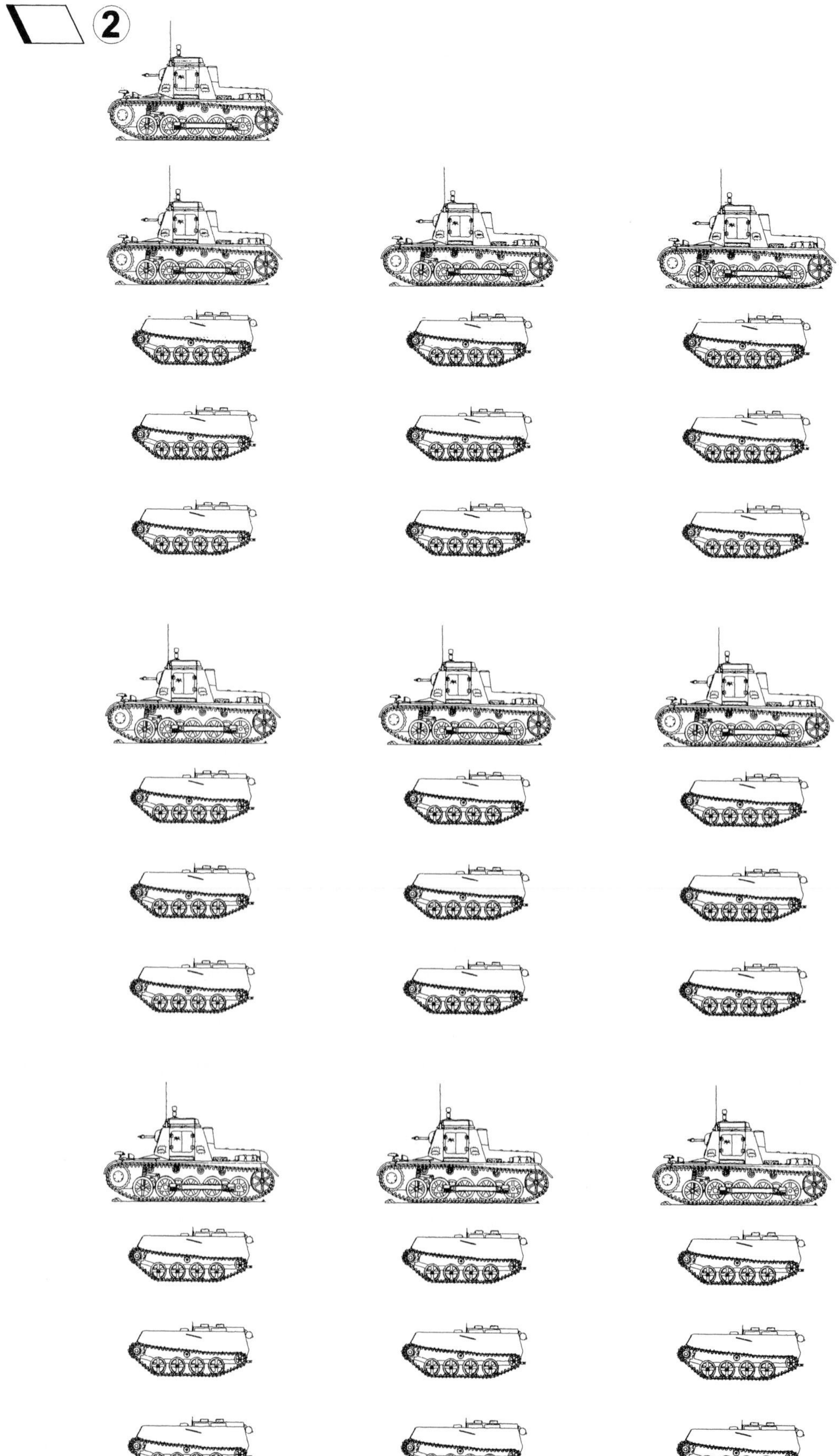

M.Jaugitz / Trojca 2000.

After moving to its new garrison at Weißenburg at Alsace, the newly formed *Minenräum-Abteilung 1* was given obstacles and fortifications within the Hagenau Fortress sector of the Maginot Line to test its vehicles on. (Hand)

In addition to the wire obstacles, there were innumerable buried mines. (Hand)

During the testing, the larger *Borgward BII* was used for the first time and destroyed. (Hand)

Opposite page, top right: *Leutnant* Senne, the adjutant of *Minenräum-Abteilung 1*, conducts an inspection. The tactical insignia of the battalion has been applied not only to the fender of the motorcycle but also to the sidecar. (Schick)

Opposite page, bottom right: Officers of *Minenräum-Abteilung 1* during a break. From left to right: *Leutnant* Senne (Adjutant); *Dipl. Ing. Major* Müler (Battalion Commander); *Oberleutnant* Blau (Company Commander of the *1./Minenräum-Abteilung 1*); and, *Hauptmann* Weicke (Company Commander of the *2./Minenräum-Abteilung 1*). (Hand)

Top left: This *Büssing-NAG 500 S* truck (Army license plate number *WH 155 644*) veered off road and rolled over. (Baumann)

Above: At the end of May 1941, the battalion started its move east. This *Borgward BII* is waiting to be loaded on the train. (Wizgall)

Bottom left: With the assistance of two *kleine Befehlswagen 265's*, the *Borgward BI*, which was thrown out of the truck, was put back on its tracks. (Baumann)

An antiaircraft vehicle (*Kfz. 4*) of the *1./Minenräum-Abteilung 1*, in the east Prussian border area. (Hand)

If needed, the battalion could only fall back on antiaircraft machine guns. Larger calibers, such as the 20 or 37 millimeter guns, were not available. (Hand)

Starting in the middle of June 1941, the *1./Minenräum-Abteilung 1*, started to occupy assembly areas under the control of the *VIII. Armee-Korps*. (Maier)

The *2./Minenräum-Abteilung 1* was attached to the *XXXIX. Armee-Korps*. (Ling)

Souvenir photo taken while it was still peaceful. This *BII* looks brand new. (Ling).

Above: The modifications made to the *Sd.Kfz. 265* are easily seen in this photograph. The all-round scope has been added to the roof hatch of the superstructure. In front of it is the antenna mounting for the radio-control antenna. (Bähr)

Left: The final hours before the start of Operation "Barbarossa." The command and control tanks were positioned in bivouac areas in the woods to facilitate camouflage. (Aichele)

Left: Just before the start of the operation, several *BII's* had to be recovered that had sunk into the soft ground in the assembly area. (Hand)

Below Left: The vehicle identification can be seen well in this photograph. Next to the parallelogram can be seen the circle with the company number. Below it is the individual vehicle number — *21*. Note how the insignia has been placed on both the superstructure and the front fender. (Hengstberger)

Below: Snapshot taken in the early-morning hours of 22 June 1941. The attack on the Soviet Union had begun. (Aichele)

Two *BII* mine-clearing vehicles follow a command and control tank into the jump-off position. The vehicle numbers are painted in white on the rear of the vehicles. (Bundesarchiv 265/5/A)

The vehicle drivers move the BII's into the jump-off positions. Note the camouflage which has been applied to help reduce the chances of enemy target acqusition. (Bundesarchiv 265/5/2A)

The drivers have to maneuver the vehicles until they reach their positions. (Bundesarchiv 265/5/3A)

The battalion commander observes the advance of the individual sections. (Bundesarchiv 265/5/10A)

Above: The command and control tank of the battalion commander has been outfitted with a frame antenna. (Bundesarchiv 265/5/14A)

Top right: The battalion commander's tank did not have the all-round optics. *Major* Müller observes the advance. (Bundesarchiv 265/5/16A)

Bottom right: *Dipl. Ing. Major* Müller, the commander of *Minenräum-Abteilung 1*. (Bundesarchiv 265/5/20)

Number *22* rolls forward. The command and control tank has already occupied its position and observes the potential target. (Bundesarchiv 265/5/17A)

Meeting in the early morning. From the left: *Major* Müller, *Hauptmann* Weicke and *Hauptfeldwebel* Pieperhoff (the *Spieß* of the *2./Minenräum-Abteilung 1*). (Bundesarchiv 265/6/3A)

To protect the columns against aerial attack, this *Sd.Kfz. 10/4* has taken up position on high ground. (Bundesarchiv 265/5/28A)

This command and control tanks belongs to the *2./Minenräum-Abteilung 1*. (Bundesarchiv 265/6/28A)

Above: A member of the crew of the battalion commander's tank. (Bundesarchiv 265/6/30A)

Top right: In contrast to the other *Sd. Kfz. 265's* of the battalion, the battalion commander's tank did not have any tactical insignia on the vehicle's superstructure. (Bundesarchiv 265/6/16A)

Bottom right: An inquiring glance to the rear: Is anyone coming? (Bundesarchiv 265/6/35A)

A command and control tank of the *2./Minenräum-Abteilung 1* (Vehicle *6*) passes advancing infantry. (Bundesarchiv 265/6/37A)

The command and control tank has advanced as far as an outbuilding. The farmer looks on with interest. (Bundesarchiv 265/7/2)

While an officer receives the reports, the commander makes notations on his map. (Bundesarchiv 265/7/21)

Because the sections have to operate widely dispersed, all message traffic is passed through messengers in order to maintain secrecy. (Bundesarchiv 265/7/18)

Above: A propaganda company soldier films the operations of *Minenräum-Abteilung 1*. (Bundesarchiv 265/7/25)

Above: The local populace approaches the leading elements of the unit. (Bundesarchiv 265/7/34)

Right: This medium staff car (*WH 405 943*) belongs to the headquarters staff. (Gradolph)

Several Balkan crosses have been painted on this command and control tank. (Hand)

Maintenance on a command and control tank. The machine gun and the all-round optics have been removed. Portions of the extra track links have also been removed. (Schott)

The vehicles of the battalion roll eastward in long columns. (Hand)

The crew of this command and control tank belonging to the *2./Minenräum-Abteilung 1* prepares for operations. A portion of the crew has already put on its steel helmets. (Fritschken)

The *2./Minenräum-Abteilung 1* reached the Lithuanian city of Wilna on 25 June 1941. (Stolz)

Rest break. Another *Ente* of the *2./Minenräum-Abteilung 1* in the background. (Bähr)

This *BII* mine-clearing vehicle was prepared for employment in the area of operations of the *20. Panzer-Division*. (Lock)

The amphibious version of the *Borgward BII* — the *Ente* (Duck) — was employed in crossing the Düna on 5 July 1941. (Aichele)

The running gear of the *Ente* was almost identical with that of the normal *Borgward BII.* (Kugler)

Two *Enten* of the 2./*Minenräum-Abteilung* 1 await employment. Neither tactical insignia nor vehicle numbers can be seen. (Kugler)

Above: The summer heat and the constant advance affect both personnel and equipment. (Schott)

Top right: Maintenance work on the engine on command and control tank *21* of the *2./Minenräum-Abteilung 1*. (Stolz)

Right: The operations of the battalion come to an end on 8 July 1941. (Stolz)

Left: The formation assembles in Wilna. To the left of *Oberleutnant* Blau is *Leutnant* von der Recke, platoon leader in the *1./Minenräum-Abteilung 1*. (Hand)

Above: Garrison duty at Wünsdorf. (Scherer)

Left: This *BII* has been outfitted with a special detonation initiating device. After moving into an obstacle, the demolition charges were supposed to be ignited through the metal frame attached to the front of the vehicle. (Fritschken)

Change of command. *Hauptmann* Weicke assumed command of *Minenräum-Abteilung 1* in September 1941. There were still a few *BI's* left in the battalion. (Fritschken)

The former commander, *Dipl. Ing. Major* Müller, was transferred to the officer reserve of *Heeresgruppe Süd*. (Fritschken).

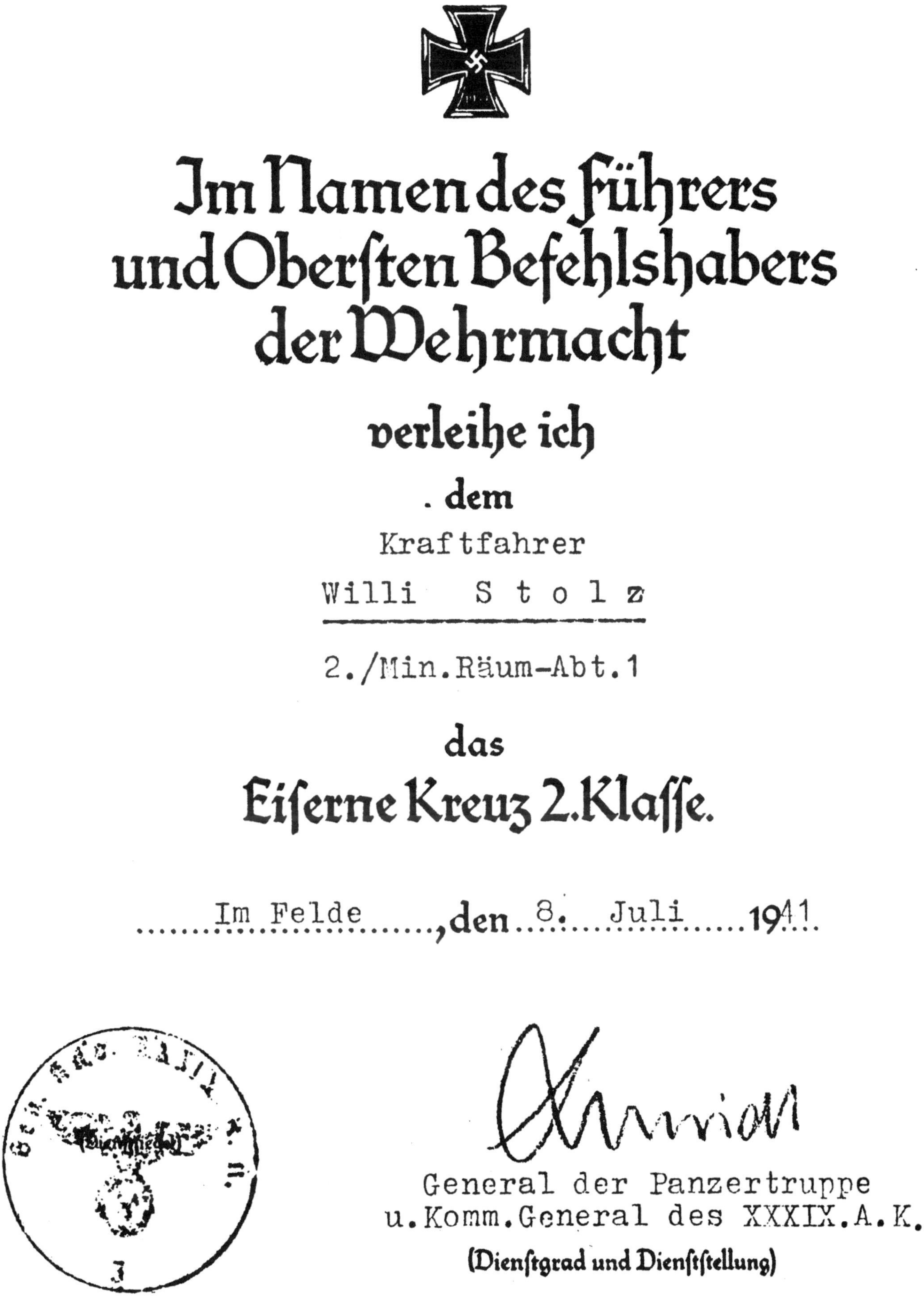

Im Namen des Führers und Obersten Befehlshabers der Wehrmacht

verleihe ich

dem

Kraftfahrer

Willi Stolz

2./Min.Räum-Abt.1

das

Eiserne Kreuz 2.Klasse.

Im Felde, den 8. Juli 1941

General der Panzertruppe
u.Komm.General des XXXIX.A.K.

(Dienstgrad und Dienststellung)

Iron Cross, Second Class award document to *Kraftfahrer* Willi Stolz of *Minenräum-Abteilung 1*.

2.2.3 Panzer-Abteilung 300 / Panzer-Abteilung 300 (Fernlenk) / Panzer-Abteilung 301

On 15 September 1941 *Minenräum-Abteilung 1* was redesignated as *Panzer-Abteilung 300* (order AHA I c 4169/41 g.K.). At first the battalion consisted of just the headquarters and two line companies.

Panzer-Abteilung 300 was commanded by *Hauptmann* Weicke. The two company commanders were *Oberleutnant* Müller (1st Company) and *Oberleutnant* Fritschken (2nd Company).

Following transfer from Wünsdorf to Cottbus in November 1941, the battalion began training on the *Panzerkampfwagen III* command and control tank. At the same time, personnel were detailed to the Hell firm in Berlin-Dahlem to assist in the building and testing of transmitter and receiver equipment. In this way it was hoped that they would become familiar with the new technology sooner.

After *Oberleutnant* Müller was killed in an accident in January 1942, *Oberleutnant* Abendroth was placed in command of the *1./Panzer-Abteilung 300.*

On 9 February 1942 the battalion was further redesignated as *Panzer-Abteilung 300 (Fernlenk)*. At the same time, an order was issued (*AHA I a II 429/42 g.K.*) for the formation of a new company to be designated as the *3./Panzer-Abteilung 300*. It was planned to equip the first two companies as *leichte Panzer-Kompanien (Fernlenk)* with radio-controlled demolition-charge carriers. The *3./Panzer-Abteilung 300 (Fernlenk)*, on the other hand, was to serve as a medium remote-control tank company equipped with wire-controlled demolition-charge carriers.

The battalion's list of officers in spring 1942 was as follows:

Battalion Headquarters
Battalion commander: *Hauptmann* Weicke
Adjutant: *Leutnant Dr.* Schlüter
Executive officer: *Leutnant Dr.* Schmidt
Officer for remote-control equipment: *Oberbaurat Dipl.Ing.* Hanke
Battalion surgeon: *Dr.* Wirth

1./Panzer-Abteilung 300 (Fernlenk): *Oberleutnant* von Abendroth
Platoon leader: *Leutnant* ?
Platoon leader: *Leutnant* Schlenzig

2./Panzer-Abteilung 300 (Fernlenk): *Oberleutnant* Fritschken
Platoon leader: *Leutnant* Fischer
Platoon leader: *Leutnant* von Rhoden

3./Panzer-Abteilung 300 (Fernlenk): *Oberleutnant* Senne
Platoon leader: *Leutnant* Dettmann
Platoon leader: *Leutnant* Sigmund

The first demonstration of the new wire- and radio-controlled demolition-charge carriers took place at the *Führer's "Wolfsschanze"* headquarters at the end of March 1942.

Jus Fischer, then a *Leutnant* and a platoon leader in the *2./Panzer-Abteilung 300 (Fernlenk)*, wrote of this in his memoirs:

In the Wolfsschanze Führer Headquarters at Rastenburg in East Prussia

It was a cold, dry day at the end of March 1942. In the outer security zone of the *Wolfsschanze* we had positioned a radio-control tank section, which consisted of a *Leitpanzer III* and two *BIV's*, and a *"Goliath"* section, with three demolition-charge carriers, for inspection by Hitler.

Hitler, commander-in-chief of the German armed forces, walked past our display with quick, long steps, his face sallow and serious. He was followed at a discrete distance by *Generalfeldmarschall* Keitel, chief of the *OKW*, and several generals of the *Wehrmacht* Operations Staff, the *Waffen-SS* and the *SS* Security Service, as well as by senior party functionaries, including Bormann.

I stepped up to Hitler and reported. He looked at me without saying a word, never once slowing his pace, and walked past the large posters and display boards we had set up. I had intended to refer to these while giving a briefing on the development of tank radio control, the test results achieved at the *Westwall*, the layout and performance characteristics of the special vehicles and radio-control equipment and, finally, their potential uses and operational requirements.

My men had worked for several days and nights making preparations. Near the end, one of the two *BIV's* broke down. A replacement had to be flown in from Bremen by the Borgward company on the last day. We only just completed putting it in on the morning of the demonstration.

And now Hitler marched past us without showing any interest whatsoever. I ran after him.

I intended to give a few quick explanations when Hitler reached the end of my small display. I never got the chance. He curtly said "Demonstration". Hitler in a bad mood! He left me with the impression of a man very conscious of his power, accessible to no one. There was not even a glimmer of friendliness.

The *"Goliath"* section started off. The soldiers leapt nimbly to the control position with three *"Goliaths"* and, I would almost like to say, in seconds the *"Goliaths"* moved off and turned skilfully toward a prepared "enemy position" about 300 meters away. The enemy fired from machine-gun positions — with blank cartridges, of course. Taking advantage of depressions in the terrain, the *"Goliaths"* reached the machine-gun positions and, once at the target, smoke charges simulated detonations. During remote-control maneuvers care had to be taken not to damage the cable running to the demolition charge carrier by driving over it, as this would result in loss of current. In real life this posed the danger of the demolition charge exploding in our own ranks, which unfortunately happened at Leningrad. The demonstration transformed Hitler. He turned to those in his immediate entourage, smiled broadly and slapped his thigh joyfully.

The engineers had constructed a massive antitank barrier for the demonstration by the *BIV* section. The objective of the demonstration was to show the *BIV* dropping its demolition charge and making the obstacle negotiable. Unlike the *"Goliaths"*, this was a live demonstration, and the demolition charge weighed 450 kilograms. Everything had to go right. On my own initiative I had disregarded a safety requirement. I was therefore nervous and my mind recalled previous unsuccessful demonstrations.

I asked Hitler to climb into the *Leitpanzer III* and I assisted him with a stepladder. I asked him to keep the hatches closed at all times. There was no room in the tank for *Feldmarschall* Keitel and I had to ask him and the other officers to step behind a safety line for the demonstration. But Keitel wanted to lie down with me behind the rear of the tank. "I can do the same as you," he said, and remained. From the tank the section leader started the *BIV* by remote control, guided it right up to the antitank barrier and jettisoned the charge. The demolition charge carrier automatically shifted into reverse gear and backed out of the blast zone at high speed.

With the explosion, Keitel and I threw ourselves down behind the rear of the tank. Immediately after the explosion I jumped onto the control tank and had it start up. We moved up to the crater with Hitler in the tank, Keitel hurrying after us on foot. He had the bad luck of having his uniform soiled by the exhaust fumes when the engine started. He hadn't stood up quickly enough and I, happy that the demonstration had gone as planned, had ordered the tank to move off too soon. At the same time, the squad leader guided a second *BIV* forward through the crater and over the wrecked barrier to demonstrate it was negotiable. The *BIV* shook as it very slowly climbed the wall of the crater. It looked as if it were about to stop, but the hydraulic drive did the job.

Hitler climbed out of the tank at the crater and walked with me toward Keitel. He spoke in a chopped staccato (something like): "I am imagining...their use in Sevastopol. First...the small *"Goliaths"*...attack on a broad front...into the infantry positions and trench systems...then...break in with the big demolition-charge carriers...! That's how we will break through this complex system of positions!"

Hitler spoke only to Keitel. Maintaining his ramrod-straight Prussian posture as before, he raised his marshal's baton at every half sentence. Whenever Hitler paused, he bowed his head slightly and said, *"Jawohl, mein Führer"* — *"Jawohl, mein Führer"* — *"Jawohl, mein Führer."*

I heard "trench systems" and "infantry positions"! That was absolutely the worst thing for remotely-controlled *BIV's*. I knew that tracked vehicles easily became hung up in trench systems, remembering the Tobruk demonstration a few days before. After Keitel's last *"Jawohl"* I said, "*Mein Führer*, the success of this first mission will depend on the terrain. We need terrain which is definitely favorable for tanks."

Hitler, almost shocked, stared straight ahead, now past Keitel too. I got the awkward feeling that I had behaved rudely. Or did Hitler have no interest in factual information which contradicted his imaginings? After a brief pause, Hitler said to Keitel: "Please look into this." He walked past Keitel, who followed at a discrete distance. Hitler's car approached and picked them up.

The next day while having lunch with officers of the general staff I heard something about Hitler's foul mood, which had been obvious from his behavior. The day before, the Chief of the General Staff, *Generaloberst* Halder, had argued strongly against Hitler's strategic plans. Hitler cast aside his recommendations, willfully redeployed the forces and, in spite of the good intelligence provided by *General* Gehlen's "Foreign Armies East," underestimated the enemy's potential. It was also for this reason that the separate demonstration had been ordered at short notice.

The security measures in the *Wolfsschanze* were already very strict at that time (March 1942). I considered them excessive.

Officers had to surrender their pistols. The *"Goliaths"* were checked to ensure that the detonators had not been fitted. There were always several *SD* officers near the armed *BIV*.

When the demonstration was discussed with the engineer *Oberst* responsible for safety, he rejected the 450 kilogram charge as totally unacceptable. My commander thereupon pointed out that the barrier could not be taken out with a smaller charge and that it would be better not to conduct the demonstration at all. It was clear to me that our charge would seem an immense amount of explosives to a engineer officer, since the combat engineers mainly dealt with blasting caps or at most a few kilo-packets of explosives for blowing up bridges. Our counterproposal, to use 300 kilograms, went unanswered. He had previously stated that 200 kilograms was the maximum amount. It was clear to both of us that such a small charge would make questionable the success achieved in the development work to date. Weicke made no decision.

I took the risk and stayed with the full charge. I had put so much time and effort into the theory and practice of remote-control that I believed I had to act. I discussed this with no one, not even afterwards. Neither was I questioned, either by the engineer *Oberst* or my commander, and I later dared to think that I had their unspoken agreement.

The same demonstration was given to the general staff the next day, but this time without detonating the charges. Interest was great and there was a lively discussion concerning details of the system and possible operational problems. An infantry general was skeptical, convinced that there was a driver hidden in the *BIV*.

Soon after returning to Cottbus, the battalion received orders to prepare for action at Sevastopol without waiting for terrain reconnaissance results. With the aid of maps and aerial photographs of the fortifications and trench systems, we were able to prepare for operations against a fortified system bolstered by very favorable terrain for the defenders.

Deliveries of the *Sd.Kfz. 301* and *Sd.Kfz. 302* began in April 1942, followed by intensive training and instruction on the new demolition-charge carriers.

At the beginning of May 1942 *Panzer-Abteilung 300 (Fernlenk)* received orders to proceed to Sevastopol in the Crimea. There it was to take part in "Operation *Störfang*", the conquest of the Fortress of Sevastopol.

The unit began rail loading in the second week of May. Several trains left Cottbus in the period between 11 and 13 May and headed east. The one-week journey to the Crimea took the battalion through Poland, White Russia and the Ukraine.

After detraining, the battalion occupied its first billets at Simferopol. The weeks leading up to the end of May were used for further training and familiarization with the radio- and remote-control vehicles. In addition, the unit scouted its future area of operations.

Generaloberst von Manstein had been ordered to take Fortress Sevastopol. His *11. Armee* was divided into two *Armee-Korps*. The *XXX. Armee-Korps* was formed from the *28. leichte Division* and the 72. and *170. Infanterie-Divisionen*. The *22.*, *24.*, *50.* and *132. Infanterie-Divisionen* belonged to the *LIV. Armee-Korps*.

Since *Panzer-Abteilung 300 (Fernlenk)* could not be committed organically, the unit was split in two. The *XXX. Armee-Korps* and the *LIX. Armee-Korps* each had one light and one half of a medium radio-control tank company attached to them. The army corps were to make radio-control platoons or sections available to the infantry divisions under their command for offensive operations. In order to provide the commanders of infantry divisions with a basic knowledge of the radio-control force, the "Instructional Pamphlet for the Employment of Remotely Guided Demolition-Charge Carriers (*Panzer-Abteilung 300*)" was issued via the army corps to their constituent formations (see Chapter 4.2.2).

The companies of *Panzer-Abteilung 300 (Fernlenk)* moved into their assembly areas between 29 and 31 May 1942.

At 0230 hours on 2 June 1942 von Manstein's artillery

began an intensive five-day barrage on the fortress installations. Then, on 7 June, the *22.* and *50. Infanterie-Divisionen* attacked from Melzer Defile and advanced through the well-defended Kamyschly Defile.

On 11 June 1942 the *XXX. Armee-Korps* attacked in the southern sector of the front and on 18 June, after breaching the enemy's defense lines, units of the *22.* and *24. Infanterie-Divisionen* reached Severnaya Bay.

Following the capture of the Fedyukiny high ground on 21 June, the *XXX. Armee-Korps* stormed the commanding Sapun high ground on 29 June and took it the next day.

On 1 July 1942 Sevastopol was in German hands. The last enemy fort was taken on 3 July and on 4 July the last fighting ceased on Khersonyes Peninsula.

In the period from 7 June to 1 July 1942, the line companies of *Panzer-Abteilung 300 (Fernlenk)* saw action with the following infantry divisions on various days:

	XXX. Armeekorps	*LIV. Armeekorps*
07 June 1942	*50. Infanterie-Division*	
10 June 1942	*22.* and *50. Infanterie-Divisionen*	
11 June 1942	*22.* and *50. Infanterie-Divisionen*	
12 June 1942	22. Infanterie-Division	
13 June 1942	*170. Infanterie-Division*	
14 June 1942	*170. Infanterie-Division*	
16 June 1942	*170. Infanterie-Division*	
17 June 1942	*72. Infanterie-Division*	
18 June 1942	*72. Infanterie-Division*	*24. Infanterie-Division*
19 June 1942	*170. Infanterie-Division*	
30 June 1942	*28. leichte Division*	*132. Infanterie-Division*

Excerpts from the diary of Heinz Prenzlin (†), a radio operator in *Panzer-Abteilung 300 (Fernlenk)*, provides details of operations in the period from 13 May to 1 July 1942:

13 May 1942: Entrained in Cottbus from 0600 to 1400 hours, then departed for Sagan. Left Sagan at 2000 hours via Glogau, Lissa.

14 May 1942: Arrived Ostrovo at 0340 hours, departed 0450 hours via Litzmannstadt to Radom, arrived 1000 hours.

15 May 1942: 0015 hours Drobkin, 0600 Lublin, 1100 Zawadowka.

16 May 1942: Departed Zawadowka at 0900, Kholm at 0915, then via Yagochin, Lumboml, arrived in Kovel at 1515 hours.

17 May 1942: 0650 hours in Mohylani, 1745 hours in Zhitomir.

18 May 1942: 0630 hours in Miranovka, 1300 hours in Bobrinskaya.

19 May 1942: 0545 hours in Pyatishatki, departed for Verkhovzevo at 1130, from there through Dnepropetrovsk. First signs of war seen there, makeshift repairs to Dnjepr bridge, factory chimneys a pile of rubble, everywhere fires and smoking rubble. Arrived Sinelnikovo at 1930 hours.

20 May 1942: At 1300 hours arrived Dshankoi, first station in the Crimea.

21 May 1942: Sarabus at 0430 hours. Detrained there and traveled by road to Simferopol, where we occupied freshly-cleaned quarters. At 1730 the city was bombed by the Russians. Four bombs fell in our billeting area but no damage was done.

22 May 1942: Checked remote-control equipment, determined strength of emanations due to different weather conditions. In the evening the "Duty Russian", an aircraft, came in very low and scattered tiny bombs without causing much damage.

23 May 1942: Same as the day before.

24 May 1942: The same as the day before. Evening in the front theater (city theater).

25 May 1942: The same as the day before. "Duty Russian" seems to be a permanent fixture.

26 May 1942: The same. We have things to do, time is short.

27 May 1942: *BIV* drivers are instructed on how to arm the special vehicles.

28 May 1942: Vehicle commanders are instructed on how to guide the special vehicles. Today the "Duty Russian" also appeared by day. Since yesterday morning our area has been guarded by two *Flak* batteries, one 88 mm one and one quad 20 mm one. The "Duty Russian" approached several times but was always driven away. Our *Flak* is on the ball!

29 May 1942: Straightened away our equipment, radio vehicles and trucks carrying spares and set crystals to new groups. "Duty Russian" as yesterday.

30 May 1942: Rest day for us. It is said that we will be moving to the attack position tomorrow. We try to figure out where we are going. Leutnant Dr. Schmidt has gone with the headquarters staff to recon the front.

31 May 1942: Wakened by the "Duty Russian", such an idiot! He is becoming more active with each passing day. Slowly, things for us are also becoming rather hectic. At 1100 hours we receive the order: "Prepare to move out!" Mission briefing at 2100 hours, assignment of march routes to the line of departure at Tole. Departure at 2200 hours.

1 June 1942: Arrived in assembly area at 0130 hours, vehicles parked on both sides of the road and camouflaged. We had major difficulties since we were on a valley road with steep slopes on both sides. "Duty Russian" is present as in previous days.

2 June 1942: Sunrise at 0330 hours, from 0400 to 2030 hours continuous attacks by *Stukas* and *Ju 88's* on Russian positions in front of us. The "Duty Russian" is also becoming more daring, however he rarely reaches his target as our fighters chase him off.

3 June 1942: Same as previous day.

4 June 1942: Same as previous day. We are now employing new rocket launchers, called *Do-Werfer*, which took up positions on both sides of us on the hills. At first we were terrified and crawled under the tanks. The things produce an ear-shattering din and howling. At first we thought the Russians were attacking.

5 June 1942: The inferno of the day before went on undiminished; now there is a battery of railway guns which began firing at 0800 hours. It consists of one 800 mm and two 600 mm long-barreled guns. It is said that they are served by engineers from the *Krupp* factory and are the biggest guns in the world. A salvo is fired every 45 minutes. We are told that the ammunition depot of Fortress Sevastopol is the target. I had to accompany *Leutnant Dr.* Schmidt to an observation post from where we were able to observe the impacts. We were told that they had planned to fire 30 shells. The second was dead on, the fifth caused the ammunition depot to blow up. The Russians attempted several counterattacks, but these were halted by *Do-Werfer* fire.

6 June 1942: Preparations for employment. Departure at 2200 hours. I was attached to the 2nd Company as radio operator as someone had fallen ill. Reported to *Oberfeldwebel* Schneider in the

2nd Platoon.

7 June 1942: Arrived at the line of departure at 0200 hours. Baptism of fire by 0530. Continuous attacks on Russian positions in the artillery gorge by *Stukas* and *Ju 88's*, also artillery and *Do-Werfer* fire. We wait for the order to be committed. Artillery fire from Ivan during the brief firing pauses by our side. Herzberg killed.

8 June 1942: We are in the Melzer defile. Under Russian artillery fire all day. The Russians have also now begun attacking us with close-support aircraft. Three men seriously wounded.

9 June 1942: At 0330 hours set out in our tanks to reconnoiter. First to the command post of *Infanterie-Regiment 47* on Railroad Hill, then to *Infanterie-Regiment 16* at the railroad viaduct, then back to Railroad Hill, as *Infanterie-Regiment 16* had moved there. From there to Hill 716 via 643, from there via 630 to 619. This is the line of departure for the attack on the barracks and beyond. Spent the night in hedgehog position.

10 June 1942: At 0730 hours summoned by radio to Point 711, where we rejoined the others. From there via 643, 642 and 641 for the attack on the White House. Three tanks knocked out; 1230 hours back to 712 to the antitank ditch. There we refueled and rearmed. 1330 hours attack on the White House from right of the rail line via 630, 639 and 664. After successful attack on the White House, in the evening back to the antitank ditch.

11 June 1942: Attack on barracks, Fort Stalin and beyond to 619, 628 and 638, from there back across 639 to 712 in the antitank ditch for the night.

12 June 1942: Our platoon moved up for attack on the defile and port of Mekensievy-Gory; march route via 712, 643, 642, 641 and 665. Attacked from there, three antitank guns put out of action and overrun. Concrete bunker put out of commission and blown up by *BIV's*. Route clear to oil tanks in the port. Five oil tanks set on fire. At this point a gunboat sailed into the harbor. Through the sight I saw a muzzle flash, a round exploded nearby. Sparks flew; we were no longer able to move straight as we had thrown our right track. We tried to keep moving to prevent the enemy from taking our range. Radioed the other tanks of the platoon for assistance and they gave us cover. We climbed out. I destroyed all radio materials, lit the demolition charge on the radio sets and was the last to get out. We then ran back through the antitank ditch to 712, all the while under heavy artillery fire. *Oberfeldwebel* Schneider was not with us. We were then strafed by low-flying aircraft but, fortunately, their aim was poor. Late in the afternoon we drove back to the Melzer defile. After darkness fell the recovery section under Paul Schäfer recovered our tank.

13 June 1942: 0800 hours pulled back to Tole rest area. Until 17 June we repaired radio and radio-control equipment there.

18 June 1942: 1630 hours departed with Schaper and Schäpler in an amphibious *Volkswagen* for the southern front. Arrived at the Biluk-Muskomja rest area at 2030 hours. At 2100 the "Duty Russian" dropped five bombs in the middle of our billeting area. The headquarters' fuel truck went up in flames. One man was killed and another seriously injured. I took the latter to the aid station at Baltawy.

19 June 1942: 0430 hours departed from the Biluk-Muskomja rest area with Lt. Dr. Schmidt and Schäpler for the Ski Jump. 1430 hours back to the *170. Division* on the Winter Path. There with the commander, *Oberfunkmeister* Martens, Richter, Güll, Ehrlichmann, Satori, Gradolph, Stolz and Götzel to observe the situation and reconnoiter in preparation for our next operation.

20 to 28 June 1942: With the *170. Division* on the Winter Path. I have been assigned as radio operator for the company commander. Lively radio traffic with the battalion and the 1st Company as well as with other division networks.

29 June 1942: 0130 hours departure from the Winter Road to Chapel Hill. From there we observed the inferno of fire on the Sapun Heights. All guns fired continuously, repeatedly pounding the heights. It was a tremendous sight. After these preparations, *Gebirgsjäger* advanced in a wedge; the *170. Division* followed up and expanded it to both sides. Only the serpentine was avoided wherever possible, as it was thought to be mined. And it was. Now our hour had come. We deployed our *BIV's*. In two hours we succeeded in clearing the road all the way to the top. Five *BIV's* without charge containers were used. The Sapun Heights were stormed at 0600 hours. There was little resistance. We secured to the sides. At 1800 hours there was a meeting with the *170. Division* on the Sapun Heights. For us the day was over.

30 June 1942: We conducted several scouting forays from the Sapun Heights. The city lay open before us. From here the terrain extends plateau-like into Sevastopol. Our mission was to determine if there were still signs of resistance. For one thing, we determined that the resistance was being directed from the water tower and the gasometer. In the evening we returned to our starting point with the *170. Division*. We had a hard day behind us.

1 July 1942: The attack on Sevastopol began at 0300 hours. By 1300 hours Sevastopol had fallen. Our flags fluttered over the city. The Russians have been driven back to Yalta and were trapped. We crisscrossed the city several times with mounted infantry to capture Russian stragglers. This action was concluded at 1500 hours and the recall order was issued. Starting at 1530 hours the city was opened to plunder. We took part until 2000 hours. We secured several containers of butter and jam. I captured a naval officer's jacket and a pair of Cossack boots. Otherwise nothing of significance. At 2030 hours we assembled and headed back to Novo-Shuti.

Additional information is provided by the diary entries of Helmut Greiner (former *Obergefreiter* in the *3./Panzer-Abteilung 300 (Fernlenk)*:

Thursday, 28 May 1942: Awoke at 0400 hours. Maintenance duties, after which the vehicles were loaded and prepared for departure. At 0500 battalion roll call, at which the commander spoke.

Friday, 29 May 1942: Awoke at 0030 hours and at 0145 we left Simferopol for Sevastopol via Batschisseray over the pass road to Yalta — Abat — Fozi — Sala. So far we had been driving on a beautiful asphalt road and the terrain was wonderful, similar to the Allgäu. Then we came to a bad road and from there it was 7 kilometers uphill. It was a difficult drive, with our vehicles moving 500 meters at a time. But then it was downhill again and outside of Urkusta a Bren went into the ditch. We got it out and another went into the ditch, but we made it. Then at 1030 hours we reached our billeting area. The *Flak* began firing just as we arrived. This went on the entire night.

Saturday, 30 May 1942: Awoke at 0600 hours. Here we are bivouacked in a fruit plantation 8 to 10 kilometers behind the front line. Today I received a second letter from my wife. The day was mainly quiet. In the evening we had a few drinks of schnapps.

Sunday, 31 May 1942: Awoke at 0600 hours. Shortly before I awoke, the 1st Company moved past us and stopped about 300 meters away. Yesterday that company suffered our first casualties: One killed and one wounded. We pulled maintenance duties and details.

Monday, 1 June 1942: Awoke at 0600 hours and then mechanical duties and details again. Today we received a lot of mail but there was none for me. But I did send an air-mail letter.

Tuesday, 2 June 1942: Awoke at 0600 hours and there were maintenance and work details again. In the morning there was a demonstration in front off all of the division's commanders and the general, during which one of the 1st Company's *Kettenkräder* broke down.

Wednesday, 3 June 1942: Awoke at 0600 hours. Maintenance and work details again. Since four this morning the Luftwaffe has been flying over us to Sevastopol, bringing death and destruction.

Thursday, 4 June 1942: Awoke at 0600 hours and, like every day, had maintenance duties and work details. At about noon the artillery began firing.

Friday, 5 June 1942: Awoke at 0600 hours. Maintenance duties and work details again. Final preparations made to all vehicles.

Saturday, 6 June 1942: Awoke at 0600 hours, after which we made preparations to depart. At noontime we packed our things prior to turning them in to the company headquarters section. At 1900 hours a 15-man-strong *Kampfgruppe*, of which I was a part, moved out for the attack position. We reached our position at 2130. The Russians were now about 800 meters away. Our position was next to a battery of 105 mm guns, which fired all night and into Sunday. We slept on a small slope.

Sunday, 7 June 1942: At about 0400 very heavy fire began falling nearby, and it went on through the morning. At about 0600 hours we shot down a Rata. And then at 1100 a round apparently struck the 105 mm battery next to us, probably landing in front of the guns, resulting in one seriously wounded and three with minor injuries. Then we lay waiting for food. When the noon meal was delivered to us our *Leutnant* brought the sad news that we had already lost two killed and several injured. *Unteroffiziere* Wimmer and Birner were dead. Those wounded were our company commander *Oberleutnant* Senne, the *Oberschirrmeister*, and *Obergefreite* Groj and Hohmann. This evening we once again moved into the forward jump-off position. Attack at 0315 hours? Before it was launched, however, we moved back.

Monday, 8 June 1942: We pulled back again, about 3 kilometers. Now we are here again, waiting for orders. This morning news reached us that while forming up prior to being committed, our 1st Platoon under *Leutnant* Dettmann suffered about 20 casualties through enemy action. We spent the entire night on the road.

Tuesday, 9 June 1942: Rested on the road to Sevastopol.

Wednesday, 10 June 1942: At about 0130 we left our assembly area and moved forward approximately 8 kilometers to wait for the attack. It began at 0520 hours, but in spite of a massive preparatory barrage by artillery and *Stukas*, there was no progress. At about 1000 four of us — myself, *Unteroffiziere* Gerstel and Schütte and *Gefreiter* Geibel — moved forward with a Utility. We had to negotiate the entire defile to the left of Chapel Hill. Unfortunately, we could not keep our vehicle in sight on account of the badly shot up terrain. We had to use every trick in the book to detonate our vehicle and prevent it from falling into enemy hands. The infantry pulled back to its former position. We were lucky to all get back in one piece. The attack was begun again at six in the evening. One of our *"Goliath"* sections took part. As soon as the attack force left the assembly area it took a direct hit which killed one infantryman and injured two of our men. Stiel was hit in the cheek and Dratius in the hand. Stiel was taken to the field hospital. But at least the advance-guard battalion got through. On the other hand, though, we lost two killed: *Unteroffizier* Vogt and *Oberschütze* Mayer, Karl-Heinz.

Friday, 12 June 1942: This morning when we tried to pull back to the ruins the Russians blanketed us with artillery fire. We never thought we would get back. I was relieved and sent to the second rest area.

Saturday, 13 June 1942: Rested all day and, in the evening, I was recalled to the front after *Unteroffiziere* Schütte, Gerstel and Frieder Noak were wounded.

Sunday, 14 June 1942: We spent the entire day on the upper-slope road. I was soon driven mad by stomach pains. Alfred was with me. He was supposed to recover a Bren from an antitank ditch. At night! He brought me something for my stomach but it did no good. Finally, *Unteroffizier* Schmid sent me back with the coffee truck.

Monday, 15 June 1942: I spent the whole day lying here alone, but at least I feel a little better, although I am not yet completely well.

Tuesday, 16 June 1942: Still at Rest Area 2 and I intended to take it easy a little longer. In the evening I was summoned by *Leutnant* Sigmund.

Wednesday, 17 June 1942: On Chapel Hill and remained there until night. Then we drove up into the departure position again taking two Brens and two Utilities.

Thursday, 18 June 1942: We moved forward into the twilight with our two tanks and four vehicles. When it became fully light our two tanks opened fire on the bunker. While this was going on I armed the vehicles in preparation for operation. The first Bren ran over a mine and failed to explode, while the second reached its target and detonated. Although somewhat slower, the Utility also functioned perfectly and reached its target, a mortar position. It also detonated on command. The second Utility could not be controlled and was fired on and set on fire. Our comrade received a jaw full of splinters. We then withdrew to the slope road, taking two captured Russians. Our first successful mission! We stopped to bathe in the valley and then returned to the departure position on the chapel road. We once again spent the night there.

Friday, 19 June 1942: This morning all our vehicles pulled back to the second rest area. Today on my daughter's birthday it was, thank God, relatively quiet.

Saturday, 20 June 1942: A great deal of Russian air activity during the night. We checked our vehicles over again and then stood down.

Sunday, 21 June 1942: Still quiet. We checked our vehicles over for the next mission.

Saturday, 27 June 1942: Still quiet.

Sunday, 28 June 1942: At 2300 hours moved forward to the Mushroom.

Monday, 29 June 1942: Once again in assembly area at the Mushroom.

Tuesday, 30 June 1942: We moved further forward to the foot of the Sapun Heights, which had meanwhile fallen. We stayed there until 1800 hours and then three combat elements launched the final assault on Sevastopol. We spent the night on the Sapun Heights, from where we moved into the infantry's front line in the morning.

Wednesday, 1 July 1942: Already under very weak fire. The *Luftwaffe* and the artillery once again pound Sevastopol with everything they have. The attack was supposed to begin at 1230 hours, but since the Romanians on our left were already in the city at 1000 hours, the order came to attack at 1045 hours. And so we took our tanks and special vehicles into the street fighting. By 0100 hours we had reached the submarine port, almost without resistance. Our 1st Company under *Feldwebel* Roncalla pressed forward from the other side. Having reached our objective, we pulled back to the line of departure. There, at 0500 hours, orders reached us to withdraw to the old rest area. The war flag had been waving over Sevastopol since 1400 hours yesterday. Sevastopol has fallen. And now comes the big question: where to next?

The commander of *Panzer-Abteilung 300 (Fernlenk)*, *Hauptmann* Weicke, summarized the battalion's experiences in the fighting for Sevastopol in two reports to the headquarters of *Armee-Oberkommando 11*: (Source: *Bundesarchiv/Militärarchiv* Freiburg):

Panzer-Abteilung 300 22 June 1942
Office 01914
1 Enclosure

To *Armee-Oberkommando 11*

The enclosure contains a report by the battalion on the elements committed to date. Summarizing, it was determined that:

1.) The terrain conditions outside of Sevastopol are very difficult, as there is a sharp rise in front of the attack area which must be negotiated over poor roads. This led to the loss of several captured vehicles.

2.) The attack area is traversed by many more trenches than revealed by maps and reconnaissance.

3.) There is an almost total lack of protected assembly areas at a favorable attack distance from the target. As a result, in almost every case, a departure point had to be selected in exposed terrain.

4.) It is the nature of fortress warfare that all approach roads, even in covered areas, are subjected to extremely heavy artillery fire.

5.) In the initial days of the offensive, the infantry was assigned distant objectives which proved impossible to reach. In order to be in position in time for these distant objectives, the vehicles were moved too far forward, as it was later determined, resulting in losses to artillery fire.

6.) In many cases, the infantry succeeded in reaching the enemy's trench system and hiding from the enemy's line of sight. This was not possible for the tanks and special vehicles and, as a rule, they attracted all of the enemy's concentrated artillery fire.

7.) The majority of the soldiers committed by the battalion had not yet seen action in Russia and had to overcome a certain shock effect caused by the massed defensive measures. Individual failures may be attributed to this shock effect.

8.) The fact that 40% of the vehicles were employed and 30% successfully concluded their missions, may undoubtedly be seen as a success for the radio-control force. It was achieved under the most difficult of conditions.

9.) In order to make an adequate comparison, one must consider the percentage of hits achieved by the artillery, *Nebelwerfer* and bombers. Inspection of enemy positions suggests that this was about 1%. Taking into consideration the losses in aircraft and artillery (worn-out barrels, etc.), the costs associated with the radio-control force are also considerably less than those of other arms.

10.) Furthermore, it must also be taken into consideration that radio-control weapons, like every new weapon, are subject to teething problems. In further considering these teething problems one should compare the German Army's attitude to the English tank arm in 1915-16.

11.) I have requested that the division and corps also prepare after-action reports. After the conclusion of the fighting I will submit another comprehensive report.

12.) The planned operation with the *1. Panzerarmee*, of which I have been informed, absolutely must be carried out, since as a (1) combined-arms operation with tanks in (2) different terrain it should provide significant new lessons, which seem vital for rounding out our experience.

13.) All affected staffs are requested to act to speed-up production of the armored *BIV* and the remote-controlled *"Goliath"* so that the advantage which Germany has over the enemy powers in remotely-controlled vehicles is not just maintained, but increased.

14.) In the northern sector the battalion lost just one tank to mines (it was later completely destroyed by artillery fire). Two other tanks were disabled: One ran over a mine, the other was disabled by artillery fire. These two tanks were recovered. When one considers that the battalion's tanks were deployed in front of the infantry and assault guns on five different days, then the losses against the formidable enemy defenses are astonishingly low. This is mainly due to the preceding remotely-controlled vehicles drawing the enemy's antitank fire, giving the tanks the chance to identify and eliminate the antitank guns. The loss of one tank and crew and one *BIV* through antitank fire are out of all proportion [to the results achieved], and this type of operation appears to be particularly promising.

15.) The type of field position employed by the Russians — deep narrow trenches in the rock — is very similar to the fortifications at Tobruk, which could only be engaged by the artillery and *Stukas* with difficulty. When detonated in the trenches, the remotely-controlled vehicles always created breathing room [for the infantry].

16.) It should be mentioned that, to date, no special-purpose vehicle has fallen into enemy hands.

signed Weicke

Hauptmann and battalion commander

Here is the enclosure to the cover memorandum of 22 June 1942 reproduced above:

36 *BIV's* were committed in the period to 21 June 1942 with the following losses:

8 through artillery fire
2 through mechanical failures
2 were hand-guided onto mines in front of friendly infantry positions
1 was lost in the explosion of a munitions truck.

20 *BIV's* were employed in a radio-control mode.

Losses while under radio control:

1 through radio-control failure
1 through artillery fire
1 moved into a shell crater
7 drew antitank fire; in each case the antitank guns were engaged and, as far as can be determined, destroyed
3 overran mines and fulfilled their role as mine detectors
1 was committed against tanks; according to the wounded radio control *Unteroffizier*, the tank, a T 26, was destroyed
1 was hit by antitank fire and disabled 80 meters in front of a bunker. The antitank gun was engaged and the Russians fled in great numbers.
1 got stuck 20 meters in front of a bunker in a mortar crater and was detonated by remote control, after which the bunker ceased firing.
2 destroyed bunkers.
1 was hit by artillery fire in front of an enemy infantry position and began to burn in the engine compartment. The Russians left their trenches and were engaged by tanks.
4 were committed against well dug-in infantry positions which were seriously hindering the friendly infantry advance and which could not be engaged by other weapon systems. In one case, use of the special-purpose vehicle enabled a position to be taken which had previously resisted three infantry assaults. In all cases the effect was considerable and helped our infantry advance.

Number of captured tanks committed: 29

Losses:

13 through artillery fire.

9 through mechanical failure.

8 were employed in a remote-control mode

Losses from the 8 employed vehicles:
1 through artillery fire after 200 meters.
2 ran over mines, creating a lane, and [others] were guided to within 40 meters of an antitank ditch (incorrect information from the infantry). It exploded under fire from our *Nebelwerfer* and, according to eye-witnesses, had a considerable effect.
3 were guided into a trench system and showed good effect. The trenches were leveled for a distance of about 20 meters; fatal effects were demonstrated to a distance of at least 100 meters. According to *Oberstleutnant* Daniel more than 50 dead Russians lay in the trenches, bleeding from the mouth and nose.
1 Bren destroyed an antitank bunker which had held up the advance of two infantry regiments for two days. The bunker had been supplying the entire surrounding trench system with ammunition (also according to statements by Oberstleutnant Daniel).

"Goliaths" lost to artillery fire: 7

1 was guided to within 20 meters of an enemy bunker, where it was stopped by a satchel charge. Nevertheless, it was detonated by remote control. Infantry opened fire on the thrower of the satchel charge. Shortly thereafter the Russians detonated an electric minefield in front of the bunker.
2 were committed against earth bunkers, collapsed them and put them out of action.

Recommendations for subsequent development:

1.) Captured tanks: The use of captured armored vehicles in difficult terrain and at great ranges usually resulted in disablement due to mechanical reasons. In the future, the remote control of captured vehicles must remain a secondary mission for the radio-control force. Research into equipping all types of captured armored vehicles with wire control must be continued. So equipped, they can be used with success in some cases, especially in Africa.

2.) Radio-controlled demolition-charge carrier must be armored to provide protection against small-arms fire and have armor protection for the driver's head, which is outside the vehicle. This will prevent (1) losses to artillery shrapnel and (2) make it easier for the special-purpose vehicle driver to follow his tank under artillery fire. Minor technical recommendations: Quicker arming and switching to radio control, higher speed (as requested by the General Staff), and further development of the radio-control system, which at present requires too much work by technically trained personnel prior to deployment.

3.) In its present form the *"Goliath"* has failed to live up to its original potential. The main reason for this is that it requires five or six men to pull it into its launch position. Such a large moving group attracts enemy fire, often resulting in casualties and, together with the very difficult terrain, this sometimes makes it impossible to follow directly behind the infantry. The absence of the launch position mentioned in the text has an extremely negative effect on *"Goliath"* operations. The infantry very much desires a vehicle, and this desire will be met if this vehicle is made self-propelled and can be moved forward by one man. The *"Goliath's"* speed must be increased from 12 to 15 kilometers and hour.

Another report by the battalion commander was written on 2 July 1942:

Panzer-Abteilung 300 2 July 1942

Report on Operations by
Panzer-Abteilung 300

The battalion saw action on 11 days while subordinated to the *XXX.* and *LIV. Armee-Korps* and served with the following divisions: *22. Infanterie-Division*, *24. Infanterie-Division*, *28. Infanterie-Division*, *50. Infanterie-Division*, *72. Infanterie-Division*, *132. Infanterie-Division* and *170. Infanterie-Division.*

1) Use of special vehicles (see report dated 22 June 1942). Also committed were:

1 *Sd.Kfz. BIV*: Ran over mine, fulfilling its role as mine detector

1 captured tank: Guided against a trench system

1 captured tank: Destroyed bunkers containing munitions

2 captured tanks: Each destroyed another bunker

1 captured tank: During street fighting eliminated flanking rifle fire

2 *"Goliaths"*: Committed against railway underpass, the crew was destroyed

2) The employment of demolition charge carriers and tanks enabled the infantry to take approximately 2,000 prisoners. In many cases the use of special-purpose vehicles and tanks made it possible for the infantry to continue to advance.

3) The tanks provided cover for the demolition-charge carriers, provided considerable support to the infantry (always in front of the assault guns) and achieved the following results:

Numerous infantry positions were engaged with tank main guns, enabling the infantry to penetrate the system of positions.
Approximately
20 bunkers and
10 small fighting positions were destroyed.
10 antitank guns,
1 76.2-mm antitank gun and
4 mortars were silenced or destroyed.
1 coastal patrol vessel sunk and
another hit and set on fire.

4) Nearly every tank sustained moderately heavy damage from artillery fire, mines, antitank guns and antitank rifles. Three tanks were written off.

5) The experiences and recommendations listed in the letter of 22 June 1942 have not changed. It is requested that every effort be made to realize the battalion's recommendations.

signed Weicke
Hauptmann and battalion commander

Following the capture of Fortress Sevastopol, the *11. Armee* was used to secure the Crimean Peninsula from 4 July 1942 until the end of August 1942.

The bulk of the armored units that had been attached to the *11. Armee* during the fighting for Sevastopol were released to *Heeresgruppe Süd*. On 7 July 1942 *Heeresgruppe Süd* was divided into *Heeresgruppe A* and *Heeresgruppe B*.

Panzer-Abteilung 300 (Fernlenk) received movement orders and on the afternoon of 4 July 1942 it entrained and was transported northeast. After arriving in the Charzyssk area, the battalion was placed under the operational control of the *1. Panzerarmee* (*III. Panzerkorps* and *LIV. Panzerkorps*).

On 9 July the *1. Panzerarmee* attacked towards the Donets on either side of Lisichansk. *Panzer-Abteilung 300 (Fernlenk)* took part in the breaching of the Soviet front line and the pursuit of enemy forces into the Donets Basin and toward the Lower Don until 24 July 1942.

Following the Battle of Rostov and Bataisk from 21 to 26 July, the battalion was assigned the task of securing the operational area.

Neither radio- nor wire-controlled demolition-charge carriers saw action during the operations of July 1942, however, the command and control tanks took part in several tank engagements.

A brief rest period near Uspenskaya was used to carry out necessary repairs and maintenance to vehicles and equipment. On 6 August the battalion entrained in Amvrosiyevka and was transported northwest in the direction of Leningrad.

The *Wehrmacht* had been trying to take the city for more than a year. After the fall of Sevastopol, units which had taken part in the reduction of that fortress were to be transferred to the Leningrad front. Hitler tasked the experienced *11. Armee* with carrying out "Operation Northern Light." The anticipated attack date was 11 September 1942.

After a ten-day journey, *Panzer-Abteilung 300 (Fernlenk)* arrived at Krasnogvardeysk on 15 August. After detraining, the battalion road marched to Tishkovitsi, 20 kilometers to the south, where it was billeted.

On 27 August 1942 Soviet forces of the Leningrad and Wolchow Fronts launched a major offensive and, by 4 September, they had achieved a major penetration against the *18. Armee*. The *11. Armee* succeeded in taking back the lost ground by 2 October, smashing the Soviet forces. The battalion, which had been redesignated *Panzer-Abteilung 301* on 9 September 1942, participated in this defensive struggle south of Lake Ladoga. Since the ammunition stockpiled for "Operation Northern Light" had been expended in the defensive fighting and as there were no reserves available, the attack on Leningrad was postponed.

In the period from 3 to 30 October *Panzer-Abteilung 301* took part in the positional fighting in the area of operations of the *11. Armee* and then in the area of the *18. Armee* until 16 November. During this time the battalion carried out an internal reorganization. The *3./Panzer-Abteilung 302* arrived from Neuruppin on 22 October 1942 and was incorporated into the unit as the *2./Panzer-Abteilung 301*. The former *2./Panzer-Abteilung 301* was taken out of action, redesignated the *3./Panzer-Abteilung 302* and transported to Neuruppin (see Chapter 2.2.5). Furthermore, *Kompanie Abendroth* was formed as a special remote-control unit. The bulk of this unit came from the *1./Panzer-Abteilung 301* with additional personnel from the *3./Panzer-Abteilung 301* and the headquarters company.

During the period from 23 October to 11 December 1942, *Panzer-Abteilung 301* remained with *Heeresgruppe Nord* in the area of the *18. Armee*, with the exception of *Kompanie Abendroth*, which left the battalion in mid-November. The onset of extremely cold weather at the beginning of November, with temperatures as low as minus 25 degrees Celsius, further restricted operations by the tanks and demolition charge-carriers. On 12 December 1942 *Panzer-Abteilung 301* began entraining for transport home to Arnswalde.

In his dairy, Heinz Prenzlin described the period from the battalion's departure from the Crimea on 2 July 1942, the fighting in the Donets Basin and the fighting at Leningrad:

2 July 1942: 0830 hours departure for Tole rest area.

3 July 1942: Tole rest area, unpacked, made brief check of the vehicles and carried out minor repairs, prepared to depart.

4 July 1942: 0100 hours departed Tole for Sarabus to entrain. Train left at 1400.

5 July 1942: 0215 hours Melitopol, 0500 continued on through Fedorovka, Saporoshje and Bologi.

6 July 1942: Detrained in Wolnowacha; 1800 hours departed for rest area.

7 to 11 July 1942: Rest area, called Charzyssk Army Baths. Rest day; we can finally get really clean again and swim. It is an open pool. We are also able to wash our clothes properly once again.

12 July 1942: Alert at 0330 hours. Prepare for operations. At 1500 hours departed for Artemovsk via Makejewka, Jassinowataya, Gorlo (paved road) and Nikitowka. There we rested, refueled, checked the vehicles and got something to eat.

13 July 1942: 0200 hours departed for Lissitchansk, arrived there at 0800. Reported to division; discussion concerning operational employment.

14 July 1942: 0200 hours briefed on mission for today. We advanced on the road through Proletarsk to the Donets; first crossing. We met little resistance, scarcely saw any Russians, only scattered groups of stragglers. These offered almost no resistance and were all captured. Then on to Schachta-Tomachta. Arrived at 2200 hours. Day's objective has been reached!

15 July 1942: Departed at 0200 hours. Today we moved into the second line [of defense], through Mikhailowka, Novo Astrakhan and on until about 2200 hours, covered about 100 kilometers.

16 July 1942: 0300 resumed advance, through Ormelko. There were some difficulties, though we were not required to intervene directly. Reached the day's objective at 2100 hours.

17 July 1942: We were at it again today. Departure at 0300, at 1300 arrived Tarassowka, onward until 1600. Already at day's objective, it's going like clockwork, have scarcely laid eyes on a Russian, they are running as fast as they can.

18 July 1942: Rested today. Checked vehicles, replaced track blocks and shock absorbers, changed oil, checked tire pressures, etc.

19 July 1942: Wakened by Ivan at 0500, low-level attack by 6 machines that had probably got off course, nothing happened. Departure at 1500 hours, made another 20 kilometers although there were a few problems.

20 July 1942: Departure at 0200 hours, drove in front again today. No enemy contact as far as Nishnij-Brochja. After that several engagements with support from army 88-mm *Flak* on self-propelled mounts, then on to Kamensk, second crossing of the Donez. Took up positions on the bank to cover the *Panzerpioniere*, who constructed a 180-meter pontoon bridge. During the night guard duty.

21 July 1942: Departure at 0300 hours via Schachty, big coal pits and mines. Combed the terrain with infantry support, boilers still under steam, must have been surprised. Didn't see a single Russian. Towards evening made camp 5 kilometers south of Schachty. Bombing raids.

22 July 1942: Ivan counterattacked at 0130 hours. Got to within 3,000 meters of us but was beaten back sucessfully. Saw Russian tanks for the first time during the advance. He attacked with ten T

34's, six of which were knocked out. They weren't even welded properly, just spot welded. Continued on at 1700 hours, bombing raids.

23 July 1942: Departed at 1100 hours, to within 18 kilometers of Nowoscherkask. Several times enemy aircraft strafed and dropped bombs. Some wounded, especially in the trains.

24 July 1942: Further air attacks against us at lengthy intervals. Ivan dropped leaflets, called us by name and declared that he would not take any prisoners from *Abteilung Abendroth*. Pleasant outlook for us. He obviously had us on his mind. At 1400 *Dr.* Schmidt was seriously injured in an air attack. At 1430 we changed positions. We moved five kilometers closer to Nowoscherkask and immediately made camp and camouflaged our position.

25 July 1942: Attack on Nowoscherkask, only limited fighting. The air attacks of the previous days were obviously supposed to hold us up in order to facilitate the Russian withdrawal. We met almost no resistance. The whole thing was over at 1530. Camped in the gardens in a suburb. They were all nice houses of the professors and lecturers from the university. We had fresh fruit again for the first time in a long while. The gardens are full of it. For the first time I ate ripe green figs, tasted great.

26 to 29 July 1942: Combed the area around Nowoscherkask all the way to Rostov looking for Russian stragglers. To date have covered approximately 1,000 kilometers.

30 July 1942: 0315 hours departed Nowoscherkask, through Rostow to Uspenskaja, there set up camp in an apple orchard.

31 July to 5 August 1942: Uspenskaja rest area. All vehicles thoroughly overhauled. The repair section has its hand full, but we can't complain about boredom either. Deadline for all work is 6 August.

6 August 1942: Departed rest area for Amwrosijewka entraining station at 0500 hours. Pulled out at 1500, reached Ilowaiskoja at 1840, Charzyssk at 1900, Chanchonkowo at 2000.

7 August 1942: Departed Chanchonkowo at 1230 hours, Jassinowatajа at 1700.

8 August 1942: 0640 hours Awdjewka, 0730 Otscheretinow, 1200 Stalino, 1435 Uljanowka, 1530 Pissmenskaja, 1745 Sinelnikowo.

9 August 1942: 0300 hours Dnjepropetrowsk, 0910 Pjatichatki, 1130 Alexandrija, 1145 Korristowka, 1430 Krjukow, 1600 Krementschuk, 1730 Glabino, 1900 Chorol, 2030 Romodan.

10 August 1942: 1000 hours Snowskaja, 1435 Gomel.

11 August 1942: Departed Shlobin at 1200 hours, 1400 Rogatschew, 1530 Byschon, 1900 Orsha.

12 August 1942: Departed Orsha at 1415 hours, 1845 Witebsk, at 2200 hours stopped 45 kilometers from Newel as the tracks had been blown up by partisans eight times.

13 August 1942: 1615 hours. Now we can proceed. 1840 Newel, 2200 Nowo-Sokolniki.

14 August 1942: 0915 hours continued onwards through Dno.

15 August 1942: 0700 hours in Krasnogwardeisk, detrained, made camp 20 kilometers south in Tischkowitzi.

16 August to 19 September 1942: Tischkowitzi rest area. It is a village which has been occupied almost exclusively by Finns for 30 years. We overhauled all our radio-control equipment, had to reestablish field-test strengths since weather conditions here are completely different. Meanwhile, the entire headquarters company has jaundice. It is a frightful situation, one always craves something that can't be had. Received packages from Papa in Greece with dried fruit, it is delicious.

20 September to 1 October 1942: Building 10-man bunkers. We went to Schlüsselburg to cut wood in the forest. From there we could see the spires of Leningrad. While cutting wood we also scouted the area. We discovered that Leningrad is teeming with German POWs.

2 to 5 October 1942: No changes.

6 October 1942: Effective today, I am transferred to the 2nd Company as communications mechanic. Had to make numerous preparations, especially have issued to me test equipment such as field-strength tester, test transmitter, test receiver, etc. Reported to the 2nd Company in the afternoon. Met my old "friend" (???) again, *Hauptfeldwebel* Mattes. Well, this will be fun. Doesn't matter, I have a thick skin, somehow it will turn out all right. He recognized me at once. New army field post number: 08914.

7 to 22 October 1942: 2nd Company is reformed as an overstrength special-purpose unit. Sonderführer, maintenance personnel and 84 drivers are transferred from the 1st Company and the headquarters company. A platoon of *"Goliaths"* comes from the 3rd Company as well. Everyone is uneasy, we don't like what's coming from the *Spieß*. He probably hasn't realized yet that we are a battle-tested field unit. We get ready for the upcoming operations. There is much discussion as to where we will be going next. The attack on Leningrad has been called off, as we hear it our plans have been betrayed. The headquarters company and the 1st and 3rd Companies move out in preparation for entraining. Supposedly back to Germany for reorganization. The headquarters company's 1st Platoon is to stay here as an instructional platoon.

In mid-November *Kompanie Abendroth* received orders to transfer into the area of *Heeresgruppe B*. The unit began entraining in Gatchina on the afternoon of 17 November and the trains departed at night. Traveling on secondary lines, *Kompanie Abendroth* arrived in Oblivskaja eight days later. There, approximately 100 kilometers southwest of Stalingrad, the company detrained on 26 November 1942.

On 19 November 1942 the Soviets launched their major offensive, dubbed "Operation Uranus". The Red Army broke through the Romanian units holding the fronts northwest and southeast of Stalingrad and encircled the *6. Armee*.

Following the Russian breakthrough at Stalingrad, *Heeresgruppe Don* was formed from *Armee-Oberkommando 11*, which had been sent to the area in mid-November. It was then inserted into the gap in the front between *Heeresgruppen A* and *B*. "Operation Winter Storm", the German relief effort, began on 12 December. After eleven days of fighting some of the lost territory was regained, however the relief force failed to break through to Stalingrad.

During this period *Kompanie Abendroth* participated in the defensive fighting at the Don and in the Kalmuck Steppe and, after a further withdrawal, in the defensive fighting in the area of operations around the Donez. Wire- and radio-controlled demolition-charge carriers were employed in various counterattacks and defensive actions.

On 30 January 1943 *Kompanie Abendroth* entrained in Lichaja and departed for Germany. After a journey which lasted almost six weeks due to numerous interruptions, the company reached its new garrison at Eisenach on 13 March.

Heinz Prenzlin described the operations of *Kompanie Abendroth/Panzer-Abteilung 301* during the period until the beginning of March 1943:

23 October 1942: We are working feverishly, the First Sergeant is screwing up royally, he is probably no longer master of the situation. I have never experienced anything like this, he immediately has problems with the newcomers no matter what their rank. The *Spieß*

still acts like we are in garrison and plays the big mouth. He will also be brought down, wait and see.

5 November 1942: The same as yesterday, frost for the first time, minus 13 degrees Celsius.

6 November 1942: It is growing steadily colder, minus 16 degrees.

7 November 1942: 20 degrees below zero.

8 November 1942: Where is the thermometer going? Minus 22 degrees.

9 November 1942: It is no longer possible to work on the vehicles. The *Spieß* still does not grasp the situation. He continues to do nothing but curse, wants to court martial everyone. I believe his brain is already frozen, he no longer has any idea what's going on. Today it was minus 26 degrees.

10 November 1942: The company commander has ordered that there will be no more morning formations. 28 degrees below zero. It can't get much colder, otherwise our thermometers will freeze up. It is now almost impossible to get the engines to start. We are receiving heat circulation equipment for the tanks. Some vehicles are kept running all the time to be used to pre-heat the others.

11 November 1942: Warmer again today, only 10 below. The *Spieß* is in his element again, but as far as we are concerned he can go to hell!!

12 and 13 November 1942: No change, 10 below.

14 November 1942: Colder again, 15 below.

15 November 1942: It is warm, minus 2, hard to believe, but in its place we receive 20 centimeters of snow and it is still snowing. Huge flakes coming down.

16 November 1942: The entire company is filled with a curious uneasiness. Scarcely an officer to be seen. As well, contrary to what we are used to, the *Spieß* is much calmer today. At 1200 hours we had to fall in. We were told to pack at once, we are being moved but no one knows where. Much debate. Some claim to have heard we were going to Germany. Some say to Kurland. Yet others say we're going to Stalingrad. This unit is frightfully strict on secrecy. It was quite different when I was with headquarters.

17 November 1942: Preparations for departure. We moved out at 1200 hours, arrived in Krasnogwardeisk at 1330, Gatchina at 1600. There we entrained. I once again had a falling-out with the Spieß, who wanted to send me on watch immediately. I outfoxed him, telling him that I had already volunteered for the *Flak* car, it worked.

18 November 1942: 2400 hours departure, 0730 Olkatsshewo, 1015 Luga, 1615 Pleskow.

19 November 1942: 0300 hours Absene, pulled out again at 1130. 1700 hours Rossitten, 2030 Dünaberg. We celebrate, we're going home.

20 November 1942: 0400 hours Wilna, 1330 Molnodetchewo, 1830 Minsk.

21 November 1942: 0515 hours Bobruisk, 0820 Schlobin, 1100 Gomel. We are all disappointed; we're not going home. There are murmurs that we are headed for Stalingrad.

22 November 1942: 0700 hours Borowasja, Sumni, 1400 Smurodewo. We are moving exclusively on side lines. Why, we cannot determine.

23 November 1942: 0500 hours Pawlograd, 1900 Sinelnikowo. There is no longer any doubt, we are heading south.

24 November 1942: 0820 hours Awdejewka, 1020 Krasnowadaja.

25 November 1942: 0300 hours Schatowo, it is bitterly cold. We often have to sweep the snow from the car, we are issued coal at every stop.

26 November 1942: 0600 hours Walkowo, 1100 Morozowskaja, 1500 Oblivskaja. We are waiting for further orders. It is said that we are bound for Tschirskaja, the last station before Stalingrad. At 1545 hours we detrain. We can go no farther, the Russians are a few kilometers in front of us.

27 November 1942: We temporarily made camp at the Oblivskaja rail station. Not a house to be seen anywhere, no trees, about 80 centimeters of snow. We receive snow chains, installing them is miserable work in this weather. The Ivans are 15 kilometers ahead of us, we are seemingly alone here, outposts are set up. Radio readiness is ordered, I have to help the company commander out as radio operator, von Abendroth is a great guy. He always falls back on his veterans. The *Spieß* grumbles.

28 November 1942: Overnight the Ivans pester us, they are now 3 kilometers away. We are looking for a better staging area.

29 November 1942: We billeted in the nearest village, or at least what passes for a village here. All vehicles are painted white for camouflage.

30 November 1942: The Russians visit us from above, the repair section is hit by two bombs.

1 December 1942: 0600 hours alert, immediate readiness called, 0900 departure back to Tornowkoi (12 kilometers).

2 to 9 December 1942: From here individual missions in platoon strength, move to Tornowkoi.

10 December 1942: Departure for Nowo-Derbinowskij.

11 December 1942: We have been assigned a company of *Infanterie-Division "Hermann Göring"* [Author's note: The troops were in fact members of the *7.* or *8. Luftwaffenfelddivision*). We deploy to attack, the Ivans are trying to force a breakthrough with tanks. We use *"Goliaths"* carried on trucks, results with them are good. "Göring's Guard" runs away, the "old man" is furious. We collect the automatic rifles, submachine-guns and machine-guns they threw away. The machine-guns are *MG 42's*, most of ours are still *MG 34's*.

12 December 1942: At 0800 we withdraw to regroup, a new line is to be formed. We pull back 60 kilometers to Perschtschewskij where we set up camp and maintain operational readiness.

13 December 1942: We occupy quarters. Each platoon places one tank on guard duty.

14 December 1942: Repair equipment and weapons as best as possible. It is very cold and we hardly ever get out of our winter things. Lice have become a plague.

15 December 1942: The same as yesterday. I have been promoted to *Gefreiter* retroactive to 1 December.

16 December 1942: Perschtschewskij rest area, vehicles and equipment are made operational.

24 December 1942: Wakened at 0400 hours, 0700 departure for Morosowskaja. We spent the night there and celebrated Christmas as best we could. There is no mail, we haven't received any in four weeks now. The kitchen sends us some mulled wine, we procure a small sheep. Heidecker is kept busy, I work all evening making preparations. By about 2400 all is ready, we all enjoy it. The *Spieß* has gotten wind of it, of course he tries to get worked up again, but we give him the cold shoulder. Abendroth is enthusiastic.

25 December 1942: 0800 hours continued on to Komschak, moved approximately 63 kilometers.

26 December 1942 to 1 January 1943: Remained in Komschak,

took up position here. The village lies in a *balka*, which is what they call a shallow valley here or, more accurately, a depression. We have dug in behind the south hill, the Russians are opposite us on the north hill, but nothing happens. All the Ivan does is fire a parachute flare now and then, strange situation. Both sides fire into the air on New Year's Eve. The Russians call New Year's greetings to us by loudspeaker, we reply.

2 January 1943: 0400 hours we pull back, 0600 departure for Chudakowski via Katzenskaja. We billet there but always remain at alert readiness. The vehicles have to be warmed up every two hours.

3 January 1943 to 12 January 1943: Remain here in readiness.

13 January 1943: 0700 hours departure to Bjelaja Kalitwa.

14 to 16 January 1943: In Bjelaja Kalitwa.

17 January 1943: 0830 hours departure to Lischaja, on the way attacks by tactical aircraft, but nothing happens. I have a terrible cold with a high fever. At 1900 I enter the field hospital in Lischaja.

18 to 28 January 1943: In the hospital.

29 January 1943: The old man visits me in hospital and takes me with him to the unit. On the way he explains why. We are supposed to pull back from the Russians and, at the first favorable opportunity, entrain for movement to Germany.

30 January 1943: Entrained in Lischaja.

31 January 1943: Load the remaining vehicles, toward midday depart to Lischaja West.

1 February 1943: Still here, no locomotive. Artillery fire is coming ever nearer, situation slowly becoming critical, finally at 1345 hours we get two locomotives. Departure at 1400 hours.

2 February 1943: 0230 hours Krasni Kolos, once again we stop, no one knows why. Towards morning we take charge of the locomotive. The engineer and stoker are Poles; the old man doesn't trust them. In the afternoon we learn that Stalingrad has fallen. It is a sad birthday for me.

3 to 5 February 1943: Still stuck here. It is said that the Russians have broken through ahead of us.

5 February 1943: 1400 hours, finally we get moving again, 1530 hours Schtschetowo.

6 to 8 February 1943: Stuck on the line 8 kilometers from Chawo-Chinow.

9 February 1943: 0300 hours continued on through Chawichin Debelzanowo.

10 February 1943: 0630 hours Jassinowataja.

11/2 to 20 February 1943: Once again unable to progress further. The Russians have broken through on a broad front between Jassinowataja and Dnjepropetrowsk. Jassinowataja is the second-largest marshalling yard in Russia with 445 spurs. The station is jam-packed with trains. Italians, Romanians, Hungarians, hospital trains, munitions trains, fuel trains and rations trains, there are also cars carrying laundry, uniforms, winter clothing and boots. The Italians have a good nose for this, they are always discovering something new.

21 February 1943: Went to the movie theater in Stalino and then to the opera together with *Feldwebel* Kallerhof and *Unteroffiziere* Lillings, Erdel, Salzwedel and Krause. It was a lovely change, it was a beautiful day.

22 to 27 February 1943: Still stuck here. It is said that the *SS "Wiking"* is supposed to get us out of here.

28 February 1943: Went to a variety show in Jassy, it was very good. The Italians, Hungarians and Romanians are beginning to mutiny. They are leaving their trains and take off on their own.

1 to 2 March 1943: Still in Jassy. No one is allowed to leave the train. The *SS "Wiking"* has done it. The tracks are now being repaired by railway pioneers.

3 March 1943: Finally, at 1530 hours, we resume our journey, at 1600 we pass Stalino West.

4 March 1943: 1015 hours Zarekonstantinowka; I am not well, have suffered a relapse.

5 March 1943: 0500 hours Saporoshje East, 0600 Saporoshje Main Station, 2000 continued on.

6 March 1943: 0700 hours Sinelnikowo, 0900 continued on, 1100 in Usel. Moved out again at 1330 hours, 1645 Werchowzewo.

7 March 1943: 0800 hours Snamenka, 1140 Kirowograd, Nowo-Ukrainka-Hobo, at 1645 hours, 1745 Pomoshnaja, 2015 the border. We move into Romanian territory, border town Bukunieno.

8 March 1943: 1330 hours Slobodoa, 1630 Popeliuki.

9 March 1943: 1215 hours Podwoloczyska, 1715 Tarnopol, 2315 Lemberg.

10 March 1943: Departed Lemberg at 0630 hours, 1015 in Przemysl, there went through delousing for the first time and was fed, continued on at 1500 hours.

11 March 1943: 0600 hours in Reichshof, 1200 Tarnow, 1800 Krakow.

12 March 1943: 0600 hours Kattowitz, 0700 Gleiwitz, 0800 Heydebeck, 1100 Neisse, 1445 Kamenz, 1730 Schweidnitz.

13 March 1943: 0530 hours Dresden-Friedrichstadt, 1115 Weißenfels, 1300 Weimar, 1445 Erfurt, 1600 Gotha, 1700 Eisenach. End of the line after six weeks on the rails. We detrained.

14 March 1943: 0700-1100 stow our gear; 1500-1800 mail call. Some of the mail is from the previous year. Most of the packages have spoiled. Beginning at 1800 hours we are given passes into the city. Everywhere people come up and ask us where we have come from and shower us with questions.

15 and 16 March 1943: Clean vehicles and uniforms. Officers and non-commissioned officers are issued new uniforms. We keep our old things.

17 March 1943: On order of the *Führer* we are given three weeks leave. This is my first home leave.

On 25 January 1943 the elements of *Panzer-Abteilung 301* which had been transferred back to Arnswalde were transported to Neuruppin for reorganization. *Panzer-Abteilung (Funklenk) 301* was to be formed there using the remnants of the disbanded *Panzer-Abteilung 302* (see Chapters 2.2.5 and 2.2.6

Gliederung der Panzer - Abteilung 300 (FL) (Stand 13.09.1942)

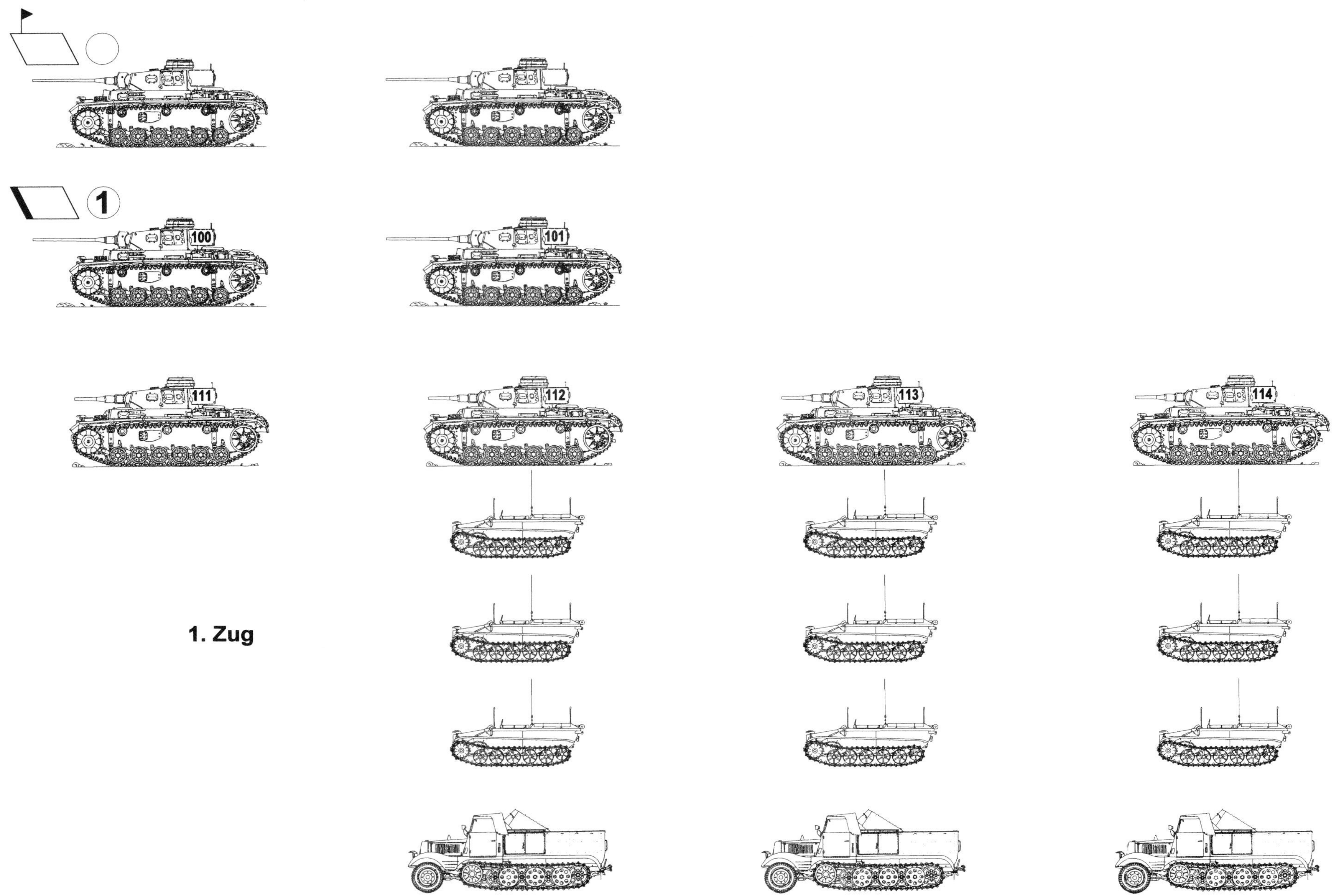

M.Jaugitz / Trojca 2000.

Gliederung der Panzer - Abteilung 300 (FL) (Stand 13.09.1942)

Sondergeräte - Reserve

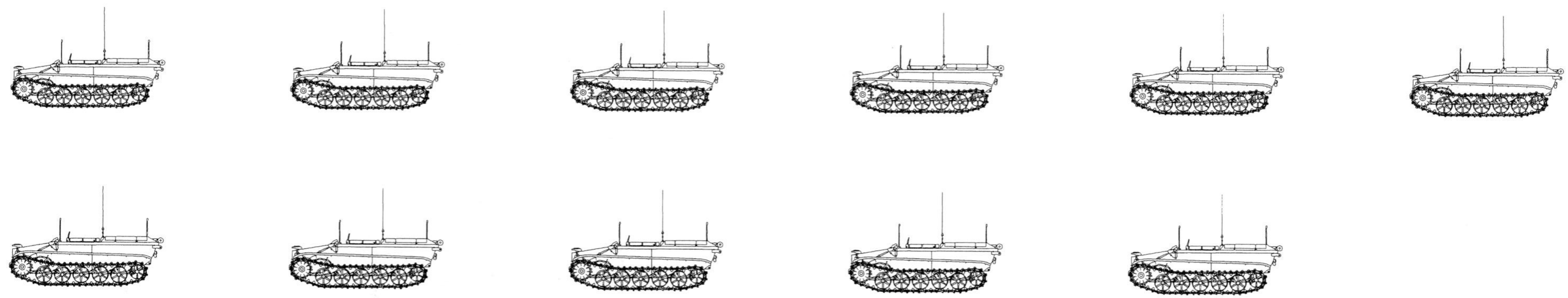

2. Zug

M.Jaugitz / Trojca 2000.

Gliederung der Panzer - Abteilung 300 (FL) (Stand 13.09.1942)

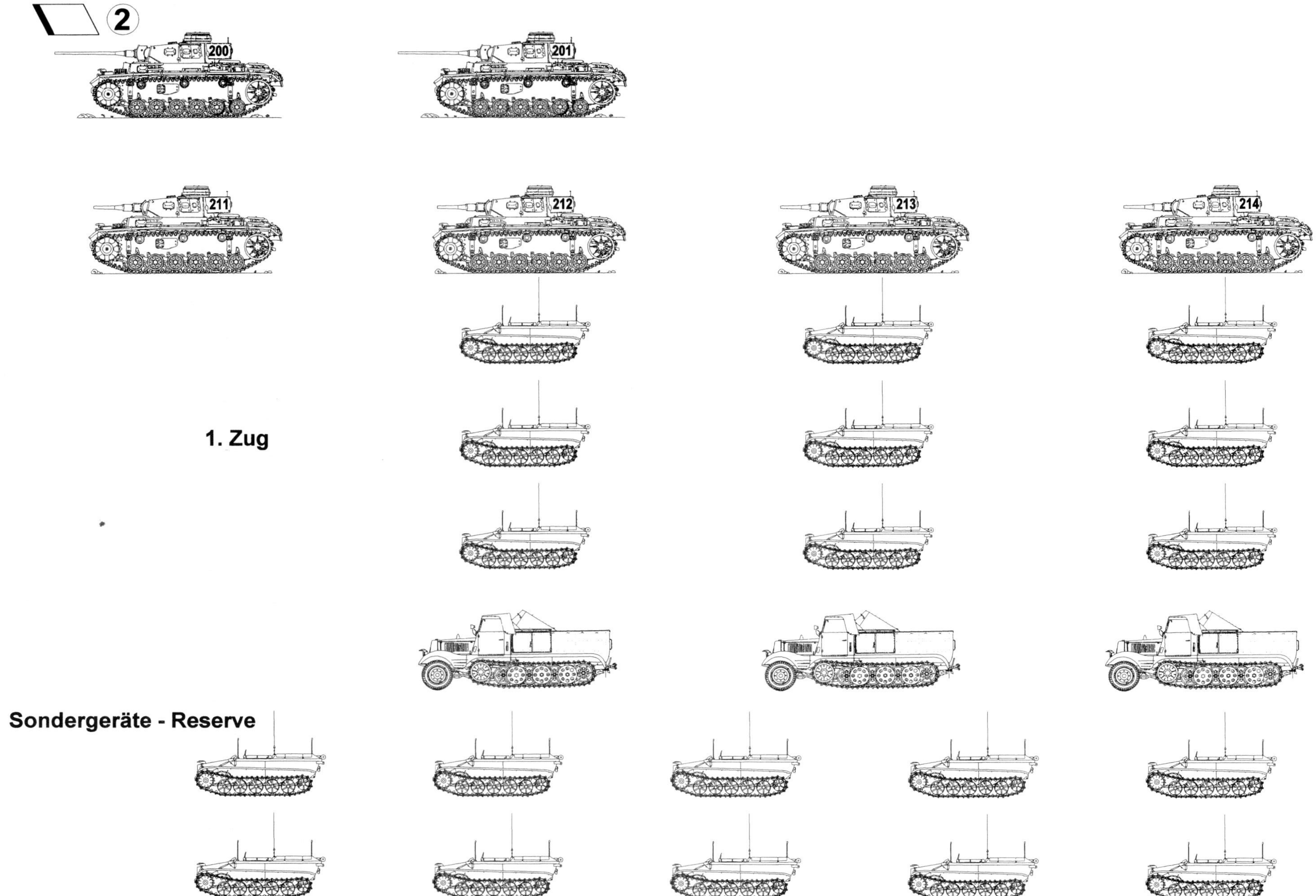

M.Jaugitz / Trojca 2000.

Gliederung der Panzer - Abteilung 300 (FL) (Stand 13.09.1942)

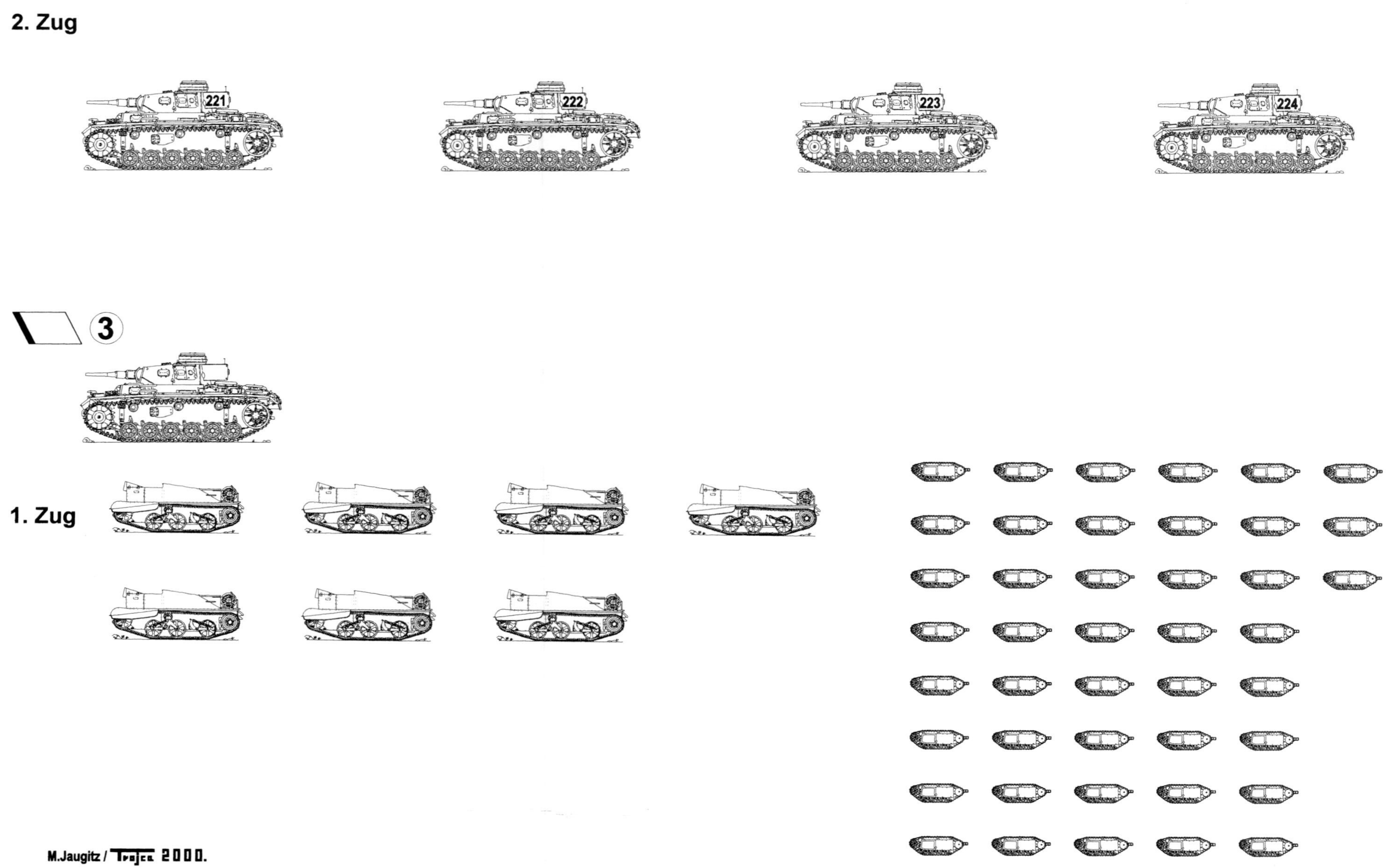

Iron Cross, Second Class award document to *Gefreiter* Hans Gradolph of *Panzer-Abteilung 300*. The designator *Funklenk* is not mentioned anywhere, probably due to security reasons or the fact that the award was prepared by another command which was not aware of the unit's proper designation.

1939

Im Namen des Führers und Obersten Befehlshabers der Wehrmacht

verleihe ich

dem

Gefreiten H a n s G r a d o l p h

Panzer-Abteilung 300

das

Eiserne Kreuz 2. Klasse

A.H.Qu., den 11. Juli 1942.

Der Oberbefehlshaber der
11. Armee
I.V.

(Dienstsiegel)

General der Kavallerie
(Dienstgrad und Dienststellung)

Besitzzeugnis

Dem Gefreiten

(Dienstgrad)

Willi Stolz

(Vor- und Zuname)

2./Pz.Abt.300

(Truppenteil)

wurde das

Panzerkampfabzeichen

— Silber —

verliehen.

Stempel

A.H. u. den 4.Juli 1942

(Ort und Datum)

Der Chef des Generalstabes

I.V.

(Unterschrift)

Oberst i.G.

(Dienstgrad und Dienststellung)

Panzerkampfabzeichen award document to *Gefreiter* Willi Stolz of the *2./Panzer-Abteilung 300*. As with the previous document, the designator *Funklenk* is not mentioned.

Besitzzeugnis

Dem Gefreiten
(Dienstgrad)

Richard W a l z
(Vor- und Zuname)

1./Pz.Abt. 300
(Truppenteil)

wurde das

Panzerkampfabzeichen
— Silber —

verliehen.

Stempel

Div.Gef.St., den 1.9.194
(Ort und Datum)

(Unterschrift)

(Dienstgrad und Dienststellung)
Generalmajor und Div.Kdr

Panzerkampfabzeichen award document to *Gefreiter* Richard Walz of the *1./Panzer-Abteilung 300*. As with the previous documents, the designator *Funklenk* is not mentioned.

Panzerkampfabzeichen award document to *Gefreiter* Hans Gradolph of the *Stab/Panzer-Abteilung 301*. As with the previous documents, the designator *Funklenk* is not mentioned.

Besitzzeugnis

Dem Gefreiten

(Dienstgrad)

Hans Gradolph

(Vor- und Zuname)

Stab / Panzer-Abteilung 301

(Truppenteil)

wurde das

Panzerkampfabzeichen

— ~~Silber~~ —

- Bronze -

verliehen.

O.U., den 20. Oktober 1942

(Ort und Datum)

(Unterschrift)

Hauptmann und Abt. Kdeur

(Dienstgrad und Dienststellung)

Krimschild award document for participation in the Crimea Campaign to *Gefreiter* Siegfried Fellhauer of the *Stab/Panzer-Abteilung 301*. As with the previous documents, the designator *Funklenk* is not mentioned.

BESITZZEUGNIS

IM

NAMEN DES FÜHRERS

WURDE DEM GEFREITEN

SIEGFRIED FELLHAUER

STAB/PANZER-ABTEILUNG 301

DER KRIMSCHILD VERLIEHEN.

GENERALFELDMARSCHALL

The *Panzer III, Ausführung J* was used as the new type of command and control tank. (Fritschken)

The company commander of the *2./Panzer-Abteilung 300* walks by his unit, which is preparing to move out. (Fritschken)

A large number of *Borgward BIV, Ausführung A's* follow the last command and control tank. (Fritschken)

The *3./Panzer-Abteilung 300* was established at the beginning of February 1942. British Bren Gun Carriers were also employed as remote-controlled demolitions carriers. (Sigmund)

The establishment of the *2./Panzer-Abteilung 300* in Cottbus in the spring of 1942. (Fritschken)

The company commander of the *1./Panzer-Abteilung 300*, *Oberleutnant* Ferdinand von Abendroth. (Gradolph)

The company commander of the *2./Panzer-Abteilung 300*, *Oberleutnant* Günther Fritschken. (Fritschken)

Above: An Opel-Blitz radio calibration truck of *Panzer-Abteilung 300*. Notice the antenna on the vehicle's roof. (Gradolph)

Right: The battalion also had a few *VW-Schwimmwagen* in its inventory. (Gradolph)

Above: A number was painted on the rear of each demolitions carrier to make section identification easier. (Lock)

Left: The entire battalion was rail-loaded in Cottbus starting 11 May 1942. (Lock)

Above: The first train is ready to depart. (Stolz)

Top right: The Bren Carriers of the *3./Panzer-Abteilung 300* are loaded. In addition to the tactical markings on the left fender, white Balkan crosses can also be seen on the superstructures. (Gradolph)

Bottom right: Bren carriers employed as demolition-charge carriers and their command and control tank, the *Panzer III*. (Gradolph)

Utility Tractors of the *3./Panzer-Abteilung 300* can be seen on the other flat cars. (Gradolph)

An antiaircraft car is coupled. Another Utility Tractor can be seen in the background. (Gradolph)

Parts of the motor pool of the battalion staff. (Gradolph)

This *Borgward BIV* has been secured against slippage by the use of wooden chock blocks. (Stolz)

This BIV carries the identifier *2 / I* with the numeral *22* beneath it. That stands for 2nd Company, 1st Platoon, Vehicle 22. (Stolz)

A few remote-controlled *NSU-Kettenkräder* were also brought along. (Schick)

Above: After arrival and off-loading in the Crimea, the equipment of the battalion was displayed for the commanders of the different infantry divisions belonging to the *11. Armee*. (Lock) **Below**: The vehicles have been camouflaged. (Bundesarchiv 230/686/24)

The *BIV* is prepared for a demonstration. (Bundesarchiv 230/686/23)

The battalion commander, *Hauptmann* Weicke, waits for the demonstration to start. (Bundesarchiv 230/686/20)

The commander of the command and control tank takes over the remote guidance of the *BIV*. (Bundesarchiv 230/686/19)

Those invited to the demonstration watch critically. (Bundesarchiv 230/686/18)

The *BIV* has reached its target. (Bundesarchiv 230/686/4)

The demolition charge has been dropped. (Bundesarchiv 230/686/06)

The commanding general observes the demonstration and thanks *Hauptmann* Weicke afterwards for its success. (Bundesarchiv 230/686/7)

The command and control tanks of the *1./Panzer-Abteilung 300* have assembled in their entirety. (Lock)

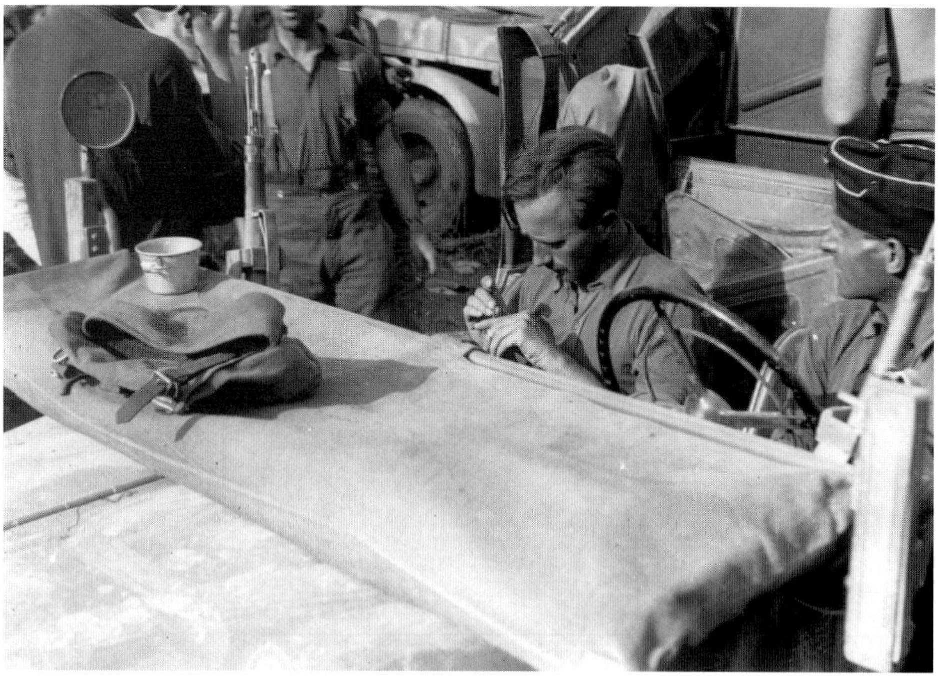

Oberleutnant Fritschken and *Leutnant* Fischer of the *2./Panzer-Abteilung 300* set out on reconnaissance. (Stolz)

Maintenance is performed on the command and control tanks. The return rollers are greased… (Stolz)

…and the roadwheels exchanged. The protective rubber cap on the roadwheel in the foreground has been damaged. (Stolz)

A snapshot taken before operations. (Lock)

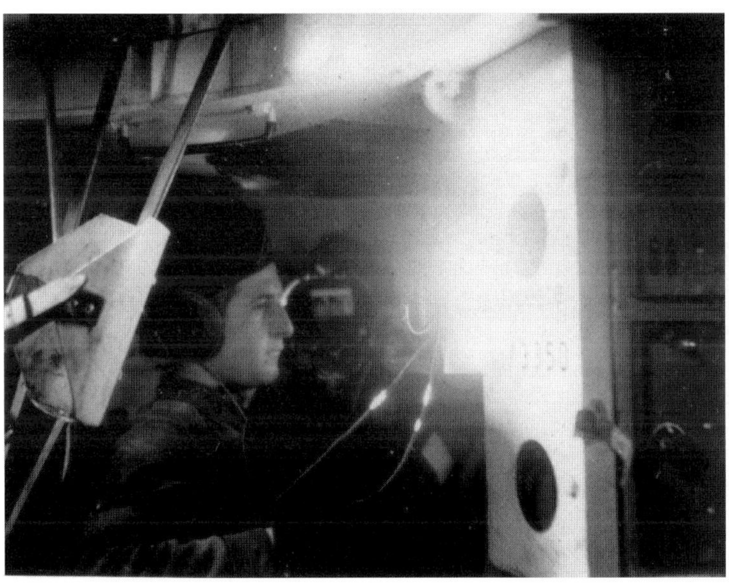

The driver of a command and control tank. The chassis number of the vehicle (73350) has been applied to the radio equipment cover. (Stolz)

A platoon leader of the *2./Panzer-Abteilung 300*, *Leutnant* Fischer, in his command and control tank. (Stolz)

Above: Spare track blocks have been affixed to the front of the command and control tank to provide additional protection. (Lock)

Above, right: This early model of the *Panzer III*, *Ausführung J* does not yet have the armored air intake covers on the rear deck. (Osterried)

Bottom right: The *Strolch* is awaiting its next operation. The radio-control antenna has already been attached. (Stolz)

Above: Due to lack of space, a replacement roadwheel has been mounted far forward on the right-hand fender. (Seeger)

Top left: The *Spieß* gets a haircut in the field. A well-camouflaged *Panzer III* command and control tank can be seen in the background. (Stolz)

Left: This tank crew is using available free time to write letters. (Stolz)

Above: One of the first *Goliaths*. (Rettig)

Right: A driver and his *BIV*. (Seeger)

Available time is used to belt machine-gun ammunition. This *Panzer III* command and control tank is also an early-version of the *Ausführung J*. (Lock)

A remote-control section moves into a jump-off position. (Rettig)

After reaching the jump-off position, the drivers wait for their orders. (Fiedler)

The release mechanism on the *BIV* is checked one more time. (Seeger)

The operational area is bleak and sparsely vegetated but heavily mined. (Fiedler)

An antiaircraft truck, a *Kfz. 4*, provides protection against surprise aerial attack. (Paust)

Three-ton prime movers with trailers bring up additional demolition charges. (Mohrhäuser)

Another section moves towards the jump-off position. A prime mover with trailer and a command and control tank follow the *Kfz. 4*. (Baisch)

They reach the jump-off position through trackless terrain...(Fiedler)

...across creek beds...(Fiedler)

...and through defiles. (Fiedler)

The demolition-charge carriers are also brought up. (Fiedler)

A Utility Tractor of the *3./Panzer-Abteilung 300* is prepared for operations. (Rettig)

Another Utility Tractor is in position. (Wizgall)

Occasionally, the *Kettenkrad* demolitions carrier is used for other purposes. In this picture, personnel of the maintenance section use this all-terrain vehicle in order to reach a disabled tank. (Off)

A few command and control tanks and *Kettenkrad* demolitions carriers are prepared for operations in a reverse-slope position. (Wizgall)

Two *Goliaths* are loaded on the rear deck of this command and control tank. The transportation cart provided for the *Goliath* was unsuitable in the rugged terrain. (Wizgall)

The majority of the tracked and wheeled vehicles on the Crimea employed this type of camouflage pattern. (Fiedler)

Opposite page, top: Maintenance work on a *BIV* of the *1./Panzer-Abteilung 300*. (Off)

Opposite page, bottom: Demolitions charges for the *BIV's* were also carried on Opel-Blitz three-ton trucks with trailers. (Paust)

Unusual camouflage patterns were seen. (Lock)

Personnel of the ammunition section wait for the next operation. Their transportation can be seen in the background: Two *Sd.Kfz. 11* with trailers. (Kornmeier)

Above: Combat engineers were also part of this crew of the *Sd.Kfz. 11*. They were responsible for ensuring the demolitions materials were handled properly. (Kornmeier)

Top right: This *Sd.Kfz. 11* of the *1./Panzer-Abteilung 300* has driven on to a mine. The right-hand running gear has been severely damaged. (Kornmeier)

Right: Personnel of the maintenance section use a *BMW* Motorcycle/Sidecar Combination to move to a disabled *BIV* and attempt to repair it. (Paust)

A *Sd.Kfz. 11* light prime mover of the *1./Panzer-Abteilung 300.* (Schick)

A Bren demolitions carrier of the *3./Panzer-Abteilung 300* and a *BIV* of the *1./Panzer-Abteilung 300* are prepared for employment. (Lock)

The main guns of the command and control tanks were provided with muzzle covers on longer road marches. (Paust)

Assembly area within the *22. Infanterie-Division* area of operations. (Paust)

The crew of this command and control tank of the *1./Panzer-Abteilung 300* supervises the employment of a remote-controlled *BIV*. (Wizgall)

Additional *BIV's* are ready to be employed in the jump-off position. (Fiedler)

The northwest outer positions of Fortress Sevastopol had been reached by 11 June 1942. (Gradolph)

Above and next two pages: A platoon of the *2./Panzer-Abteilung 300* moves into a new area of operations. Additional infantry rides mounted on the command and control tanks. (Bundesarchiv 230/688/24 and 25 230/686/21 and 22)

Top and middle: The rest area at Tole. The majority of the *Goliaths* were staged there as well as damaged *Leitpanzer III's* and some *BIV's*. (Seeger)

Opposite page, top: The command and control tanks were also used to provide the infantry with cover.

Opposite page, bottom: Proxy marriage. This was seen occasionally. (Stolz)

The 2nd Platoon of the *1./Panzer-Abteilung 300* prepares for the final assault on Sevastopol. (Seeger)

Right: *Leitpanzer III 123* negotiates the streets of the destroyed inner city and provides the infantry with cover. (Wizgall)

Opposite page, top: *Leitpanzer III 122* negotiates the streets of the destroyed inner city and provided the infantry mopping up with cover. (Münch)

This *BIV* of the *2./Panzer-Abteilung 300* ran on to a mine. The left-hand running gear was destroyed in the process. (Müller)

Opposite page, bottom: The city is almost completely destroyed. (Wizgall)

In the evening of 1 July 1942, the individual sections of the battalion assembled. (Wizgall)

The return march of the battalion to Nowo-Schuti starts. (Off)

Sevastopol on 1 July 1942. (Paust)

The company commander of the *1./Panzer-Abteilung 300*, *Oberleutnant* von Abendroth, was wounded during the final fighting for Sevastopol. (Wizgall)

The battalion remained in the Sevastopol area until 3 July 1942. Personnel of the maintenance section of the *1./Panzer-Abteilung 300* at Yalta. (Paust)

Leitpanzer III's at Nowo-Schulti. Note the placement of the additional track links. (Paust)

The fortifications of Sevastopol were toured after the capture of the city. (Baisch)

Left: The departure to the railhead started on 4 July 1942. Innumerable Russian prisoners can be seen. (Baisch)

Above and left: The battalion was attached to the *1. Panzer-Armee* after its arrival in the area of Charzyssk. (Paust)

Above: The attack was launched on 9 July 1942. *Panzer-Abteilung 300*, as part of the *1. Panzer-Armee*, advances to the Donez. (Off)

Top right: The attackers encountered stiff resistance. *Panzer-Abteilung 300* also suffered losses. (Wizgall)

Right: Due to a lack of heavy prime movers, the command and control tanks also had to be employed to recover disabled tracked vehicles. (Stolz)

The advance continues relentlessly to the east. (Scherer)

By the end of July 1942 the remaining Utility Tractors of the *3./Panzer-Abteilung 300* were lost due to mechanical problems. Repairs were impossible due to a lack of spare parts. (Rettig)

The *Luftwaffe* supported the operations of the armored formations. (Stolz)

The *1. Panzer-Armee* advanced on a wide front. (Scherer)

The command and control tanks were employed in several armored engagements in the fighting for Rostow and Bataisk. (Mohrhauser)

The *Leitpanzer III's* took the lead. (Jenckel)

A few Borgward BIV's were among them. (Scherer)

There were no wire- or radio-controlled demolition-charge carriers employed, however. (Bürck)

Short break on the steppes. A *Sd. Kfz. 4* protects against unexpected aerial attack. (Sigmund)

Complete loss. The running gear and the superstructure were severely damaged on this *Leitpanzer III*. (Mohrhauser)

Above and below: Starting 26 July 1942 the battalion received the mission to provide security in the rear area. Break at Uspenskaja. (Stolz)

Above and opposite page, top left: The vehicles were serviced and loaded with ammunition. (Jenckel)

This *NSU* demolitions carrier continued to be used within the maintenance section as a support vehicle. Its cross-country ability was appreciated. (Off)

Above and top next page: At the beginning of August the tracked and wheeled vehicles moved to the railhead at Amwrossijewka. (Paust)

Opposite page bottom and both photos this page: The entire battalion was loaded on to rail cars. A 10-day rail journey lay ahead of it. (Paust)

Above and top right: One attempted to make it as comfortable as possible. Food was purchased from the local populace at halts. (Fiedler)

Top left and right: On 15 August the trains reached Krasnogwardeisk. (Wizgall)

Right: After unloading, the battalion moved to Tischkowizi, 20 kilometers away. (Lock)

Services were performed on the vehicles. They had been overdue for weeks. (Stolz)

New numbers were painted in white on the *Leitpanzer III's* of the *1./Panzer-Abteilung 300.* (Lock)

BMW motorcycle/sidecar combination of the *2./Panzer-Abteilung 300.* (Hofmann)

The battalion received orders to establish its winter quarters. All helped to construct bunkers. (Off)

Vehicle motor pool of the 2nd Platoon of the *1./Panzer-Abteilung 300* in Tischkowizi. (Lock)

The *Borgward BIV's* were driven on to wooden blocks to avoid freezing to the ground and then covered with shelter quarters. (Scherer)

This *Leitpanzer III* belonged to the redesignated *2./Panzer-Abteilung 301*. (Stolz)

In the middle of September 1942 six new *Leitpanzer III's* with the long-barreled 5 cm *Kwk 39 L60* were delivered to the battalion. (Schick)

Left and top next page: The spaced armor is easily identifiable. (Paust)

The wooden containers used for storing personal gear have not yet been mounted. These were usually found on the right fenders. (Schick)

The horseshoe was intended to bring luck. (Schick)

Leitpanzer III 213 has sunk into the soft ground. (Ling)

The wear-out rate on roadwheels was large. The rubber on the road wheels wore down quickly. A damaged roadwheel has been mounted on the vehicle. (Wizgall)

The weather worsened daily at the end of October 1942. *Leitpanzer III 122* and a Sd. Kfz. 250 await new orders. (Lock)

On 17 November 1942 *Kompanie Abendroth* began to entrain at Gatschina. This disabled *Borgward BIV* is towed by a *Sd.Kfz. 250.* (Rettig)

Before the loading of the *Leitpanzer III's* began, the stability of the wooden ramp was tested. (Lock)

The command and control tanks are carefully moved on to the flat cars. (Lock)

The remainder of the battalion remained attached to the *18. Armee* under *Heeresgruppe Nord* for an additional four weeks. The *3./Panzer-Abteilung 300* still had several Bren demolitions carriers. (Rettig)

Due to the severe weather, this open-topped *Sd.Kfz. 250* has been covered with a tarp. (Rettig)

The command and control tanks of the *2./Panzer-Abteilung 301* have received a coat of whitewash. The turret number — *213* — has not been covered. (Mohrhauser)

The company commander of the *3./Panzer-Abteilung 300*, *Oberleutnant* Dettmann. (Rettig)

The rest of *Panzer-Abteilung 301* was rail loaded at Gatschina starting 12 December 1942. (Fiedler)

The vehicles were off-loaded at Arnswalde and temporarily staged there. (Rettig)

Leitpanzer III 121 was also brought back. (Rettig)

Below and next page: In the meantime, *Kompanie Abendroth* occupied quarters in the area of the newly formed *Heeresgruppe Don*. The demolitions carriers had to be staged in the open. (Jenckel)

Above: Due to the cold, the employment opportunities for demolitions carriers was restricted. (Jenckel)

Left, top and bottom: These *Leitpanzer III's* had the short-barreled 5 cm main gun. (Jenckel)

The command and control tank of the company commander, *Oberleutnant* von Abendroth, was outfitted with the long-barreled 5 cm main gun. Note the special winter tracks (Fiedler)

When the weather permitted it, operations were again conducted with the *Borgward BIV's*. (Paust)

A break in the action. (Fiedler)

Vehicles were occasionally disabled due to mechanical failure. The outer drive sprocket is broken; the track has been thrown. (Lock)

Kompanie Abendroth was loaded at Lischaja at the end of January 1943. A 6-week odyssey started which did not end until the unit reached Eisenach on 13 March 1943. (Lock)

2.2.4 Versuchs-Kommando (F.L.) Tropen

On 24 June 1942 the *AHA* issued an order (*Nr. 3240/42 g.Kdos*) for the formation of *Versuchs-Kommando (F.L.) Tropen* (Radio-Guidance Test Detachment — Tropical). The unit had an authorized strength of 78 men and consisted of two platoons plus the vehicle maintenance and field test sections and the trains. The 1st Platoon consisted of two sections, each equipped with five *Borgward BIV's*. The 2nd Platoon was equipped with wire-controlled *Goliaths*. In addition, the detachment was issued three *Panzerkampfwagen III's, Ausführung J* with crews. These vehicles, which were used as command and control tanks, did not appear in the table of organization.

The unit was formed by *Panzer-Ersatz-Abteilung 5* at the See Garrison at Neuruppin and consisted mainly of personnel released by the replacement battalion plus members of *Panzer-Versuchs- und Ausbildungs-Kompanie (FL) 300.* The unit was assigned field post number 23514. Commander of the test detachment was *Oberleutnant* J. Fischer. The *BIV* platoon was led by *Leutnant* Knoblauch, the *Goliath* platoon by *Leutnant* Hecker.

The purpose of *Versuchs-Kommando (F.L.) Tropen* was to gain experience in the use of remotely-controlled vehicles in the North African theater.

After the fall of Tobruk in the summer of 1942, the advance by German and Italian forces had come to a halt between El Alamein and the Qattara Depression. This area of mainly flat desert terrain seemed ideal for the use of remotely-controlled vehicles against fortified positions.

After several practice missions had been carried out and the members of the detachment had been outfitted with tropical clothing, the unit was loaded aboard trains on 12 August 1942 and transported to Italy. There the detachment was temporarily housed in the Instituto Constancea Ciano near Naples.

In Naples the unit's wheeled and tracked vehicles were loaded on ships for transport by sea to Tarent and Benghasi. The only personnel to accompany the vehicles were the drivers. The remaining personnel were transported by train to Brindisi on 8 September and, from there, were flown via Crete to Tobruk on 15 September.

Once the transports had been unloaded, the tracked vehicles were transported on the still-functioning railway along the Via Balbia to El Daba west of El Alamein. The wheeled vehicles were driven to the new quartering area.

Tests with radio- and wire-controlled demolition charge carriers were resumed. At this time the technical officer, *Leutnant* Dr. Otto, joined the unit. In addition to trials, the unit also gave a demonstration to *Generaloberst* Rommel and senior German and Italian officers.

During the course of further exercises it was determined that the radio-controlled demolition charge carriers were not working satisfactorily.

Kurt Hecker, former *Leutnant* and *Goliath* platoon leader, described the situation:

> Depending on the time of day and temperature, the *BIV's* receivers suffered from interference at longer ranges which prevented them from receiving further commands. In such cases a motorcycle/sidecar combination had to trail the "runaway". In order to "capture" the *BIV* the man in the sidecar had to jump onto the moving vehicle and remove the ignition key.
>
> During the day visibility was reduced by strong reflections and shimmer caused by the bright sunlight. Additionally, depressions in the ground not visible from a distance caused problems which ended the control process prematurely.
>
> There were no "live-fire" operations during the entire test period.

Versuchs-Kommando (F.L.) Tropen also conducted experiments with captured tanks, and one Bren Carrier was converted into a wire-controlled demolition-charge carrier.

The expected British offensive at El Alamein began on 23 October 1943. As a result, an operation by the *Goliath* platoon planned for 26 October was not executed.

The test detachment's three command and control tanks were detached to the *90. leichte Afrika-Division (mot.)*, which was deployed on the coast road north of El Alamein. The three *Panzerkampfwagen III's* were tasked to escort supply trains to the front.

Beginning on 3 November 1942, the remainder of *Versuchs-Kommando (F.L.) Tropen* were led by *Oberleutnant* Fischer back to Tripoli by way of the Halfaya Pass, Tobruk, Derna and Benghasi. The three command and control tanks rejoined the unit later in the Tripoli area.

On 27 November 1942 the bulk of the unit's tracked and wheeled vehicles were attached to the *90. leichte Afrika-Division (mot.)* with one driver each. Following removal of their radio command receiver equipment, the *BIV's* were blown up in the desert, as were the wire-controlled *Goliaths*.

Beginning on 11 January 1943 the remaining members of *Versuchs-Kommando (F.L.) Tropen* were flown by *He 111* to Italy and Greece. From there they returned to *Panzer-Versuchs- und Ersatz-Abteilung 300.* At the beginning of February 1943 *Versuchs-Kommando (F.L.) Tropen* was disbanded and the members of the detachment were divided among the companies of *Panzer-Versuchs-und Ersatz-Abteilung 300.*

Gliederung des Versuchskommando (FL) Tropen (Stand: Oktober 1942)

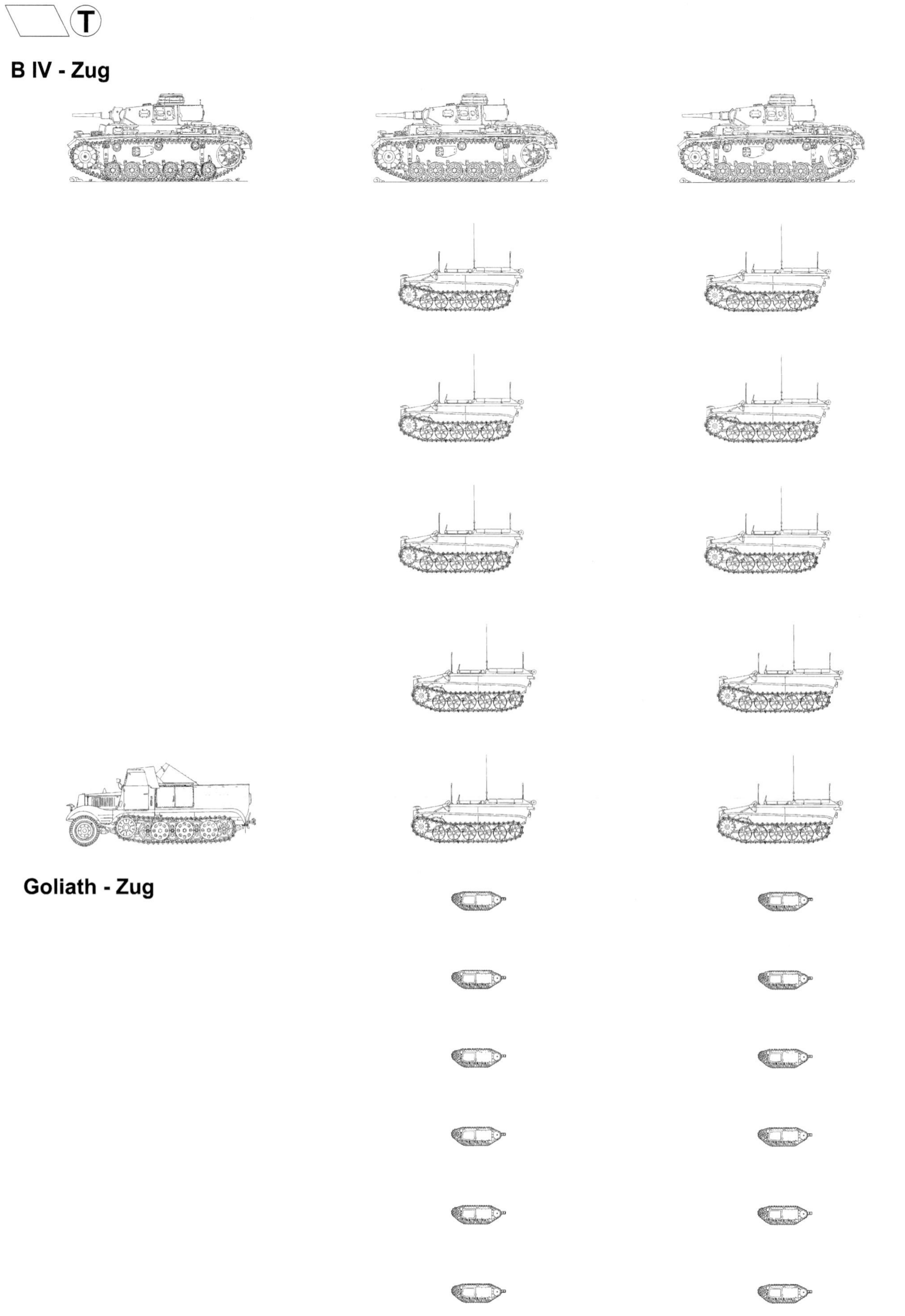

M.Jaugitz / Trojca 2000.

After being loaded by rail at Neuruppin, the unit was sent south. Note the tactical insignia of the *Versuchs-Kommando (FL) Tropen* on the left fender of the *Opel-Blitz* truck. (Lehmann)

One of the command and control tanks — a *Panzer III, Ausführung J* with the short-barreled main gun. (Lehmann)

All of the vehicles have the sand-brown camouflage scheme, some with irregular patches of gray. This scheme was prescribed in *Heeresmitteilung 1942, Nummer 315*. (Lehmann)

The command and control *Panzer III's* could be differentiated from other *Panzer III's* by the rectangular box affixed to the rear of the turret. (Lehmann).

Right: All of the wheeled and tracked vehicles were staged around the Instituto Constancea Ciano at Bagnoli (near Naples) after their arrival. (Lehmann)

Below: The left-hand command and control tank has received additional armor. (Lehmann)

Guard duty! (Lehmann)

The unit assembled in the area around El Daba after its arrival in North Africa. This *Panzer III* command and control vehicle was christened *Strolch II* (rascal). (Lehmann)

An additional mounting for a radio-control antenna has been mounted on this *Opel-Blitz* truck. (Susenbeth)

The same vehicle seen previously has been partially dug in. (Hecker)

The *Goliath* demolition-charge carriers (*Sd.Kfz. 302*) have been loaded on trucks. (Susenbeth)

Well-camouflaged truck with *Goliaths* in the North African desert. (Susenbeth)

A *Goliath* is blown up during a demonstration. The demolition-charge carrier had been loaded with a smaller charge. The normal practice demolition charge of 60 kilograms would have been too dangerous. (Hecker)

The command and control tanks of the *Versuchs-Kommando (FL) Tropen* were occasionally detached to the *90. Leichte Afrika-Division* to support it. (Lehmann)

The withdrawal of the *Versuchs-Kommando (FL) Tropen* started in November 1942. This *Borgward BIV* with additional water cans is rolling westward. (Lehmann)

Again and again the three command and control tanks were detached to other units to support them. (Lehmann)

The command and control tanks were also damaged during these engagements. The left-front track guard is missing on the *Leitpanzer III*. (Hinzmann)

During the withdrawal the command and control tanks were also festooned with all sorts of additional equipment. As a result, they were often referred to a "gypsy caravans". (Hinzmann)

The *Strolch II* covers the withdrawal movements of Italian formations. (Lehmann)

Personnel of the *Goliath* platoon assemble the transport carriers for the *Goliaths*. (Susenbeth)

Many soldiers of the *Versuchs-Kommando (FL) Tropen)* wear light-brown, one-piece coveralls. (Susenbeth)

The lead vehicle pulls a *Goliath* loaded on its transport carrier. (Susenbeth)

Withdrawal over the Halfaya Pass. (Susenbeth)

Two *Borgward BIV's* follow. (Lehmann)

This Bren Carrier has been converted to a wire-guided demolition-charge carrier. It had been captured in North Africa. (Susenbeth)

A *BIV* in the desert. (Susenbeth)

A *BIV* passes by a military cemetery at Capuzzo. (Susenbeth)

A *Sd.Kfz. 10* of the *Versuchs-Kommando (FL) Tropen* in a war-torn North African village. It is not known what purpose the cylindrical container on the half-track served. (Susenbeth)

Leutnant Hecker of the *Goliath* platoon next to his *VW-Kübelwagen*. (Susenbeth)

The *Versuchs-Kommando (FL) Tropen* reaches Tripoli. A few *BIV's* have been loaded on the trailers. (Susenbeth)

Personnel of the *Versuchs-Kommando (FL) Tropen* await the withdrawal to Germany from Tripoli. (Susenbeth)

The majority of the vehicles were transferred to the *90. Leichte Afrika-Division*. (Susenbeth)

The *Borgward BIV's* were later blown up to prevent their capture by the enemy. (Susenbeth)

2.2.5 Panzer-Abteilung 302

On 10 September 1942 the *AHA* ordered the formation of another remote-control unit on 15 October 1942. The new unit was designated *Panzer-Abteilung 302* and was intended to consist of headquarters, headquarters company, three tank companies and a maintenance platoon. The battalion's home garrison was to be Neuruppin. Cadre for the battalion was to be drawn from existing units.

The bulk of the new unit came from the *III./Panzer-Regiment 10*. After taking heavy losses on the Eastern Front, the battalion was detached from the regiment and, after being brought back up to strength by *Panzer-Ersatz-Abteilung 10* in Großglienecke, it was sent to Neuruppin.

In addition, on 29 September 1942, the *1. leichte Panzer-Kompanie "f"*, which had been stationed at Paderborn with *Panzer-Regiment 11*, was moved to Neuruppin. After arriving in its new billets, the Friedrich-Franz Garrison, it was renamed the *3./Panzer-Abteilung 302* on the same date.

Since *Panzer-Abteilung 302* had no experience in wire- or radio-control operations at this time, a further reorganization was carried out. Then, in the second half of October, the *3./Panzer-Abteilung 302* was placed on alert and, following the issuing of winter clothing, was shipped out for Leningrad.

There it was incorporated into *Panzer-Abteilung 301* as its new *2. Kompanie*. At the same time the former *2./Panzer-Abteilung 301* was withdrawn from the unit, redesignated the *3./Panzer-Abteilung 302* and transferred back to Neuruppin.

After the formation of *Panzer-Ersatz-Abteilung 300* on 1 November 1942 and the organizational changes in *Panzer-Abteilung 301*, which were inspired by frontline operations, *Panzer-Abteilung 302* was again reorganized.

By order of the *AHA* (*I a II Nr. 721/43*, dated 9 January 1943), the *1./Panzer-Abteilung 302* was renamed *Panzer-Kompanie (Fkl) 311*, the *2./Panzer-Abteilung 302* became *Panzer-Kompanie (Fkl) 312*, and the *3./Panzer-Abteilung 302* was renamed *Panzer-Kompanie (Fkl) 313* (see Chapter subsections 2.2.7 to 2.2.9). These changes were effected on 25 January 1943.

The headquarters, headquarters company and maintenance platoon of *Panzer-Abteilung 302* were released for the formation of the new *Panzer-Abteilung (Fkl) 301* (see Chapter subsection 2.2.6).

Once the redesignation process had been completed, *Panzer-Abteilung 302* was officially disbanded.

2.2.6 Panzer-Abteilung (Fkl) 301

On 20 November 1942 the army called for the formation of 10 radio-control armor units. At the same time, it proposed that existing personnel and radio-control equipment be brought together in a radio-control instructional unit of the *Panzertruppen-Schule* to be based at Neuruppin.

This demand led to the reorganization of the remote-control force, the related formation of separate radio-control tank companies, a replacement battalion for the radio-control force, and the reorganization of *Panzer-Abteilung (Fkl) 301*.

Panzer-Abteilung (Fkl) 301 was formed in Neuruppin on 25 January 1943 from elements of the disbanded *Panzer-Abteilung 302* (see Chapter subsection 2.2.5) and the elements from *Panzer-Abteilung 301* brought in from Arnswalde in Pomerania (see Chapter subsection 2.2.3).

Panzer-Abteilung (Fkl) 301 consisted of a headquarters, a headquarters company, a maintenance platoon and three tank companies (*2.* to *4. Kompanie*). *Kompanie Abendroth*, which was still employed in southern Russia, was to be incorporated as the *1. Panzer-Kompanie* at a later date.

When the formation process was over, personnel not incorporated into the new battalion were transferred to *Panzer-Ersatz-Abteilung (Fkl) 300* in Eisenach.

In spring 1943 *Panzer-Abteilung (Fkl) 301* consisted of the following officers:

Battalion Commander: *Major* Reinel
Adjutant: *Oberleutnant* Guckel
Special Staff Officer: *Leutnant* Sickendick
Radio-control officer: *Oberleutnant* Voss
Headquarters Company Commander: *Oberleutnant* Stein
Battalion Surgeon: *Oberarzt Dr.* Wick
2./Panzer-Abteilung (Fkl) 301: *Oberleutnant* Hoyer
3./Panzer-Abteilung (Fkl) 301: *Oberleutnant* Krämer
4./Panzer-Abteilung (Fkl) 301: *Oberleutnant* Busse
Maintenance Platoon: *Leutnant* Meyer

When Kompanie Abendroth returned to Germany in mid-March 1943 it was not sent to Neuruppin. Instead, it was moved to the replacement battalion at Eisenach. Therefore, the plan to incorporate the company as the *1./Panzer-Abteilung (Fkl) 301* was not carried out. Until redesignated as *Panzer-Kompanie 315 (Fkl)*, *Kompanie Abendroth* had the status of a separate radio-control company.

At this time the newly-formed separate tank companies — *Panzer-Kompanien (Fkl) 311*, *312*, *313* and *314* — were also attached to *Panzer-Abteilung (Fkl) 301*, however, *Panzer-Kompanie (Fkl) 312* had already been moved to the garrison of *Panzer-Ersatz-Abteilung (Fkl) 300* at Eisenach on 29 January 1943.

Following the allocation of the first command and control assault guns and demolition-charge carriers, training on vehicles and equipment began in spring 1943 with emphasis on armor tactics and radio-control procedures. In spite of fuel rationing, the necessary training exercises were successfully carried out on the nearby Neuruppin training grounds. In July 1943 *Major* Reinel and *Oberleutnant* Guckel were on the Eastern Front as observers, with orders to file a report on operations by the remote-control companies (see Chapter 4.2.5).

At the beginning of September 1943 *Panzer-Abteilung (Fkl)* 301 reported that it was ready for operations and, on 11 September, received orders to transfer into the area of the *OB West* (Commander-in-Chief West). On its arrival the unit was to be placed under the operational control of the *14. Panzer-Division*, which was being reformed in the Angers area in the summer of 1943 after having been destroyed at Stalingrad.

Several trains were required to transport *Panzer-Abteilung (Fkl) 301* and it traveled via Troyes, Orleans and Tours to Mirebeau, not far from Potiers, where the unit was assigned quarters in several towns. In mid-October an advance party moved to Nimes via Bordeaux, Toulouse and Montpellier to prepare for the arrival of the main body of *Panzer-Abteilung (Fkl) 301* at the Cote d'Azur.

Since the steep cliffs of the Mediterranean coast were totally unsuited to the use of armor and demolition-charge carriers, the order was rescinded and the unit was moved into the Amiens area. There it occupied quarters approximately 40 kilometers to the east in the Peronne and Roisel area.

There *Panzer-Abteilung (Fkl) 301* was temporarily attached to *Panzer-Grenadier-Division "Feldherrnhalle"*, however that division was already in the midst of preparations for transfer to *Heeregruppe Mitte* on the Eastern Front. Tactically, the battalion was attached to the reorganized *9. SS-Panzer-Division "Hohenstaufen"*.

According to a status report dated 1 November 1943, *Panzer-Abteilung (Fkl) 301* possessed a full complement of 32 assault guns and 108 demolition-charge carriers as per the table of organization.

In November 1943 the *9. SS-Panzer-Division "Hohenstaufen"* conducted a large-scale training exercise which was observed by the Inspector-General of Armored Forces, Guderian, and *General der Panzertruppen* Geyr von Schweppenburg. *Panzer-Abteilung (Fkl) 301* also took part in the exercise.

Effective December 1943, *Panzer-Abteilung (Fkl) 301* was given administrative control over *Panzer-Kompanie 315 (Fkl)*, a separate radio-control tank company.

Early on the morning of 22 January 1944 Allied forces landed on the Italian coast between Anzio and Nettuno southwest of Rome and established a beachhead. In the days that followed, German units succeeded in sealing off the landing zone with the objective of launching a counterattack with the forces under *Armee-Oberkommando* 14 at the end of January.

The *Wehrmacht* Operations Staff was ordered to release the strongest possible forces for the planned attack. As a result, instructions were issued for *Panzer-Abteilung (Fkl) 301* to be brought in from the west as quickly as possible.

The order for the attack never took effect. On 30 January 1944 the enemy attacked from the bridgehead with three divisions. The Anglo-American forces succeeded in driving as far as Aprilia.

On 31 January 1944 *Panzer-Abteilung (Fkl) 301* received orders to proceed immediately to Italy. The unit was to be ready for departure by 1 February. After the battalion had entrained at Roisel on 31 January, the trains rolled through Karlsruhe, Augsburg, Munich, the Brenner Pass and Florence to Ficulle, north of Orvieto. After detraining, the unit traveled overland toward the south. At dawn on 5 February the unit passed through Rome and was subsequently quartered in the Instituto Agraria Roma (an agricultural college) five kilometers south of Rome.

On 7 February the commander-in-chief of the *14. Armee*, *Generaloberst* von Mackensen, inspected *Panzer-Abteilung (Fkl) 301* and was shown the unit's remote-control technology. He was followed on 11 February by the *Oberbefehlshaber Südwest*, *Generalfeldmarschall* Kesselring.

To enable the *Oberbefehlshaber Südwest* to assemble the various armor units under his command, the Inspector-General of Armored Forces placed at his disposal *Panzer-Regiment-Stab 69*.

By 10 February 1944 the German formations had succeeded in holding most of their own positions while establishing blocking positions around the Allied beachhead. Preparations immediately began for the friendly attack, however, these were complicated by bad weather.

On 15 February *Panzer-Abteilung (Fkl) 301* moved into the assembly area in the Genzano — Velletri — Albano area. From there the line companies moved into departure positions near the front.

After the buildup was completed, the German attack, code-named "Operation Red Sky", was to begin on 16 February. The main burden was to be borne by the *LXXVI. Panzer-Korps* under the command of *General der Panzertruppen* Herr. In addition to various other armored units, the first wave included *Panzer-Abteilung (Fkl) 301*. The *4./Panzer-Abteilung (Fkl) 301* was placed under the operational control of *schwere Panzer-Abteilung 508*. The *2.* and *3./Panzer-Abteilung (Fkl) 301* were deployed on both sides of Route No. 42 with *Infanterie-Regiment 735* in the sector of the *715. Infanterie-Division.*

The German units achieved a minor penetration into the enemy's main line or resistance, however, they were pinned down by naval gunfire and the attack was halted on 20 February. There followed extensive regroupings as well as preparations for another attack, which began on 29 February 1944.

The *LXXVI. Panzer-Korps* had been deployed between the Mussolini Canal and the Spaccasassi Creek. Positioned there was the *114. Jäger-Division* with the attached *Panzer-Grenadier-Regiment 1028*. To its east was the *362. Infanterie-Division* bolstered by the *8./Panzer-Regiment 4*, a *Goliath* company and a company of *Pioneer-Bataillon 60*. The main effort consisted of the *26. Panzer-Division* (reinforced by the *I./Panzer-Regiment 4*, a *BIV* company and a battery of *Sturmpanzer-Abteilung 216*) and *Panzer-Division "Hermann Göring"* (reinforced by *schwere Panzer-Abteilung 508*, and the *1./schwere Panzerjäger-Abteilung 653*).

The attack began at 0445 hours. By noon enemy artillery fire had again brought the attack to a halt with heavy casualties. Prevailing weather conditions made the attack sector unsuitable for armored vehicles. On 29 February the Commander-in-Chief Southwest ordered the attack halted.

At the beginning of March 1944 preparations for "Operation Margarethe," the German occupation of Hungary, made necessary a redistribution of forces in the area of the *Armee-Oberkommando 14*. However, with the exception of *Panzer-Abteilung (Fkl) 301*, all of the armored units dispatched in February remained with the Commander-in-Chief Southwest.

The casualties for *Panzer-Abteilung (Fkl) 301* during the fighting in Italy were 25 killed, 31 wounded and 2 missing plus 15% losses in vehicles. The formation was withdrawn from the front and, on 10 March 1944, was moved back to Orvieto. After entraining in Ficulle on 14 March, the battalion was transported through central Italy and the Po Plain to Vienna, where it detrained on 17 March. Its stay there was a brief one. On the evening of 22 March the unit entrained once again, this time at Vienna-Pötzleinsdorf.

Panzer-Abteilung (Fkl) 301 detrained in Arras in northern France between 26 and 28 March. The unit, which was scheduled for a complete rest and refit, was placed under the operational control of the *2. Panzer-Division* stationed in the Arras — Cambrai — Bapaume area.

Since the *1./Panzer-Abteilung (Fkl) 301* still had not been formed, the Army General Staff ordered on 2 April 1944 that separate *Panzer-Kompanie (Fkl) 312* be incorporated into the battalion organization (see Chapter subsection 2.2.8).

On 28 and 29 April 1944 *Panzer-Abteilung (Fkl) 301* took part in a division exercise by the *2. Panzer-Division* in the Arras — Mareuil — Béthune area. The battalion was subsequently transferred into a new quartering area near Corbie east of Amiens on 30 April 1944. During the first half of May, members of the battalion and French prisoners were used to install post obstacles, so-called "Rommel's Asparagus".

After yet another transfer, the elements of the battalion occupied quarters on 15 May in the Frelles — Bertangles area, six kilometers north of Amiens. During this period there was a noticeable increase in Allied bombing raids on northern French cities. Twice, on 19 and 28 May, the battalion's *Flak* platoon contributed to the defensive effort against Allied raids.

On 20 May 1944 there was a change of command. *Major* Reinel left the battalion and was replaced by *Hauptmann* von Abendroth. By the end of May *Panzer-Abteilung (Fkl) 301* had been brought back up to authorized strength in personnel.

On 1 June 1944 the Army General Staff ordered the transfer of *Panzer-Abteilung (Fkl) 301* to *Heeresgruppe Nordukraine* in Lemberg. Beginning on 5 June the battalion was transported east, minus its *4. Kompanie.*

It was intended to incorporate the *4./Panzer-Abteilung 301 (Funklenk)* under the command of *Oberleutnant* Witter into the newly-formed *Panzer-Abteilung (Fkl) 302* at a later date (see Chapter subsection 2.2.15).

In mid-June 1944 *Panzer-Abteilung (Fkl) 301* had the following officers:

Battalion Commander: *Hauptmann* von Abendroth
Adjutant: *Leutnant* Schreiner
Special Staff Officer: *Leutnant* Griewe
Radio-Control Officer: *Oberleutnant* Voss
Headquarters Company: *Oberleutnant* Meyer
Battalion Surgeon: *Stabsarzt Dr.* Wick
1./Panzer-Abteilung (Fkl) 301: *Oberleutnant* Nolte
2./Panzer-Abteilung (Fkl) 301: *Oberleutnant* Scheer
3./Panzer-Abteilung (Fkl) 301: *Oberleutnant* Krämer
Maintenance Platoon: *Oberleutnant* Bayerl

After arriving on the Eastern Front in the Zloczow area between 9 and 10 June 1944, the battalion was placed under the operational control of the *8. Panzer-Division* on 12 June. Until the end of June *Panzer-Abteilung (Fkl) 301* trained with *schwere Panzer-Abteilung 507* in joint operations involving *Tiger* tanks and demolition-charge carriers.

At the beginning of July, the battalion increased the number of its scouting missions to the front. The battalion was taken by surprise when, on 7 June 1944, it received an order that all demolition-charge carriers were to be sent back to Jaroslau. The battalion was subsequently ordered to a heightened level of readiness.

The Soviet offensive began on 14 June. *Panzer-Abteilung (Fkl) 301* was immediately committed as a pure tank unit with the *8. Panzer-Division.*

On 15 July, the battalion commander, *Hauptmann* von Abendroth, was badly wounded and put out of action. Acting command was passed to *Hauptmann* Nolte on 17 June and, after he was wounded, to *Hauptmann* Krämer.

In the weeks that followed — until 18 August 1944 — the battalion was committed at various hot spots on the front and suffered considerable losses in men and equipment.

The battalion's war diary for the period from 29 March to 21 August 1944 has survived and it provides details of the unit's time in northern France, its transfer to the Eastern Front, and its participation in the defensive fighting and withdrawals in Eastern Galicia, the Carpathians and the Dukla Pass.

(Source: Ludwig Schreiner, *Leutnant* and adjutant in *Panzer-Abteilung (Fkl)* 301 during the period covered by the report. The **boldface** text indicates the format followed)

Day Time Location	**Description of Events** **Important elements: Assessment of the situation [friendly and enemy], receipt and transmission times of reports and orders.** **Weather situation; state of health; replacements; ammunition consumption; arrival and departure of attached elements; and, lessons learned**
29 March 1944	The battalion occupied its new quartering area Avesnes, southeast of Arras. Quarters in Avesnes le Comte — le Comte Novelle — Vion — Wanquetin — Harbareq — Gomy en Artois area.
Avesnes le Comte	Command and control: Under operational control of the *2. Panzer-Division*, reporting directly to *Panzer-Regiment 3*.
30 March 1944	The *4./Panzer-Abteilung (Funklenk) 301* was assigned as a ready force and in attached to the *Kampfruppe* for operational deployment.
Avesnes	
5 April 1944	*Panzer-Kompanie (Fkl) 312* has been consolidated with the battalion and redesignated as the *1./Panzer-Abteilung (Fkl) 301.*
Avesnes	
6 April 1944	The battalion immediately began work on its quartering area to make it less vulnerable to air attacks as well as acts of sabotage and espionage; formed anti-paratroop detachments.
Avesnes	
17 April 1944	The *1./Panzer-Abteilung (Fkl) 301* arrived in France from Holland and occupied quarters in Conde-Folie (25 kilometers west of Amiens).
Avesnes	Arrival and departure of attached elements: *1./Panzer-Abteilung (Fkl) 301* from Holland
30 April 1944	Battalion transferred into new quartering area east of Amiens. Departure during night of 2-3 May. Quartered in Querrieu — Bussy les Daours — Allenville — Lahoussoye area.
3 May 1944	Transfer of battalion command post to Corbie (10 kilometers east of Corbie Amiens).
4 May 1944	For purposes of administration and supply the battalion is placed under the operational control of the *2. Panzer-Division*; for training it reports to *Panzer-Regiment 3.*
Corbie	
06 May 1944	By order of the *2. Panzer-Division*, post obstacles were placed in the area on either side of the Somme to guard against an enemy airborne landing.
Corbie	
15 May 1944	Battalion again transferred within the area of the 2. Panzer-Division; quartered due north of Amiens.
Corbie	The battalion is in a state of constant alert readiness; one company has to be ready to move at all times. Billets are being prepared for all-round defense. Battalion needs to be operational within 2.5 hours; the ready company within 1.5 hours.
20 May 1944	By order of the *OKH* the battalion commander, *Major* Reinel, has been transferred to the *OKH.* His replacement as battalion commander is *Hauptmann* von Abendroth.
Bertangles	
24-26 May 1944	Replacement personnel for the various companies arrived from *Panzer-Versuchs- und Ersatz-Abteilung 300 (Fkl)* in Eisenach: *Leutnant* Wulff plus four noncommissioned officers and 69 enlisted men. As soon as they arrived they were assigned to the companies and briefed on their duties.
Bertangles	Replacements: *Leutnant* Wulff plus noncommissioned officers and 69 enlisted men.
24 May 1944	Monitoring lines set up in case of an invasion. Their purpose is to prevent able soldiers from staying in the rear area.
25 – 30 May 1944	1) The companies used all the means at their disposal to carry out repair work internally and restore their fighting strength after operations in Italy. Replacements for missing vehicles were ordered in Germany and those available were delivered to the battalion.
Arras/Amiens	2) Individual unit training was intensified in conjunction with *Panzer-Regiment 3* and *Panzer-Grenadier-Regiment 304.* Exercises were carried out, some with live ammunition. Repeated alert exercises allowed the battalion to satisfy itself that the companies were ready for operations. 3) At the end of May the companies received replacements, allowing them to reach the strengths authorized in the table of organization. 4) Replacement radio-control equipment was also received; now at 100% 5) Work on fortifying billets pursued vigorously. Installation of post obstacles also achieved excellent results. The battalion emplaced 8,547 posts.

3 – 18 June 1944	Operations with *Heeresgruppe Nordukraine*
3 June 1944 (1100)	*2. Panzer-Division* alerted the battalion by telephone; orders for express rail movement to the east, without the *4. Kompanie.*
(1700)	Transfer postponed by 24 hours.
4 June 1944	Efforts by battalion to leave the *1. Kompanie* (*Oberleutnant* Nolte) behind instead of the *4. Kompanie* (*Oberleutnant* Witter) appeared to have been successful, however, in the afternoon a telex was received from the *XXXVII. Panzer-Korps* stating that the *4. Kompanie* was going to remain in the west.
Bertangles	Arrival and departure of attached elements: *4./Panzer-Abteilung (Fkl) 301* remains in the west.
5 June 1944 (2000)	Battalion placed under the operational control of *Heeresgruppe Nordukraine*. Destination Lemberg. (Telex *XXXVII. Panzer-Korps*). Order received from *Heeresgruppe D* via the *2. Panzer-Division* that the battalion is to leave 4 motorcycles, 10 cars and 11 trucks in France. The vehicles are to be used in the reorganization of *Panzer-Abteilung (Fkl) 302*. Rear detachment to remain behind in France.
Bertangles	
5 – 6 June 1944	Battalion loaded aboard six trains in Amiens (St. Roch) and Boves (8 kilometers southeast of Amiens). Little air activity on day of entraining. No harassing attacks.
(1700 – 1300)	
9-10 June 1944	Battalion detrained in Zloczow (70 kilometers east of Lemberg). Assigned quartering area north and southeast of Zloczow. Attached to the *II. SS-Panzer-Korps*.
12 June 1944	*II. SS-Panzer-Korps* moved out. New command relationship: *8. Panzer-Division / XXXXVIII. Panzer-Korps.*
Zloczow	
13 June 1944	Under the operational control of the II. Panzer-Korps after its arrival. Oherwise nothing out of the ordinary. Continuous alert readiness; ability to move out within two hours of notification assured.
Zloczow	
14 June 1944	*Schwere Panzer-Abteilung 507*, the *4./Panzer-Regiment 10*, and *Panzer-Abteilung (Fkl) 301* placed under the command and control of *Panzer-Regiment 9* (Overall subordination to the *8. Panzer-Division* remains in effect).
Zloczow	
16 June 1944	Blocking lines in the rear area ordered by the army group and the army.
Zloczow	
17 June 1944	The *1./Panzer-Abteilung (Fkl) 301* practiced cooperation with a *Tiger* battalion: the *3./schwere Panzer-Abteilung 507* Committed: Three *BIV's*; two failed to detonate, the third after jettisoning (as it is supposed to).
Zloczow	
18 June 1944	New alert plan released by the division. Battalion to be ready to move in two hours. One company ordered to act as a ready company and be prepared to move out at a moment's notice. Maps for approach routes issued at the same time.
Zloczow	
20 June 1944	Leaders' reconnaissance south of Jasionow, attended by the regimental commander, *Major* Schmidt. Subsequent viewing of the enemy's main battle area from Hill 261 (6 kilometers south of Brody).
Zloczow	
22 June 1944	Leaders' reconnaissance east of Kruhow. Carried out by Major Schmidt. Briefing on probable operational sector and future missions.
Zloczow	
24 June 1944	Exercise with the *2./Panzer-Abteilung (Fkl) 301* with the *2./schwere Panzer-Abteilung 507* east of Jasiono. Purpose: Cooperation between *Tiger* tanks and remote-control vehicles. One live *BIV* moved into a ditch, detonated on its own and was destroyed. The exercise was not a complete success because of heavy rain and muddying of the exercise area.
Zloczow	
26 June 1944	The 2./Panzer-Abteilung (Fkl) 301 again exercised with schwere Panzer-Abteilung 507. Purpose: Forest fighting — together with grenadiers. At the same time range firing.
Zloczow	
29 June 1944	Rear detachment arrived from France. *Schwere Panzer-Abteilung 507* was withdrawn and rushed away for employment elsewhere. At the same time, *Major* Schmidt gave up command of the regiment, replaced by *Hauptmann* Lekschat (commander of the *I./Panzer-Regiment 11*).

Zloczow

1 July 1944 — *Oberst* von Hake took over the former *Panzer-Regiment Schmidt*. Regiment renamed: *Panzer-Regiment von Hake*. *Panzer-Abteilung (Fkl) 301* attached to *Panzer-Regiment von Hake*.

Zloczow

4 July 1944 — *Hauptmann* von Abendroth conducted a terrain briefing for his officers on Hills 375 and 386 (east of Manajo and Harbuzow). Visual inspection of our main line of resistance and enemy's main battle area.

Zloczow

7 July 1944 — Battalion moved to alternate quarters by telephoned order from the 8. Panzer-Division; threat of bombardment exists in villages on the main road. New quartering area in Zazule (5 kilometers east of Zloczow) and in the woods to the east of it. At same time division ordered to heightened state of alert readiness.

(2200) — *Hauptmann* von Abendroth ordered to division with all of the battalion's operational plans.

Zloczow

8 July 1944 (0030) — Order received by telephone from *Heeresgruppe Nordukraine*: Battalion to turn in all of its *BIV's* to a collection point at Jaroslau. Detail led by *Oberleutnant* Voss. Effective immediately the battalion will only be employed with assault guns.

Zazule

8 July 1944 (0700) — Commander announces the new order to the battalion. Sharp increase in friendly air activity.

9 July 1944 (1000) — Battalion visited by new regimental commander *Oberst* von Hake. Presentation of officer corps.

10 July 1944 (0200) — Battalion command post and specialists also moved into the woods. All built-up areas evacuated by order of the division. Continued heightened level of alert.

(0600) — New quartering area occupied; all elements brought to alert status. The *4./Panzer-Regiment 10* attached to the battalion, effective immediately (Acting commander: Oberleutnant Pressberger).

Arrival and departure of attached elements: The *4./Panzer-Regiment 10* placed under the operational control of the battalion.

11 July 1944 — Heightened level of alert maintained by order of the division.

Zazule — Intelligence suggests that enemy offensive will begin in the period 10-12 July.

Oberleutnant Voss and detachment entrained for Jaroslau.

12 July 1944 — Stand-by for all elements. Enemy attack expected on morning of 13 July 1944, otherwise nothing out of the ordinary.

Zazule

13 July 1944 (0100) — Harassing fire (artillery) increased in Hnidawa sector. Periods of barrage fire. Not yet apparent if this is prelude to attack.

Zazule (0700) — No enemy attack. Stand-by cancelled.

14 July 1944 (0100) — Alerted by regimental adjutant, *Panzer-Regiment von Hake*. Enemy barrage fire begun two hours ago, concentrated on sector of the *349. Infanterie-Division* (Hnidawa).

Zazule

(1100) — Battalion attached to *Panzer-Grenadier-Regiment 28*. March route: Zazule, Strutyn, Pluhow, Podlipce, Iwaczow, Wolczkowce. After personal reconnoitering by the commander, occupied readiness position east of Wolczkowce. Close-support aircraft very active.

(1800) — Divisional commander personally gave the order to attack at Hill 426 (east of Monitovka). Attack did not proceed as *Panzer-Grenadier-Regiment 28* not yet ready. At 2400 hours the battalion again readied itself northeast of Jaroslawice. The *1./Panzer-Abteilung (Fkl) 301* was attached to the *II./Panzer-Grenadier-Regiment 28*. Assembly area due west of Monitowka.

Weather situation: Clear, sunny, warm

Arrival and departure of attached elements: The 4./Panzer-Regiment 10 deployed by the division on its own.

Ammunition consumption:

High-explosive rounds: 300

Armor-piercing rounds: 80

Illumination rounds: 50

Machine-gun rounds: 4,000

15 July 1944 (0300)	The battalion (minus the *1. Kompanie*) placed under the operational control of the *Kampfgruppe* of the *8. Panzer-Division.*
(0530)	Start of attack on Hill 426. Heavy artillery and rocket fire.
(0800)	Attack objective reached, the *1. Kompanie* returned to battalion. Enemy infantry driven away from the area of Hallezyna, Dollina. because a *Panther* battalion was attacking farther southwest, the commander had the battalion guard the division's left flank, with *1.* and *2. Kompanien* near Hill 416. The *3. Kompanie* formed the deep flank on the hills southeast of Hukatowce. (1400)
(1800)	The *1. Kompanie* counterattacked and eliminated enemy penetration north of Hill 416. Two assault guns knocked out. The *1.* and *2. Kompanien* each knocked out an enemy assault gun (150 mm) which had broken through; the 1. Kompanie also destroyed two T34's. The *3. Kompanie*, with no infantry support, prevented the enemy from driving in the direction of Wolczkowce. Broken-through enemy attacking from north of Iwaczow. The supply units there had to clear out, making it impossible to supply the combat elements with ammunition and fuel. While making contact with the *Panzer-Grenadier-Regiment*, the battalion commander, *Hauptmann* von Abendroth, was seriously wounded.
(1830)	Radio contact with the *3. Kompanie* lost. *Hauptmann* Nolte led the main body of the battalion (*1.* and *2. Kompanien*) and, by 2000 hours, made himself available to the fusilier battalion of the *357. Infanterie-Division* on Hill 416. With no resupply of fuel and ammunition, these elements withdrew toward Meteniow and sought to link up with the supply units.
	Weather situation: Sunny, dry, warm
	High-explosive rounds: 300
	Armor-piercing rounds: 60
	Illumination rounds: 50
	Machine-gun rounds: 3,000
16 July 1944 (0600)	The *1.* and *2. Kompanien* were attached to the armored *Kampfgruppe* sent to join it on Hill 401, north of Kazimierzowka.
(0700)	After checking with division, the adjutant established contact with the supply elements, which had pulled back toward the hill south of Zarwanica.
	The previous evening, the *3. Kompanie*, with 17 infantry, had withdrawn toward Hill 410 east of Welczkowce as it had been outflanked on both sides. Fired on by friendly *Flak* unaware of the situation.
(1900)	After expending all its ammunition and with fuel reserves for only 15 kilometers, the *3. Kompanie* and the remaining infantry withdrew to Kabarowce and, after Meteniow, received supplies.
	Maintenance units and the field trains (which had already been fired on by enemy patrols in Zezule on 15 July 1944) withdrew in the direction of Przemyslany.
	Weather situation: Clear all day, rain in the evening.
	Ammunition consumption:
	High-explosive rounds: 300
	Armor-piercing rounds: 40
	Illumination rounds: 30
	Machine-gun rounds: 2,000
17 July 1944	The *3. Kompanie* — with 11 *Sturmgeschütze* — rejoined the battalion under the command of *Hauptmann* Nolte, and then *Hauptmann* Krämer after the former was wounded. It formed up on Hill 401 north of Kazimierowka and then attacked toward Iwaczow as part of the armored *Kampfgruppe*. After the capture of Iwaczow these secured north of it. Four antitank guns, three T34's and enemy infantry in company strength were destroyed. One of our assault guns destroyed by antitank fire. *Hauptmann* Nolte and *Leutnant* Radicke were wounded. *Hauptmann* Krämer took over temporary command of the battalion. While en route to the maintenance facility with three semi-serviceable assault guns, *Leutnant* Endemann was ordered by *Oberst* Dollmann of the *III. Panzer-Korps* to secure the Zloczow — Lemberg road near Lackie and subsequently knocked out a T34.
(0900)	Weather situation: Clear by day, rain in the evening
	Ammunition consumption:
	High-explosive rounds: 200
	Armor-piercing rounds: 30
	Illumination rounds: 20
	Machine-gun rounds: 2,000
18 July 1944	The planned attack toward Nuszcze by the armored *Kampfgruppe* was halted after enemy troops with tank support

attacked the division's exposed right flank and elements of the *Kampfgruppe* were detached to eliminate the threat. The assault guns and *Tigers* destroyed one 150 mm assault gun and four T34's.

Great difficulty in supplying the battalion because of very bad road and weather conditions and heavy enemy artillery fire. Nevertheless, *Leutnant* Zywietz managed to make it happen. The prime movers employed by the supply elements distinguished themselves, delivering supplies to the combat elements in spite of the artillery fire.

Weather situation: Rain all day, cloudy.

Ammunition consumption:

High-explosive rounds: 150

Armor-piercing rounds: 40

Illumination rounds: 30

Machine-gun rounds: 2,000

19 July 1944 (2130) Battalion continued to secure north of Iwaczow. The *8. Panzer-Division* was pulled out of its previous sector. Together with the *1. Panzer-Division* it was supposed to break open the ring around the *XIII. Armee-Korps* north of Iwaczow. The withdrawal movements were carried out in total darkness and heavy rain showers and contact with the *Kampfgruppe* was lost. One assault gun became stuck in the difficult terrain and, with the enemy pressing hard and no prime movers available, it had to be blown up. All of the assault guns that were employed developed serious mechanical problems during the night march and were sent to the maintenance section.

Weather situation: Steady rain, cloudy

20 July 1944 *Leutnant* Endemann and five assault guns were placed under the operational control of *Panzer-Grenadier-Regiment 28* and supported the attack on Woroniaki. Heavily-fortified enemy positions south of Woroniaki were taken out; Woroniaki was captured. In the process six antitank guns, two trucks, two troop carriers and enemy infantry in company strength were destroyed.

Weather situation: cloudless, clear

Ammunition consumption:

High-explosive rounds: 100

Armor-piercing rounds: 50

Illumination rounds: 50

Machine-gun rounds: 2,000

21 July 1944 Three assault guns (*Leutnant* Endemann) attached to *Major* Lange's armored *Kampfgruppe*. Attack toward Jasionowce. After reaching the hills south of Jasionowce the attack force turned to Woroniaki to relieve the hard-pressed infantry. Infantry assault team brought in two antitank rifles. Many of the enemy killed. Withdrawal toward Szpiklosy during the night. Two assault guns pulled security at the northern outskirts of Woroniaki and then also withdrew. Losses: One assault gun blown up after running out of fuel.

Weather situation: Periods of rain, some sun

Ammunition consumption:

High-explosive rounds: 250

Armor-piercing rounds: 40

Illumination rounds: 40

Machine-gun rounds: 2,000

22 July 1944 Division withdrew via Wicyn and Wipiski into the area north of Ostalowice. Attached to *Panzer-Grenadier-Regiment 28*. Secured Szpiklos. Counterattack toward Remizowce with *Panzergrenadiere*. Destroyed an antitank gun and a mortar. Withdrew toward Zukow. Enemy tanks appeared west of the town. Two antitank guns knocked out and infantry in company strength destroyed. During the course of the subsequent withdrawal, blocking positions set up east of Wisniowskyk and friendly infantry supported in fending off enemy attacks.

Two assault guns disabled by mechanical problems and towed back to the battalion's supply elements. The last assault gun followed at 1830 hours and was assigned outpost duty with two *Panzer IV's*. These three smashed an enemy attack near Lipowce as it was beginning, destroying two antitank guns and two heavy machine-guns. The assault gun was hit four times by antitank fire and began to burn. The crew made their way back to our lines. Battalion out of contact with the division. It fought its way back through Wipiski to Statowisce. In spite of strong oncoming traffic, *Oberleutnant* Meyer personally delivered fuel at the last minute. Once again the battalion's medium-wave radios proved their worth, making it possible to avoid destroying six assault guns and a *Tiger* for lack of fuel. Fuel was even supplied to other units and to the division headquarters.

Weather situation: Sunny, warm.

Ammunition consumption:

High-explosive rounds: 150

Armor-piercing rounds: 30

Machine-gun rounds: 2,000

Lesson learned: In this case, because of the rain-soaked, softened terrain, it was not possible to deliver supplies even with prime movers, and the use of trucks was out of the question.

The medium-frequency radios proved invaluable, especially because of the very great distances between the combat elements and the trains resulting from the withdrawal. The bad roads, the congestion — with three columns on the road side by side at times — and the very heavy air activity in the first days of the attack, made it almost impossible for dispatch riders to operate. The frequent rain also limited the use of solo motorcycles.

Overlapping and jamming of the frequency by the Russians was a frequent problem, however, this was mastered by the radio operators in spite of the difficult conditions.

23 July 1944 The Commanding General of the *XXXXVIII. Panzer-Korps*, *General der Panzertruppen Balck*, ordered the commitment of six conditionally-serviceable *Sturmgeschütze* northeast of Ostalowice to seal off the enemy breakthrough. *Hauptmann* Krämer was placed under the operational control of the *100. Jägerdivision* and secured to the east on Hill 305 in the sector of *Feld-Ersatz-Bataillon 380.*

Weather situation: Sunny, warm.

Ammunition consumption:

High-explosive rounds: 150

Armor-piercing rounds: None

Illumination rounds: 10

Machine-gun rounds: 2,000

24 July 1944 In the course of further withdrawal movements, the combat elements of the battalion, together with five self-propelled guns, provided the mobile firepower of the *100. Jäger-Division.* One antitank gun and one light artillery piece were destroyed at Zedowice.

Weather situation: Sunny, cloudless

25 July 1944 Counterattack toward the hills southeast of Dobrzanika against enemy penetration at Korzalico. With the arrival of darkness withdrew through Dobrazanika, Tuczna and Ploka to Lubecza.

Weather situation: Brief rain showers, otherwise dry.

Ammunition consumption:

High-explosive rounds: 180

Machine-gun rounds: 2,000

26 July 1944 Battalion's command and control relationship lifted by order of the *XXXXVIII. Panzer-Korps.* Battalion returned to the *8. Panzer-Division. Leutnant* Werthmann and *Leutnant* Hellwig were wounded but stayed with the unit. One assault gun could not be recovered and had to be blown up.

Weather situation: Sunny, dry

Ammunition consumption:

High-explosive rounds: 100

Armor-piercing rounds: None

Illumination rounds: 10

Machine-gun rounds: 1,000

27 July 1944 The *8. Panzer-Division* moved to a new area of operations north of Kalusz. Battalion assembled west of Zurawno and marched via Sokolow, Lisowice, Bolechow, direction of Dolina to Hoszow.

Weather situation: Sunny, dry

Ammunition consumption:

High-explosive rounds: 300

Armor-piercing rounds: 50

Machine-gun rounds: 3,000

28 July 1944 By personal order of the Commander-in-Chief of *Panzer-Armee-Oberkommando 1*, *Generaloberst* Rauß, eleven assault guns under *Oberleutnant* Scheer were attached to the *Feldersatz-Bataillon* of the *101. Jägerdivision.* They had been en

route to Moszisca (the divisional command post). The enemy had infiltrated from the southeast on the Bolechow — Stanislau road and was southeast of Dolina. Combat elements secured east of Dolina. Retreating infantry were stopped and a counterattack threw back the pursuing enemy.

Weather situation: Sunny, no clouds.

29 July 1944 (0430) Enemy continued to press into the outpost line of the *Feldersatz-Bataillon*. During a counterattack, *Oberleutnant* Scheer was killed while leading his men. Enemy attacked Dolina from the south, making it necessary to withdraw the main line of resistance toward the outskirts of town. Our assault guns covered the withdrawal movements. The Hungarians in the neighboring sector withdrew further west, as a result of which the main line of resistance had to be built up due west of Dolina. The assault guns and a small force of German infantry formed the backbone of the defense there. Two mortars were destroyed in a counterattack.

Weather situation: Sunny, no clouds.

Ammunition consumption:

High-explosive rounds: 300

Armor-piercing rounds: 30

Illumination rounds: 10

Machine-gun rounds: 3,000

30 July 1944 The enemy continued to press hard. As a result, the main line of resistance had to be moved back to the railway embankment west of Dolina. While resisting enemy attacks, two antitank guns and two mortars were destroyed in counterattacks. One enemy battery of light artillery was silenced. An assault gun (*Leutnant* Nebe) was put out of action with a burst barrel. Crew suffered only minor injuries.

(2200) The battalion's combat elements attached to *Kampfgruppe Zahn*.

Weather situation: Sunny, some cloud.

Ammunition consumption:

High-explosive rounds: 400

Armor-piercing rounds: 40

Machine-gun rounds: 3,000

31 July 1944 Six assault guns under Lt. Möser deployed on the Kampfgruppe's left wing took Rachin then turned toward Dolina, which was taken from the north and east. One assault gun ran over mines in Dolina. Two antitank guns were destroyed.

1400 hours The combat elements were attached to *Bataillon Heger* of the *Panzer-Grenadier-Regiment* of the *8. Panzer-Division* and advanced toward Turza Ml. Two antitank guns on the commanding hill were destroyed. Turza Ml was fired on and set alight, and then cleared of the enemy with the assistance of *Tigers*. Subsequent advance to the pass road was halted because the enemy antitank belt could not be silenced. One of our assault guns was hit by antitank fire but remained operational. As per orders from the *XXIV. Panzer-Korps* (via the *8. Panzer-Division*), the battalion detached some cadre — the assault gun crews of the *1. Kompanie* — to Krakow to receive new equipment. The company's assault guns were issued to the *2.* and *3. Kompanien*.

Weather situation: Sunny during the day, rain in the evening.

Ammunition consumption:

High-explosive rounds: 400

Armor-piercing rounds: 40

Machine-gun rounds: 3,000

Arrival and departure of attached elements: the *1. Kompanie* left the battalion to receive new equipment.

1 August 1944 Riflemen cleared the road and *Kampfgruppe Zahn* advanced on Wygoda. Fierce enemy resistance (artillery and antitank guns) broken after several hours. Wygoda taken, however the enemy still holds the surrounding hills. Combat elements securing Wygoda while the *Tigers* clear the flanks together with the riflemen.

Weather situation: Rainy.

Ammunition consumption:

High-explosive rounds: 300

Armor-piercing rounds: 50

Illumination rounds: 50

Machine-gun rounds: 3,000

2 August 1944 — Assault guns and 20 riflemen set out to screen the right flank toward Mizun. Fierce enemy resistance. Three antitank guns, two light artillery pieces and numerous mortar positions were destroyed. Two trains (approximately 20 to 30 Panje wagons) were captured; Mizun was taken. After reaching *Bataillon Erlenkampf* the assault guns followed the attack on the pass road to Weldzirz. They then secured Weldzirz and supported *Bataillon Heger* in clearing the surrounding hills. One assault gun was put out of action after its cannon was hit, however, it was able to withdraw under its own power.

(1830) — Moved out again toward Zakle to establish contact with the Hungarian 20th Light Division. Two kilometers south of Weldzirz, heavy fire from an antitank belt in Maksynowka and Lolin made any further advance impossible. Sudden cloudburst began and, after 1.5 hours of fighting, everyone was pulled back. Three antitank guns destroyed.

(2100) — Assault guns withdrawn to the forester's house north of Wygoda to resupply.

Weather situation: Heavy rain all day.

Ammunition consumption:

High-explosive rounds: 250

Armor-piercing rounds: 30

Illumination rounds: 30

Machine-gun rounds: 2,000

3-4 August 1944 — All elements of the *8. Panzer-Division* pulled out of the line and deployed on the left wing of the *1. Armee* with the *III. Panzer-Korps*. March route: Bolechow — Stryi — Drohobycz — Borislaw — Opaka — Turze — Topoluica — Busowisko. March severely hampered by the onset of rain and poor road conditions.

Weather situation: Rain, cloudy.

Ammunition consumption:

High-explosive rounds: 200

Armor-piercing rounds: 60

Illumination rounds: 20

Machine-gun rounds: 3,000

Lessons learned: This major withdrawal movement, with a movement of 200 kilometers, once again showed that during rainy weather in the east it is impossible to return to action quickly without prime movers.

5 August 1944 — Marching via Stary-Sambor — Stary-Sol — Chyrow — Krosienko, the battalion reached the Jasien quartering area south of Ustrzki Dln. with five serviceable assault guns.

(1145) — Planned continuation of march by the division postponed as a result of enemy penetration on the Woitkowa — Krosienko road.

Five assault guns under *Hauptmann* Krämer attached to *Panzer-Grenadier-Regiment 28* with orders to drive back an enemy force situated at Woitkowa. From Woitkowa the assault guns attacked with mounted infantry north through Trzieniec, Leszczawa Grn., Leszczawa Dln. to Bircz road. In spite of difficult terrain they reached their objective at 1900 hours unhindered by the enemy, who fled left and right into the forests. While withdrawing, one assault gun became stuck in a ford and, despite much effort, could not be freed. Weapons and equipment were removed and the assault gun was abandoned by its crew. Recovery was planned for 6 August 1944 as the enemy was no longer present. Combat elements reached Woitkowa at 0300 hours on 6 August 1944 and waited there for repairs.

Weather situation: Dry, somewhat cloudy.

Ammunition consumption:

High-explosive rounds: 100

Machine-gun rounds: 3,000

6 August 1944 — One assault gun was attached to *Panzer-Grenadier-Regiment Pretzl*. Secured the northern outskirts of Woitkowa and repulsed attacks by enemy infantry.

Weather situation: Sunny, no clouds

Ammunition consumption:

High-explosive rounds: 100

Machine-gun rounds: 500

7 August 1944 — Two assault guns (*Leutnant* Hellwig) guarded Woitkowa against enemy attack. Counterattack from Woitkowa over Hill 488 to the wood line. Objective reached.

(2030) — Withdrawal to new main line of resistance by way of Krosienko, Berehy Dln., Ropienka.

(2100) The armored combat elements of the *8. Panzer-Division* consisting of: *Panzer-Abteilung (Fkl) 301*, the *4./Panzer-Regiment 10* and the *I./Panzer-Regiment 11*, were combined into *Panzer-Einsatz-Abteilung Krämer*.

Weather situation: Sunny, no clouds

Ammunition consumption:

High-explosive rounds: 80

Machine-gun rounds: 500

8 August 1944 Battalion marched via Ustrzyki into new quartering area in Olszanica.

(0900) At Ropienka, two of the battalion's assault guns under *Leutnant* Möser were placed under the operational control of *Panzer-Grenadier-Regiment Pröll* and provided cover due east of Ropienka.

Weather situation: Sunny and warm.

Arrival and departure of attached elements: The *4./Panzer-Regiment 10* and *Einsatz-Kompanie/Panzer-Regiment 11* were combined with *Einsatz-Kompanie/Panzer-Abteilung (Fkl) 301* to form *Einsatz-Abteilung Krämer*.

9 August 1944 (1100) *Leutnant* Möser relieved from attachment to *Panzer-Grenadier-Regiment Pröll*. (*8. Panzer-Division*). Division placed under the operational control of the *XXIV. Panzer-Korps*.

(1930) At 1930 hours the battalion marched through Lesko and Tarnawa into the Kuleszne area and, due to congested roads, arrived there at 0400 hours on 10 August 1944. Three of the five *Panthers* broke down en route.

Weather situation: Sunny, warm.

10 Aug 1944 (0930) Issuing of orders by the Operations Officer of the *8. Panzer-Division*. *Kampfgruppe Bradel* (*1. Panzer-Division*) to be relieved by the *8. Panzer-Division* in the Odrzcechowa area. Battalion marched via Plonna, Karlikowx and Bokosko to Nagorzany and was attached to *Panzer-Grenadier-Regiment Pröll*.

(1930) Battalion secured along the Pisarowce — Bukosko road due south of Bielnia with three *Sturmgeschütze* and two *Panzer IV's*. Battalion command post plus battalion reserve of two *Panzer IV's* and one assault gun south of Hill 474, 200 meters west of Nowotaniec. At dawn the next day the enemy tried to break through with tank support. Enemy was repulsed.

Weather situation: Afternoon rain, cloudy.

Ammunition consumption:

High-explosive rounds: 80

Armor-piercing rounds: 20

Machine-gun rounds: 500

11 Aug 1944 (1000) Mission briefing held by *Oberst* Bradel. The armored combat elements of the *8. Panzer-Division* were placed under the operational control of the battalion and the battalion's outposts south of Pielnia were relieved. The armored *Kampfgruppen* of the *1.* and *2. Panzer-Divisionen* were given the task of establishing contact with the *208. Infanterie-Division* on the left. In the process Odrzcechowa was to be taken.

(1100) Attachment to *Panzer-Grenadier-Regiment Pröll* lifted; now attached to *Kampfgruppe Bradel* (*1. Panzer-Division*).

(1430) Moved into assembly area north of Wola-Sekowa with one *Panzer V*, three *Panzer IV's* and four assault guns.

(1530) Set out toward Hill 410 northwest of Wynianka. Antitank fire from the direction of Hill 452 (west of Pielna) while forming up. From Hill 410 the assault guns opened fire on Odrzechowa. Tanks quickly reached the crossroads 1 kilometer south of Odrzechowa, then came under heavy antitank fire (antitank-gun belt). Battalion attempted to advance west along the road. No contact with the *Panzergrenadier* elements, heavy flanking fire from enemy antitank-gun belt (14 armored antitank guns); consequently, the attack bogged down. Battalion moved into a reverse-slope position 800 meters south of the western part of Odrzechowa.

(2030) Planned resumption of the attack cancelled.

(2230) By order of the *Kampfgruppe*, the battalion, which was not in contact with the enemy, returned to its starting point and secured there until about 0130 hours. One *Panzer IV* was lost to antitank fire, another was damaged by mortar fire; one assault gun was damaged by antitank fire. One *Panzer IV* and three assault guns were disabled through mechanical malfunction. Two enemy tanks, one of them probably an assault gun, and two to three antitank guns were destroyed.

Weather situation: Sunny, no clouds.

Ammunition consumption:

High-explosive rounds: 80

Armor-piercing rounds: 30

Machine-gun rounds: 3,000

Arrival and departure of attached elements: All armored elements of the division placed under the operational control of *Einsatz-Abteilung Krämer.*

12 Aug 1944 (0400) Battalion and attached armored elements placed under the command of the *Panzer-Grenadier-Regiment.*

(0700) Battalion's remaining tanks secured on Hill 445 south of Wola Sekowa.

(0900) Battalion and *Panzergrenadiere* placed under the direct operational control of the division. The division's commander, *Oberst* Schmidhuber, personally expressed his appreciation to the battalion commander for the armored attack of the *Kampfgruppe* on Odrzechowa on 11 August 1944 and praised the actions of the tank crews.

(1200) One *Sturmgeschütz*, one *Panzer IV* and *Leutnant* Zywietz were attached to *Kampfgruppe Meschkat* (commander of *Aufklärungs-Abteilung 8*). They once again attacked Odrzechowa. Odrzechowa was taken towards evening after 18 batteries and two infantry regiments had been committed. The latter took possession of the hills northwest of Odrzechowa and eliminated the enemy threat from there. The two armored vehicles destroyed four antitank guns and one antitank rifle, two mortars and five supply vehicles.

Weather situation: Sunny, warm.

Ammunition consumption:

High-explosive rounds: 40

Armor-piercing rounds: 20

Machine-gun rounds: 500

Arrival and departure of attached elements: Armored elements departed again with the exception of the *4./Panzer-Regiment 10* and the *Einsatz-Kompanie/Panzer-Regiment 11.*

13-14 August 1944 Continuation of attack to the west. Contact established with the *208. Infanterie-Division.*

Leutnant Zywietz still attached to *Kampfgruppe Meschkat.* Attack on Zarszyn, which was taken. One tank, five antitank guns, four mortars and three supply vehicles were destroyed. One artillery position was silenced and subsequently captured by our infantry.

Weather situation: Sunny during the day, rain in the evening.

Ammunition consumption:

High-explosive rounds: 70

Machine-gun rounds: 500

15 Aug 1944 (0600) During the night of 15-16 August, the division moved west into the area south of Jaslo after establishing contact with unit on the left and forcing the enemy back north of the Sanok — Krosno road. The battalion moved through Kulaszne, Komancza, Jaskiska and Dukla to Zmigrod (18 kilometers south of Jaslo). Planned continuation of the move cancelled as the 8. Panzer-Division was committed east of Jaslo in response to renewed enemy pressure against the main road.

(1000) As per telex from the Panzer-Armee-Oberkommando 1 the battalion was released from its attachment to the 8. Panzer-Division and ordered to Mielau, East Prussia, to rest and refit. In turn, the units attached to the battalion (*Einsatz-Kompanie/Panzer-Regiment 11* and *Einsatz-Kompanie/Panzer-Regiment 10*) were released to return to their parent formations.

Weather situation: Sunny, no clouds.

Arrival and departure of attached elements: *Einsatz-Kompanie/Panzer-Regiment 11* and *Einsatz-Kompanie/Panzer-Regiment 10* left and were placed under the direct command of the division as *Einsatz-Kompanie Buch* (commander of the *3./Panzer-Regiment 11*).

16-17 August 1944 The support base detachment (formerly in Medyka) was sent from Krakow to rejoin the battalion. The battalion made itself ready to move and entrain. All companies entrained in Jaslo. The Maintenance Platoon and the unserviceable assault guns were loaded at Humenné (Slovakia).

Weather situation: Cloudy, warm, brief rain shower in the afternoon.

18 Aug 1944 (1600) Battalion loaded aboard three trains at Jaslo station. The commander of the division, *Oberst* Schmidhuber, again personally thanked the battalion commander for the battalion's efforts and expressed his wish that it return to the *8. Panzer-Division* after its rest and refit.

19 August 1944 While still en route, the *OKH/In 6* sent orders that the battalion was to be redirected to the training grounds at Grafenwöhr.

21 August 1944 The battalion reached its destination and was directly subordinated to the Inspector-General of Armored Forces, for the purpose of rest and refit to *Arbeitsstab Waldeck.* It was planned to reform the battalion as *schwere Panzer-Abteilung (Fkl) 301* equipped with *Tiger* tanks.

On 19 August 1944 the Inspector-General ordered the unit reorganized into a radio-control armor battalion with *Tiger* command and control tanks. For this purpose the battalion was to be moved to the training grounds at Mielau. The movement order was amended, however, and the battalion was redirected to Grafenwöhr, where it arrived in the second half of August.

By order of the *AHA Stab I* (*No. 40843/44* dated 2 September 1944) the unit was reorganized and renamed as *schwere Panzer-Abteilung (Tiger/Fkl) 301*.

The organigrams for this battalion are found with the section covering the *4./Panzer-Abteilung (Funklenk) 301* later in the book.

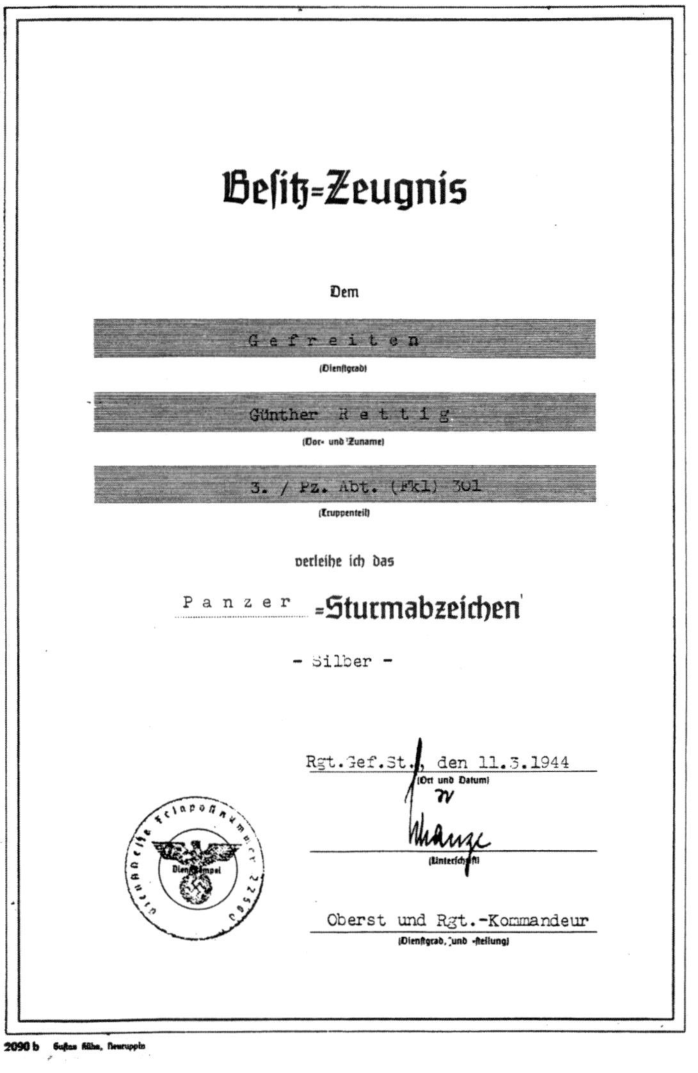

Besitz-Zeugnis

Dem

Gefreiten
(Dienstgrad)

Günther Rettig
(Vor- und Zuname)

3. / Pz. Abt. (Fkl) 301
(Truppenteil)

verleihe ich das

Panzer =Sturmabzeichen

- Silber -

Rgt.Gef.St., den 11.3.1944
(Ort und Datum)

(Unterschrift)

Oberst und Rgt.-Kommandeur
(Dienstgrad und -stellung)

2090 b

Im Namen

des

Führers

und Obersten Befehlshabers der Wehrmacht

verleihe ich

dem

Gefreiten Günther Rettig
3./Pz.Abt. (Fkl) 301

das

Kriegsverdienstkreuz 2. Klasse

mit Schwertern

Div.Stabsqu., den 20. April 1944

Dienstgrad und Dienststellung
Generalleutnant und stellv.Kdr.2.Pz.Div.

Award documents to *Gefreiter* Günther Rettig of the *3./Panzer-Abteilung (Funklenk) 301* for the *Panzerkampfabzeichen* in Silver (Above. Also note the regiment awarding him the badge apparently did not have the proper form and used one for the General or Infantry Assault Badge) and the War Service Cross, Second Class.

Left: Award document to *Oberfeldwebel* Max Palitz of the *3./Panzer-Abteilung (Funklenk) 301* for the *Panzerkampfabzeichen* in Silver.

Below: Award document to *Gefreiter* Dieter Höland of the *1./Panzer-Abteilung (Funklenk) 301* for the Iron Cross, Second Class.

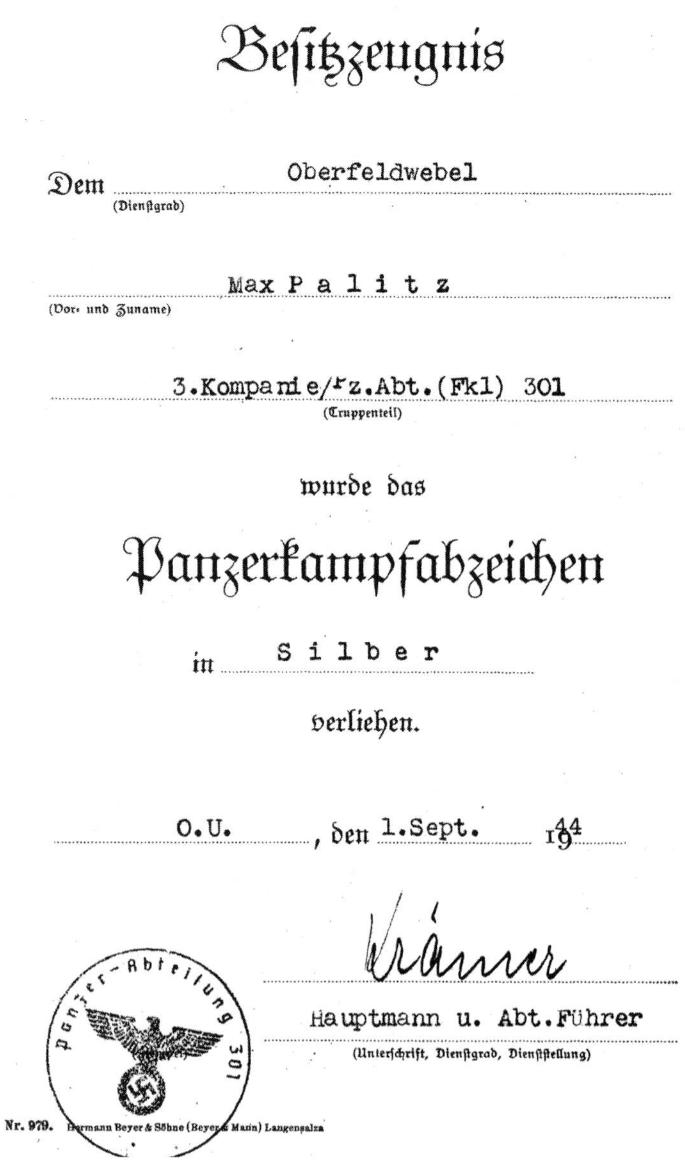
Besitzzeugnis

Dem Oberfeldwebel
(Dienstgrad)

Max Palitz
(Vor- und Zuname)

3.Kompanie/Pz.Abt.(Fkl) 301
(Truppenteil)

wurde das

Panzerkampfabzeichen

in Silber

verliehen.

O.U., den 1.Sept. 1944

Krämer

Hauptmann u. Abt.Führer
(Unterschrift, Dienstgrad, Dienststellung)

Panzer-Abteilung 301

Nr. 979. Hermann Beyer & Söhne (Beyer & Mann) Langensalza

IM NAMEN DES FÜHRERS
VERLEIHE ICH
DEM

Gefreiten
Dieter Höland
1./Pz.Abt.(Fkl)301

DAS
EISERNE KREUZ
2. KLASSE

Div.Gef.Stand, 30.Juli 1944

(DIENSTSIEGEL)

Oberst u.Fhr.d.8.Pz.Div.
(DIENSTGRAD UND DIENSTSTELLUNG)

Training personnel of *Panzer-Abteilung (Funklenk) 301* started in the spring of 1943 at Neuruppin. *Panzerschütze* Hoffmann with his *BIV*, *Ausführung A*. (Hoffmann)

The windshield is cleaned. The numeral *314* means 3rd company, 1st section, 4th vehicle. (Hoffmann)

Leutnant Schreiner inspects a few demolition-charge carriers. (Hoffmann)

The first few *Sturmgeschütz III's* of the battalion attract the attention of the younger crowd. (Rettig)

Moving out to the training area. (Weiß)

Staging the command and control vehicles. (Weiß)

The radio sets of the assault guns are dismounted. Mechanics check out the radio-control receivers of a *BIV* in the background. (Rettig)

During live-fire exercises the assault gun of the company commander loses its barrel. The battalion commander, *Major* Reinel, is in the middle. (Schreiner)

A heavy motorcycle/sidecar combination of the maintenance section of the *2./Panzer-Abteilung (Funklenk) 301*. (Fender)

The battalion was sent to France in the middle of September 1943. This *Sturmgeschütz* belongs to the *3./Panzer-Abteilung (Funklenk) 301*. The chassis number (95004) can be seen painted on the driver's vision slot. (Rettig)

Preparing a meal during the rail movement next to a *BIV*. (Rettig)

Battalion-level training was reinstated after another move — this time to the area around Amiens — in October 1943. A radio-control section is positioned behind a command and control vehicle. Note the vehicle identification number (*452*) painted on the rear of the demolition-charge carrier. (Beer)

Most of the radio-control equipment consisted of new *BIV's*, *Ausführung B*. These demolition-charge carriers have already been painted a sand color. (Bodsch)

Opposite page top right: A *Büssing-NAG* 4.5-ton truck with a *Bilstein* rotating crane. This vehicle belonged to the maintenance platoon. (Franzen)

Opposite page bottom right and this page left: Demolition-charge carriers operations were conducted in November and December 1943. The command and control *Sturmgeschütze* and the *BIV's* of the *3./Panzer-Abteilung (Funklenk) 301* have a striking camouflage pattern. (Rettig)

Below: This *Sturmgeschütz III* command and control vehicle belonged to the *2./Panzer-Abteilung (Funklenk) 301*. (Franz)

Above and right: This assault gun caught on fire while it was being welded. The follow-on explosion destroyed the vehicle. This *Sturmgeschütz III* had been built using a *Panzer III, Ausführung M* chassis. (Franzen)

Maintenance work being conducted on a *Sturmgeschütz III* command and control vehicle. The main gun has been removed as well as the superstructure roof. The chassis number (92933) can be seen on the driver's vision slot. This vehicle was also based on a *Panzer III 8/ZW* hull. Note the one-piece, forward-opening hull-access covers. (Franzen)

A command and control vehicle of the *2./Panzer-Abteilung (Funklenk) 301* at Comples in November 1943...(Franz)

...and again on 19 December 1943 at Roitel. (Franzen)

Left: After the Allied landings at Anzio/Nettuno at the end of January 1944, *Panzer-Abteilung (Funklenk) 301* was moved rapidly by rail to Italy on 31 January 1944. (Rettig)

Opposite page, left: The fighting compartments of the assault guns have been covered with tarpaulins. The soldiers have set up a small field stove in front of their vehicle. (Rettig)

Opposite page, right: The demolition-charge prime movers have also been stuffed full. (Franz)

Left: After being unloaded at Ficulle, the battalion continued south towards Rome. (Franz)

Below left: A photograph taken just after the first commitment of the battalion at Aprilia on 16 February 1944 (around 1600). This is the radio-control section of *Feldwebel* Palitz of the *3./Panzer-Abteilung (Funklenk) 301*. (Palitz)

Below: This *BIV* attempted to climb a set of stairs after its arrival at the Insituto Agraria Roma. (Franz)

Above right: Personnel of the maintenance platoon work on running-gear damage. This *Panzer III, Ausführung H* belonged to the battalion commander. Additional skirts have been added to the turret. The mounting brackets for the skirts can also be seen on the hull and fenders. (Franzen)

Opposite page: The German attack was stopped on 20 February 1944. The demolition-charge carriers were staged near the front and camouflaged. (Bischlepp)

Right: Short break for the *BIV* drivers. The tactical insignia and vehicle identification number (*453*) have been added to the rear of the demolition-charge carrier. (Beer)

The second German offensive started on 29 February 1944. One company of *Panzer-Abteilung (Funklenk) 301* was attached to the attack force. Grenadiers from the *Luftwaffen-Panzer-Division "Hermann Göring"* are standing next to this command and control vehicle. (Bundesarchiv 311/940/32)

Panthers of the *I./Panzer-Regiment 4* have gone into position behind this *BIV*. The attack was called off in the evening without success. (Bundesarchiv 311/940/14)

Panzer-Abteilung (Funklenk) 301 was loaded on trains again in the middle of March 1944. Another view of the *Büssing-NAG* crane truck of the maintenance platoon (*WH 1310239*). The vehicle has been painted in a camouflage scheme since the last picture of it. (Franzen)

Training of new recruits was initiated after the battalion returned to France in the spring of 1944. Here: *Leutnant* Glässel of the *4./Panzer-Abteilung (Funklenk) 301* with *BIV* drivers and combat engineers. (Weiß)

One of the most important supply vehicles — the mess truck! Note the *K* above the rear axle (*K* = *Küche* = kitchen). (Weiß)

In the middle of April 1944 *Panzer-Kompanie (Funklenk) 312* — previously a separate company — is integrated into *Panzer-Abteilung (Funklenk) 301* as its *1. Kompanie*. The gun commander of this command and control assault gun is *Leutnant* Endemann. (Höland)

This *BIV*, *Ausführung B* carries the tactical insignia of the *1./Panzer-Abteilung (Funklenk) 301*. (Höland)

Maintenance work on a demolition-charge carrier. In the foreground a portion of the removed gearbox can be seen. (Franzen)

After being transferred to the Eastern front, the battalion was attached to the *8. Panzer-Division.* The *Sturmgeschütz III* command and control vehicles have been staged, dispersed, in the open. (Weiß)

Several terrain walks are conducted with officers of the *schwere Panzer-Abteilung 507*, *Panzer-Abteilung (Funklenk) 301* and the *8. Panzer-Division* in the second half of June 1944. (Höland)

The assault guns were staged near the front and camouflaged. (Geissel)

The Soviet offensive started on 14 July 1944 in this sector. *Panzer-Abteilung (Funklenk) 301* was immediately committed. This *Sd.Kfz. 251, Ausführung C* belonged to the battalion staff. (Weiß)

Above and top photo next page: Even at this late date, this *Panzer III, Ausführung N* was employed with the battalion. It was committed to the defensive fighting with gun commander Höland. (Franzen)

A rarity! Only two examples of the *Maultier* based on the *Borgward* 3-ton truck were produced. This prime mover was used by the maintenance platoon. (Franzen)

The streets and roads at the front were softened up at the beginning of August 1944 by heavy downpours and were not readily negotiable. This *Büssing-NAG* truck has put chains on its front wheels. (Franzen)

The *Borgward Maultier* has become disabled and is being towed by the crane truck. (Franzen)

Maintenance work on the *Borgward Maultier*. The running gear (roadwheels and track) consists of elements used on the 3-ton prime movers. (Franzen)

Panzer-Abteilung (Funklenk) 301 was released from the operational control of the *8. Panzer-Division* in the middle of August 1944. A maintenance truck on the way to the railhead at Humenné. (Franzen)

The move to the rear begins. Here: A rear view of the *Borgward Maultier*. (Franzen)

2.2.7: Panzer-Kompanie (Fkl) 311

As part of the reorganization of the remote- and radio-control force, the *AHA* ordered on 25 January 1943 that the *1./Panzer-Abteilung 302* was to be renamed *Panzer-Kompanie (Fkl) 311* (see Chapter Subsection 2.2.5).

As per *Kriegsstärkennachweisung 1171 f*, *Panzer-Kompanie (Fkl) 311* was to be equipped with the *Sturmgeschütz III, Ausführung G* command and control assault guns (*KStN 1171 f Ausf. A*). The unit was assigned army field post number 17589 and its company emblem was a slightly modified devil figure from an illustrated story by Wilhelm Busch.

The company commander was *Oberleutnant* Bachmann, while platoon leaders and radio control officer were *Leutnants* Epke, Hitzdeld and Werth.

While in training, the company was attached to *Panzer-Abteilung (Fkl)* 301 then being formed in Neuruppin. The company began receiving its assault guns and demolition charge-carriers in March. After several weeks of instruction on its new equipment *Panzer-Kompanie (Fkl) 311* was moved from Neuruppin to Eisenach in May.

On 11 July 1943 the Army General Staff Organizational Department (*Generalstab des Heeres/Organizations-Abteilung*) ordered that the company was to be placed under the command of the *III./Panzer-Regiment "Großdeutschland"* for more extensive trials involving working with a *schwere Panzer-Abteilung (Tiger)*.

On 18 July 1943 *Panzer-Kompanie (Fkl) 311* entrained in Eisenach for transport to Kharkov on the Eastern Front. During "Operation *Zitadelle*" *Panzer-Kompanie (Fkl) 312* was attached to a *Tiger* unit, *schwere Panzer-Abteilung 505*. *Panzer-Kompanie (Fkl) 311* was to take part in a further test to determine if remote-control vehicles and *Tiger* tanks could operate efficiently in a single formation.

After detraining, *Panzer-Kompanie (Fkl) 311* was sent into the area of operations of *Panzer-Grenadier-Division "Großdeutschland"* and assembled in the Achtyrka forest. The company was committed on 14 August, taking part in an advance by units of *Panzer-Grenadier-Division "Großdeutschland"* in a northeasterly direction to Belsk.

The subsequent report filed by the company commander, *Oberleutnant* Bachmann, has survived (Source: Bundesarchiv/Militärarchiv Freiburg):

Panzer-Kompanie (Fkl) 311 25 October 1943

Report on the use of a *BIV* during the reconnaissance-in-force toward Belsk on 14 August and the destruction of a bridge using a *BIV* on 14 August 1943.

1.) Mission: Reconnaissance-in-force toward Belsk with one remote-control platoon and a platoon of grenadiers.

After the terrain had been scouted, the two platoons deployed in a depression west of the town. The radio-control platoon took up a reverse-slope position. Under covering fire from the assault guns, the grenadier platoon advanced to within 600 meters of the town. At the same time three *BIV's* were dispatched to reconnoiter and one reached the edge of the town. The enemy's heavy rifle and machine-gun fire diminished noticeably with the approach of the *BIV's*. The enemy appeared perplexed. After resuming their advance, the grenadiers came under heavy mortar and artillery fire from the direction of the northern outskirts of Kotelwa. After withdrawing briefly from the town, the *BIV* was moved to the edge of the town again and its demolition charge was dropped. The *BIV* was recalled. After the charge detonated the enemy ceased firing for a time and the grenadiers took advantage of this to pull back. The pause in firing lasted 5 to 7 minutes.

Results: The reconnaissance mission was completed. The enemy lost numerous killed. Several machine guns and three mortars were destroyed.

Losses: None.

Lessons learned: The use of the BIV was possible because the command and control assault guns and the BIV's had been carefully synchronized in the rest area. The time necessary for each mission is five to six hours. If the vehicles must travel (10 kilometrs) or there is a period of more than twelve hours between checks and a mission, then the equipment must be checked again.

The terrain was also favorable for the use of the BIV. It proved possible to carry out another necessary check and ready the BIV in the depression west of the town out of sight of the enemy (duration 10 minutes). The steadily rising terrain made it possible to control the vehicle out to 1,200 meters. There were no hindrances such as shell craters, ravines or woods.

The loss of two *BIV's* was due to problems that developed during the radio-control procedure. On one *BIV* a relay became stuck.

2.) Mission: Destruction of a bridge near Cholowdewchia through the use of a *BIV*.

The target bridge lay under enemy fire. As a result, it was necessary to use a *BIV* to destroy it. The *BIV* was moved toward the bridge from a reverse-slope position. The enemy's response was cautious and hesitant, allowing the *BIV* to reach the bridge. The command to jettison the charge failed; consequently, it had to be detonated in place, destroying the *BIV* as well.

Result: Mission accomplished.

Losses: None.

Lessons learned: The BIV was moved without difficulty from a reverse-slope position to the bridge, which was about 500 meters distant. The enemy seemed surprised and remained quiet. The 18-ton bridge was completely destroyed. All of the pilings were destroyed; the destruction reaching to beneath the surface of the water. The bridge was 25 meters long.

signed Bachmann
Oberleutnant and company commander

The defensive fighting west of Kharkov, which lasted until the middle of September 1943, was followed by a fighting withdrawal through Poltava in the direction of Kremenchug on the Dnjepr.

Panzer-Grenadier-Division "Großdeutschland" took part in the defensive fighting in the Dnjepr bend between Mishurin Rog and Kriviy Rog and Kirowograd until the end of 1943. In spring 1944 the German forces made a fighting retreat across the Bug, followed by defensive fighting around Targul Frumos in northern Bessarabia and the Carpathian foothills That lasted until the end of April 1944.

Demolition-charge carriers saw action during the course of various defensive movements by *Panzer-Grenadier-Division "Großdeutschland"* in the period from September to May 1944. The radio-control company's command and control assault guns also operated in an antitank role and provided support to the fusilier and grenadier regiments of the division.

On 18 May 1944 the *AHA* issued instructions for *Panzer-Kompanie (Fkl) 311* to proceed to its garrison at Eisenach as soon as possible. At that time the unit was in the Bakau area in Romania.

Following transport to Germany, the company was garrisoned at Eisenach and other bases and brought back up to strength. On 25 July 1944 the *AHA* ordered *Panzer-Kompanie (Fkl) 311* incorporated into *Panzer-Abteilung (Fkl) 302* as its *4. Kompanie*.

Gliederung der Panzerkompanie (Fkl) 311 (Stand 01.08.1943)

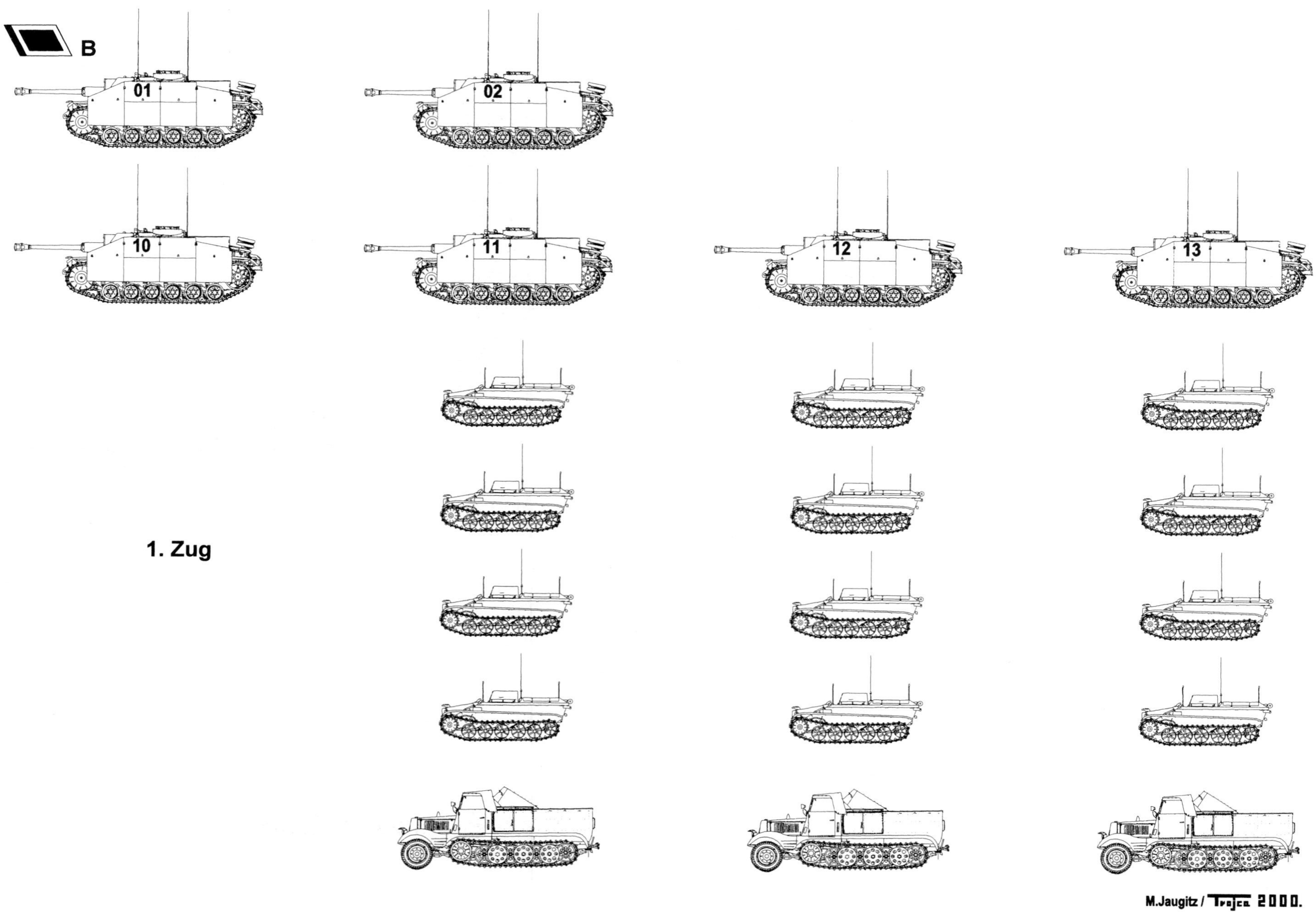

Gliederung der Panzerkompanie (Fkl) 311 (Stand 01.08.1943)

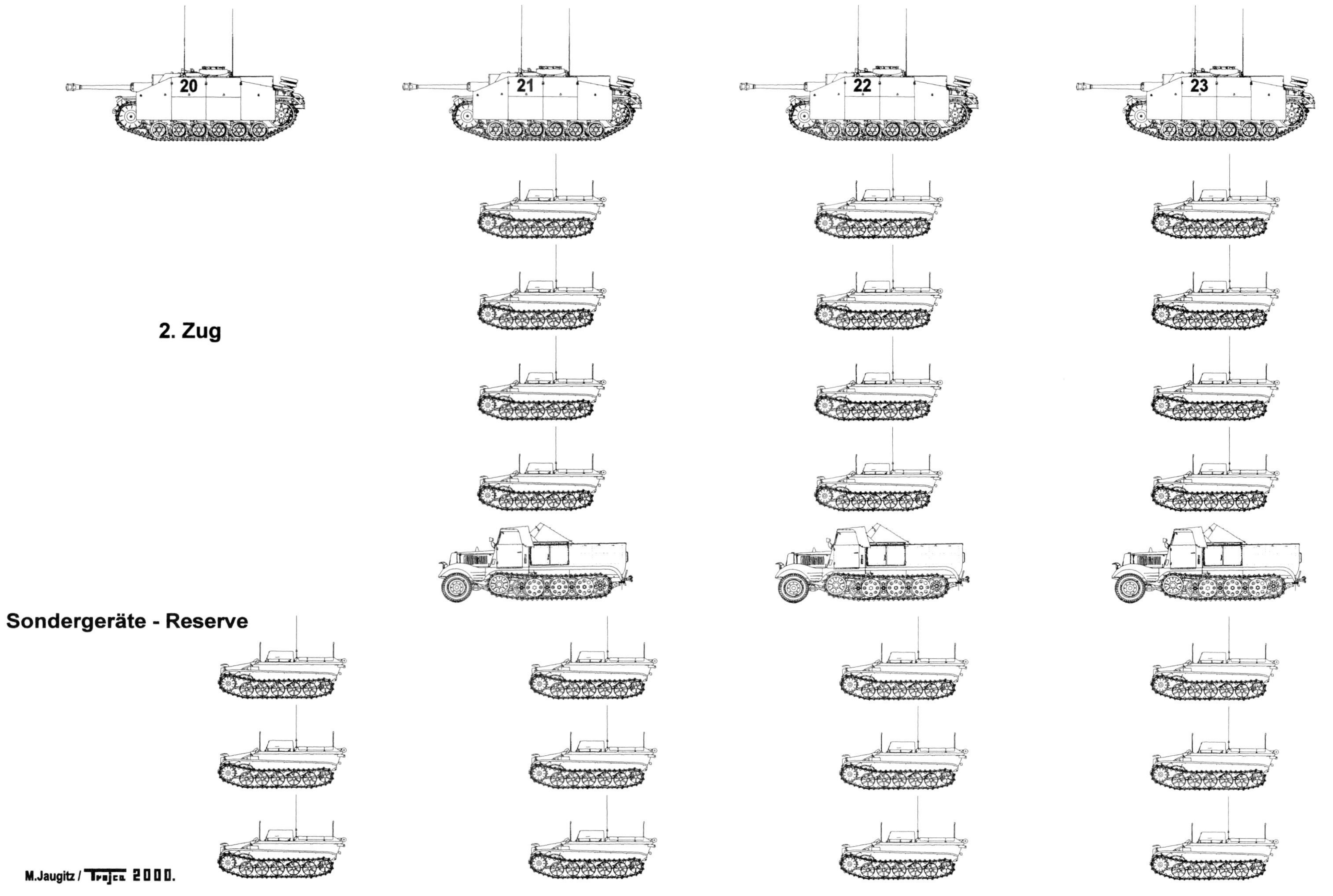

The formation of *Panzer-Kompanie (Fkl) 311* in the spring of 1943 at Neuruppin. (Geuss)

The company has been equipped with 10 command and control assault guns — the *Sturmgeschütz III, Ausführung G.* (Hitzfeld)

Some of the assault guns were built on the chassis of the *Panzer III, Ausführung M.* Hitzfeld)

This attempt at radio-control guidance has failed. The *BIV* has "broken loose" and bottomed out on the slope. (Hitzfeld)

The rail load of the company starts on 18 July 1943 at Eisenach. (Susenbeth)

The motorcycle/sidecar is firmly tied down. (Susenbeth)

The tactical insignia — a parallelogram with a *B* for *Bachmann* — and the company insignia are painted on all vehicles. (Susenbeth)

The side skirts were removed before rail transport and stowed in front of the vehicle. (Susenbeth)

Additional smoke-grenade launchers have been mounted on this *Sturmgeschütz*. (Susenbeth)

The demolition-charge carriers have been secured with cable to prevent slippage during the rail movement. (Susenbeth)

Vehicle numbers have been painted on the rear of the wooden boxes affixed to the fenders. The rear *Sturmgeschütz* has the number *20*. (Susenbeth)

Above and top left next page: The uploading of the vehicles is done. *Oberleutnant* Bachmann conducts a briefing in front of the assembled company prior to departure. (Susenbeth)

Right: Even during the rail movement the 3-ton prime movers are used in their capacity as haulers for replacement demolition-charge containers. (Kurnik)

Below: Assembly area in the woods at Achtyrka. Note the company insignia on the driver-side door. (Kurnik)

Above and left: During the defensive fighting west of Kharkov, which lasted until the middle of September, numerous radio-control operations were conducted. (Kurnik)

Panzer-Kompanie (Fkl) 311 occupies billets at Marijefka in January 1944. Here: The maintenance of a 3-ton prime mover during extremely cold weather. (Kurnik)

Above: Snow-covered terrain poses no problem for the *BMW* heavy motorcycle/sidecar; on the other hand, the *BIV's* can hardly be employed due to problems with the oil-steering boxes. (Niepenburg)

Above and below: The *Sturmgeschütze* of *Panzer-Kompanie (Fkl) 311* were frequently employed with the tanks of *Panzer-Regiment "Großdeutschland."* During some of the fighting, the assault gun of *Leutnant* Hitzfeld, the platoon leader of the 1st Platoon, received a hit on the gun mantlet. (Hitzfeld)

The next five photographs show the complete removal of the main-gun assembly on *Sturmgeschütz 10. Leutnant* Hitzfeld and the crew pitch in on the repair work. The *Sd.Kfz. 9/1* (an 18-ton prime mover with a 6-ton-capacity rotating crane) belongs to a maintenance unit of *Panzergrenadier-Division "Großdeutschland."* (Hitzfeld)

Above: It frequently happened that when using the new-style dead track on the running gear, the track would be thrown and wind up perched on the drive sprocket. (Hitzfeld)

Opposite page, top: Deserving crews are decorated. (Hitzfeld)

Opposite page, bottom: Occasionally radio-controlled demolition-charge carriers are again employed in the spring of 1944. This *BIV* still has the cross-shaped drive sprocket. (Hitzfeld)

Right: The company commander, *Oberleutnant* Bachmann, and the platoon leader of the 1st Platoon, *Leutnant* Hitzfeld, in the Bakau area of Rumania in the spring of 1944. (Hitzfeld)

2.2.8: Panzer-Kompanie (Fkl) 312

On 9 January 1943 the *AHA* issued a directive calling for the formation of *separate Panzer-Kompanie (Fkl) 312*. The formation process was to take place at the garrison at the newly-formed *Panzer-Abteilung (Fkl) 301* in Neuruppin.

The new company was formed from a small cadre of former members of the renamed and then disbanded *2./Panzer-Abteilung 302*. Then, in the second half of January 1943, the main body of the *2./Panzer-Abteilung 301* arrived in Neuruppin from Arnswalde. On 29 January 1943 *Panzer-Kompanie (Fkl) 312* was moved to Eisenach.

Panzer-Kompanie (Fkl) 312 was issued army field post number 04717. *Oberleutnant* Nolte remained as company commander, while the 1st Platoon was led by *Leutnant* Gottschalk and the 2nd Platoon by *Leutnant* Endemann. For its company emblem *Panzer-Kompanie (Fkl) 312* selected the "Uncle Nolte Head", another figure from the drawings of Wilhelm Busch.

In addition to theoretical training in garrison, much time was spent in practical training on the newly delivered *Sturmgeschütz III's* and *BIV* demolition-charge carriers. Bivouacs and exercises on the Berka and Ohrdruf training grounds were the order of the day, as were long marches in the area around the base.

The company received orders to be deployed at the end of April 1943. The unit moved to the railhead on 30 April to board the two transports to the east. During entraining there was an accident that cost the company its first fatality, *Panzerschütze* Neubauer.

After traveling through Germany and Poland, the first train under the command of *Leutnant* Gottschalk struck a mine laid by partisans on the morning of 3 May 1943. Two cars were derailed, but there were no casualties. Members of the company and personnel of a *Reichsbahn* repair train repaired the section of track, allowing the train to continue after a 26-hour delay.

The company detrained at Orel station on 6 May 1943 in the middle of a fierce air attack. The Soviet bombers were driven away by the German *Flak* and fighters before they could inflict any damage. Widely dispersed, *Panzer-Kompanie (Fkl) 312* set up its first field camp near the village of Sokolowka.

On 19 May the unit moved into its new rest area in some woods near Filosovo. The move there was exceptionally difficult because rain had made the roads almost impassable. On 26 May *Gefreiter* Neuhaus and *Panzerschütze* Dahmen were killed in an accident with an unexploded bomb.

In the weeks that followed, *Panzer-Kompanie (Fkl) 312* was visited several times by senior German commanders. Both *Generaloberst* Model and *Generalfeldmarschall* von Kluge came to investigate the operational possibilities of radio-controlled demolition-charge carriers.

On 3 and 4 July the combat elements and combat trains I and II left the rest area separately for their respective assembly areas. At that point in time the company was attached to *schwere Panzer-Abteilung 505* which, together with the *20. Panzer-Division* and the *6. Infanterie-Division*, had been combined into *Panzer-Brigade 21*. This armored formation under the command of *Oberst* Burmeister was intended to drive an attack wedge south into the front.

The attack began at 0320 hours on 5 July 1943 after a heavy bombardment by artillery, dive-bombers and close-support aircraft.

In the days that followed, *Panzer-Kompanie (Fkl) 312* saw action near Podolyan, Verkh-Tagino, Step, Saborowka, Glasunowka, Trosna, Krasnikowo, Mish-Mukhonowka and the high ground near these towns. Fifty-six enemy tanks, 28 guns and antitank guns and 18 vehicles were destroyed.

The number of wounded sustained by *Panzer-Kompanie (Fkl) 312* was astonishingly low and the unit lost just three killed.

The company commander, *Oberleutnant* Nolte, compiled a detailed after-action report on the unit's operations during the period 5 to 8 July 1943 which provides excellent insight into the operations during this time. (Source: National Archives, USA)

Nolte Location classified, 13 July 1943

Oberleutnant and Company Commander

Pz.Kp.(Fkl) 312

8 Copies

3rd copy of 8

Secret!

Report on Operations from 5 – 8 July 1943

For the attack by the *6. Infanterie-Division* on 5 July 1943 against the Tagino main line of resistance, the 2nd Platoon with three assault guns and twelve *BIV's* was attached to *schwere Panzer-Abteilung 505*. After friendly infantry had reached Yasnaya Polyana and the Wiedehopf Woods as well as Hill 240.4, *schwere Panzer-Abteilung 505*, with the attached 2nd Platoon, *Panzer-Kompanie (Fkl) 312*, overran the infantry's frontline and attacked in the direction of Hill 230.7.

I rode with the commander of *schwere Panzer-Abteilung 505*. The unit was formed into *Kampfgruppe Oberst Burmeister*. The platoon maintained continuous combat reconnaissance in front of the *1./schwere Panzer-Abteilung 505*. Three *BIV's* were in front of the *Tiger* company at all times, their principal task being to locate mines. None were found. The enemy had installed defensive weapons on Hill 230.7 and northwest of Step and linked them with weak field positions. One *BIV* was sent against a group of 2 to 3 antitank guns. There was a small ditch in front of the position and the *BIV* drove into it. 50 to 60 infantry immediately gathered round the stationary *BIV*. The controller gave the command to detonate and destroyed the antitank position and the 50 to 60 infantry. The range was 700-800 meters. The remaining targets which presented themselves were engaged with fire and largely destroyed. Apart from the *BIV* which was detonated, there were no losses.

At about 0800 hours on 6 July 1943 the company received orders to scout out from Sabarowka in the direction of Hill 230.4 for employment opportunities. At the southern outskirts of Sabarowka, the *20. Panzer-Division* was already engaged with enemy tanks on Hill 230.4 and both sides of it, mainly T34's. The town itself lay under heavy artillery fire. I deployed the 1st Platoon (three assault guns and twelve *BIV's*) from the depression at the southern edge of Sabarowka and held the 2nd Platoon (one assault gun, two *BIV's*; the other two assault guns disabled mechanically) on standby at the northern edge of Sabarowka. The 1st Platoon directed two *BIV's* against the T34's on the hill, which were at least 1,800 to 2,000 meters away. Both *BIV's* were destroyed before they could reach their target. At the same time the platoon joined the firefight.

At 1300 hours *Kampfgruppe Burmeister* — consisting of the *II./Panzer-Regiment 3*, the *I./Panzer-Grenadier-Regiment 304*, the *3./Pionier-Bataillon 38* and *Panzer-Kompanie (Fkl) 312* — passed through Sabarowka and attacked Hill 230.4. The 1st Platoon was interspersed with the lead company of *the II./Panzer-Regiment 3.* The 2nd Platoon was given the mission to support the advance by the *I./Panzer-Grenadier-Regiment 304*, however, it was subsequently also incorporated into the point unit of the *II./Panzer-Regiment 3*.

In front of *II./Panzer-Regiment 3* was a large number of T34's and KVII's, especially to the south and southeast. The firefight was opened at a range of 2,000 to 2,500 meters. Only to the southeast were the ranges somewhat less, on the order of 400 meters. I myself moved in the middle of the armor battalion and first committed the elements of the company against the closer targets to the southeast. A T34 was approached successfully. The T34 tried to ram the *BIV*. The controller sent the command to detonate and the T 34 was destroyed. In addition to the T 34, a large number of Russian infantry was also destroyed.

Hills 230.4 and 238.5 were guarded by heavy field fortifications. Three *BIV's* were successfully maneuvered against three heavy gun bunkers at a range of 400 to 600 meters and detonated, destroying the targets. During this operation one *BIV* was hit shortly after leaving friendly lines and burned out. A second *BIV* was attacked with gasoline bombs in the Russian position. It was subsequently detonated and, as far as could be seen, inflicted considerable damage on the Russian position.

Then I took the company south on the right wing. There were no opportunities to employ the *BIV's* there and the company's only action was taking part in the firefight. The firefight was conducted at a range of about 1,800 to 2,500 meters. During this attack one assault gun was disabled by a hit to the drive sprocket. It was recovered, however, and was operational the next morning.

On 7 July 1943 the company strength: five assault guns and ten *BIV's*) was attached to the *2. Panzer-Division* as division reserve. Together with the *I./Panzer-Grenadier-Regiment 304*, it screened toward the south while the remaining elements of the *2. Panzer-Division* launched an attack southeast against the Aier Woods.

On 8 July the company (Strength: five assault guns, ten *BIV's*) participated in an attack by *Kampfgruppe Burmeister* (consisting of two armor battalions and a *Panzergrenadier* battalion) against Hill 238.1, Teploye and Hill 253.5. The company was deployed with the leading elements of *Panzer-Abteilung 35* and took part in the fighting from there.

It was committed against Hill 238.1. One *BIV* was sent against a bunker (probably an antitank bunker) situated on the reverse slope. The *BIV* caught fire about 10 meters from the bunker, reached its target and was detonated. The bunker was destroyed. The range was approximately 400 meters. A second *BIV* was sent against a gun bunker, which was destroyed. A third *BIV* was committed against an infantry position that was located on the reverse slope and could not be reached by the main guns of the assault guns. The position housed three guns. The *BIV* was detonated, destroying the three guns and the gun crews. Range was approximately 800 meters.

The attack was then carried forward against Teploye, which was taken by the grenadiers. One assault gun was knocked out by enemy action. The company received orders to conduct zone reconnaissance — looking for mines in particular — in preparation for the attack on Hill 253.5, which had previously been bombed by *Stukas*. The armored attack enveloped Teploye on the east side and rolled into the depression between Hills 253.5 and 272.9, direction southwest. Zone reconnaissance was carried out by two *BIV's*. No mines were found. Four of the company's assault guns formed part of the first wave of the attack. One of the *BIV's* engaged in zone reconnaissance was knocked out and destroyed by fire. The second *BIV* was used to blow up a gun bunker.

The attack bogged down in the depression. One assault gun was knocked out, but was recovered. The company took part in the firefight against Hill 253.5. In the further course of the attack the company's remaining four assault guns occupied a firing position east-southeast of Teploye, placing fire on Hill 253.5. After the fighting had died down somewhat, the company came under continuous flanking fire from the direction of Hill 334.5. No muzzle flashes were visible from that direction, neither was there any sign of the enemy. It was determined that the enemy weapons had to be on the reverse slope. I made the decision to commit one *BIV*. The *BIV* was driven onto the hill and suddenly disappeared. In order to prevent the *BIV* from falling into enemy hands intact, I ordered *Leutnant* Endemann to secure the *BIV* while a second assault gun provided covering fire. It turned out that the *BIV* had fallen into a bomb crater.

Under cover of smoke, in the most difficult of conditions, the equipment was removed in the midst of a grain field occupied by enemy infantry. The *BIV* was destroyed with hand grenades. In the evening, when darkness fell, the company was ordered to pull back with *Panzer-Abteilung 35*. It reached the passage position of the *2. Panzer-Division* without leaving any vehicles on the battlefield.

Lessons Learned

During the course of the fighting it became very obvious that there had been insufficient training in working with armor units. In most cases the company committed the *BIV's* on its own. During the attack by the armor units there was no real cooperation which could have exploited the success which the *BIV* was undoubtedly capable of. One of the most serious hindrances was the fact that the equipment used by the units with which we were supposed to work was completely foreign, making it impossible for us to coordinate our efforts. The missions were conducted entirely on our own initiative, and these showed that a radio-control unit could only be completely effective if the command and control vehicles are incorporated into the lead elements of the armor units.

The firefight was generally opened at 2,000 to 2,500 meters, which was easily within the capabilities of our weapons. It is obvious then, that the potential for successful use of the *BIV* was great when the range was at least 2,000 meters and not 700 – 800 meters as was often the case. Because of radio silence, the company could only check the equipment using a test antenna before an attack and thus could not determine the range accurately. Calibrating the sets in Germany had not been possible, because the company only received its remote-control equipment two days before its departure. Furthermore, it was found that the sensitivity of the equipment suffered somewhat as a result of road movement. A check should be made to determine whether improvements could be made in the stability of the equipment itself or its mounts. Jettisoning of the charge has so far not been possible since, to date, the *BIV* has been exposed to such heavy fire that recovery has been out of the question.

Use of the *BIV* with its low profile in cornfields and fields with heavy vegetation has shown that the *BIV* immediately disappears from sight in cornfields and even in shallow depressions. Since the company had realized this after several exercises prior to the attack, a solution was found: 1.) A small pennant was attached to the antenna and 2.) Rods, roughly 2 meters long, were affixed on the right and left sides of the *BIV*. The rods had the additional advantage of making it easier for the controller to determine the vehicle's attitude and identify changes in direction at great distances. With the pennants it was not always possible to determine the *BIV's* direction, especially as range increased. Consideration should be given to providing the *BIV* with two folding rods as standard equipment.

During the guidance phase it was shown that the only relatively certain method of avoiding enemy fire is for the *BIV* to maintain a sharp zigzag course. It must be stressed in training that following a straight path will surely lead to the destruction of the BIV and that only a constant zigzag control procedure will allow it to reach its target in most cases.

Experience has also shown that the *BIV* must follow quite close behind the command and control vehicle. It is requested that these findings be evaluated in training: 1.) The assault gun offers the *BIV* some protection, 2.) Communication between the assault gun and the *BIV* is impossible at great distances. Remaining in close proximity is especially important should the *BIV* drive into a ditch, for example, and require the assault gun's assistance.

During zone reconnaissance with the *Tiger* battalion it was determined that the *BIV's* speed off roads is inadequate. The *BIV* cannot keep pace with the *Tiger* tanks for which it is supposed to be conducting reconnaissance.

The side entrance hatch has not proven itself. It has been shown that the driver can get out more quickly through the top, and is not exposed to fire. Crawling out the side hatch is slower and he can more easily be shot at from foxholes.

The five-pin plug must be mounted in such a way that it is not dislodged when the *BIV* bumps into something or when the demolition charge is raised, as this makes it impossible to detonate the *BIV*. It is therefore recommended that the five-pin plug be located near the driver's seat.

The detonation relays were disconnected on my order — but also mainly because of a directive issued by the Commander-in-Chief — for the following reasons: It is apparent when a *BIV* drives over a mine, even without the detonation relay. If a *BIV* is detonated, this results in a crater through which no vehicles can pass. This point is especially important near roads and bridges. Further, the explosion of any bomb or shell threatens the *BIV* through sympathetic detonation.

It has been found that the distance counter should only be set to 100 meters, since beyond that targets can suddenly appear which must be engaged.

The engine of the *BIV* has proved trouble-free.

Just prior to the start of the attack, *Panzer-Kompanie (Fkl) 314* discovered other equipment was using its radio-control frequencies. It proved a disadvantage that 1.) The company was unaware of the frequencies, 2.) No measures were taken to prevent other equipment from using the same frequencies. Tests have shown that the radios of the assault guns interfere with the radio-control frequencies. It is requested that higher command channels again check to ensure that the radio-control frequencies are available. Due to the intervention of the army signals officer an interim solution was found in the final days before the attack, but this proved an inconvenience to all units. It also turned out that several companies were committed in the same area and had the same frequency bands. For this reason, two days before the company moved into its assembly area it had to exchange equipment with *Panzer-Kmpanie (Fkl) 313*, which was a major burden for the company at that time.

The company refused to employ unarmored *BIV's* and suggested that these *BIV's* be sent back to the replacement battalion since it was impossible to manually operate an unarmored *BIV* if it were in the lead. Such a practice would surely result in the driver being incapacitated by artillery shrapnel and result in the loss of the vehicle.

The use of assault guns as command and control vehicles proved a success. In the company's view the only shortcomings were that not all of the *Sturmgeschütze* were equipped with side skirts and, more importantly, the company did not receive the ten assault guns it was supposed to. Use of the assault gun in operations where it was employed as part of the leading armor units revealed that the absence of a machine-gun is a disadvantage. Given the limited amount of high-explosive ammunition carried by the assault gun, the absence of a machine-gun makes it impossible to engage enemy infantry in broken terrain. While negotiating cornfields the commanders can defend themselves through submachine-guns and hand grenades.

In fighting such as took place in this area, it is difficult to move forward prime movers, even through our own infantry. The greatest problem remains picking up the driver who has left his *BIV* on the battlefield. As per orders, the assault guns picked up the *BIV* drivers. Consideration should be given to *SPW's* instead of prime movers. Combat engineers and communications personnel were not needed on the battlefield during the attack, as the *BIV's* had been prepared for operations before the fighting.

To summarize, in four days of fighting 20 *BIV's* were committed. Destroyed by *BIV* detonation were: one T 34, one antitank position with 2 to 3 antitank guns and 50 to 60 infantry, nine bunkers (gun or antitank gun bunkers), one infantry gun position with three guns. Two *BIV's* were set on fire by gasoline bombs in the Russian position, resulting in the destruction of a number of infantry. Furthermore, detonation resulted in each case in the destruction of an unknown number of enemy infantry. In my view, better cooperation with the armor units could result in even greater success.

signed Nolte
Oberleutnant and company commander

Distribution:
Panzer-Armeeoberkommando IX: 2
Brigadestab 21: 2
2. Panzer-Division: 1
Panzer-Ersatz- und Versuchs-Abteilung 300: 1
Internal Use: 2

The start of the Russian counteroffensive on 11 July 1943 forced the German units to break off their attack and withdraw. After a successful withdrawal, the company received orders to proceed to Mogilew.

The company proceeded to Shlobin, where it remained for a week, and then to Mogilew. The unit entrained in Briansk on 1 August 1943; destination Eisenach. This was changed at short notice, however, and *Panzer-Kompanie (Fkl) 312* was transferred to the Grafenwöhr training grounds, where it arrived on 13 August. The company's personnel were then sent on leave.

After the return of the personnel many of the tank crews took part in a *Panther* course at Erlangen. The reason for this: Plans to employ the *Panther* as a command and control tank. At the same time, *Panzer-Versuchs- und Ersatz-Abteilung (Fkl) 300* was conducting trials at Eisenach. The results were less than satisfactory because of problems housing the command transmitter equipment in the *Panther*.

On 18 October 1943 the company was ordered to move again. On 23 October it departed Grafenwöhr by train for Epe in Holland. There the company was temporarily housed in a trades school. Later *Panzer-Kompanie (Fkl) 312* moved to the nearby training grounds at Oldebroek/Wezep near Zwolle. On 27 November 1943 the *AHA* ordered the unit's equipment and personnel brought up to authorized strength.

On 20 December 1943 — in response to the enemy's growing use of mines — the Inspector-General of Armored Forces ordered the *3. Kompanie* of a *schwere Panzer-Abteilung (Tiger)* equipped with radio-control vehicles. It was decided that *Panzer-Kompanie (Fkl) 312* would be incorporated into *schwere Panzer-Abteilung 508* as its *3. Kompanie*, however, this order was nullified soon afterwards.

New orders were issued on 25 January 1944: The company was now to be incorporated into *schwere Panzer-Abteilung 507*. The company was ordered to be ready for its new role as the *3./schwere Panzer-Abteilung 507* by 31 March 1944. The *Tiger* battalion was transferred to the Eastern Front sooner than expected, however, and the consolidation never took place.

Finally, on 2 April 1944, the *Generalstab des Heeres/Organizations-Abteilung* ordered the company integrated into *Panzer-Abteilung (Fkl) 301*. On 17 April it was renamed the *1./Panzer-Abteilung (Fkl) 301*.

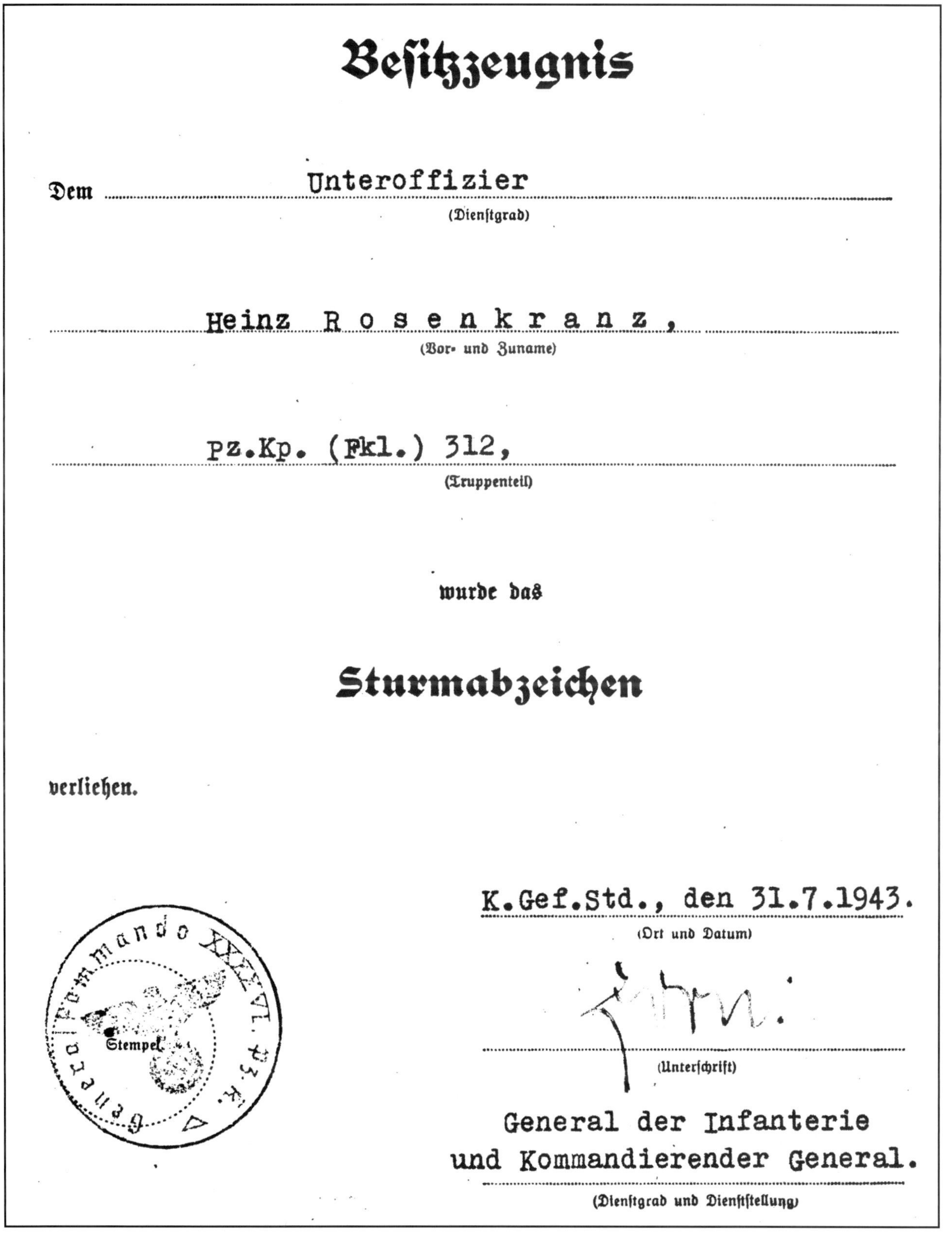

Besitzzeugnis

Dem Unteroffizier
(Dienstgrad)

Heinz Rosenkranz,
(Vor- und Zuname)

Pz.Kp. (Fkl.) 312,
(Truppenteil)

wurde das

Sturmabzeichen

verliehen.

K.Gef.Std., den 31.7.1943.
(Ort und Datum)

Stempel

(Unterschrift)

General der Infanterie
und Kommandierender General.
(Dienstgrad und Dienststellung)

Award document for the General Assault Badge to Unteroffizier Heinz Rosenkranz of Panzer-Kompanie (Funklenk) 312.

Gliederung der Panzerkompanie (Fkl) 312 (Stand 04.07.1943)

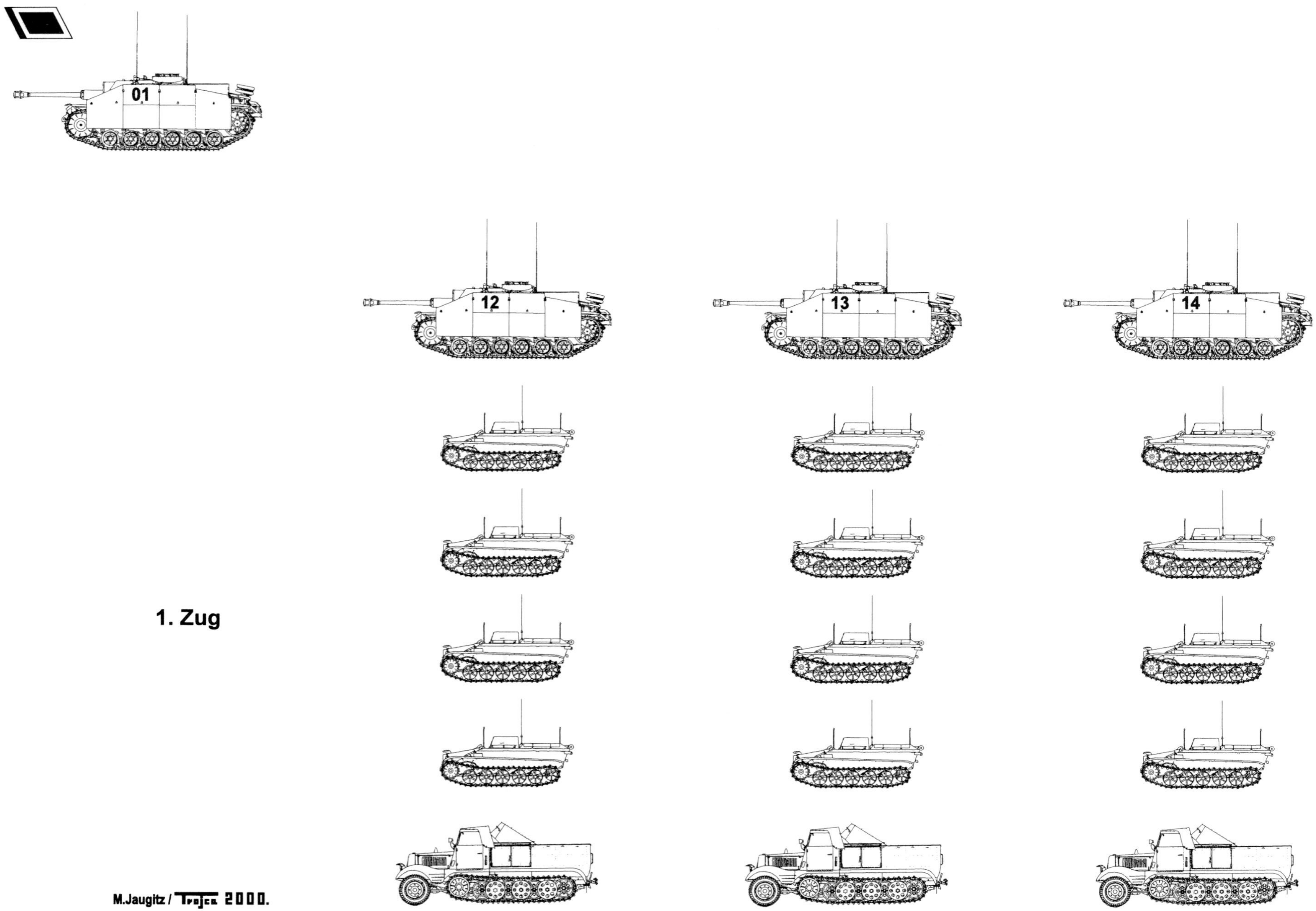

Gliederung der Panzerkompanie (Fkl) 312 (Stand 04.07.1943)

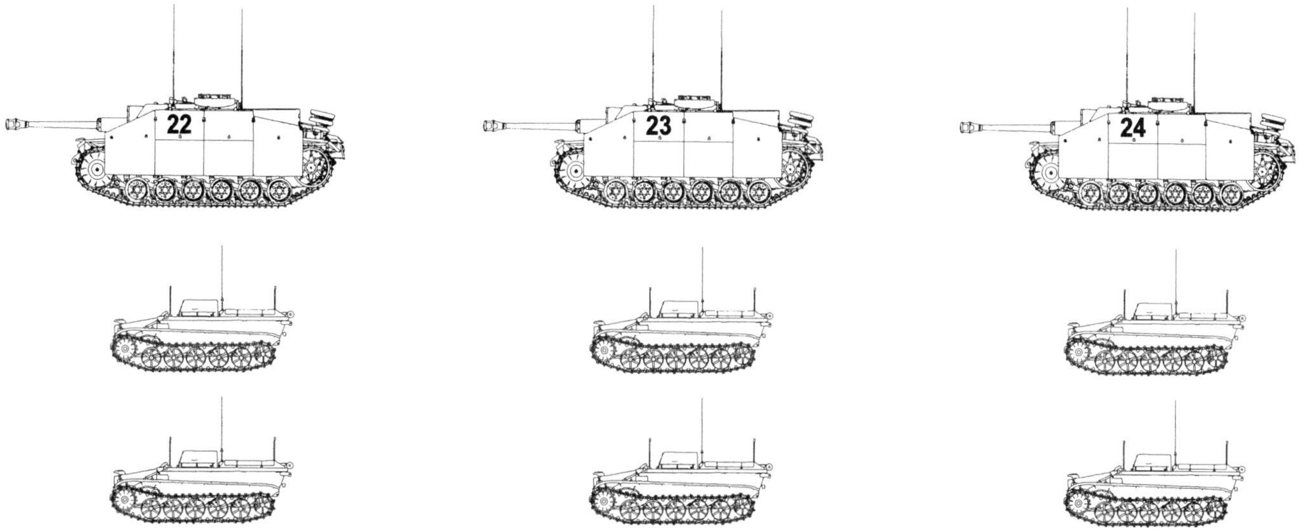

2. Zug

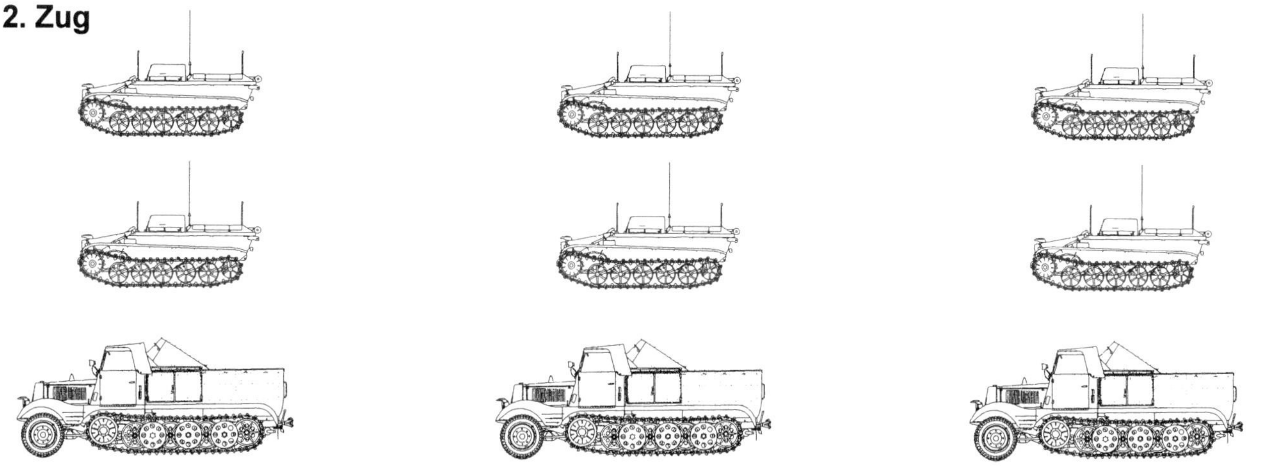

Sondergeräte - Reserve

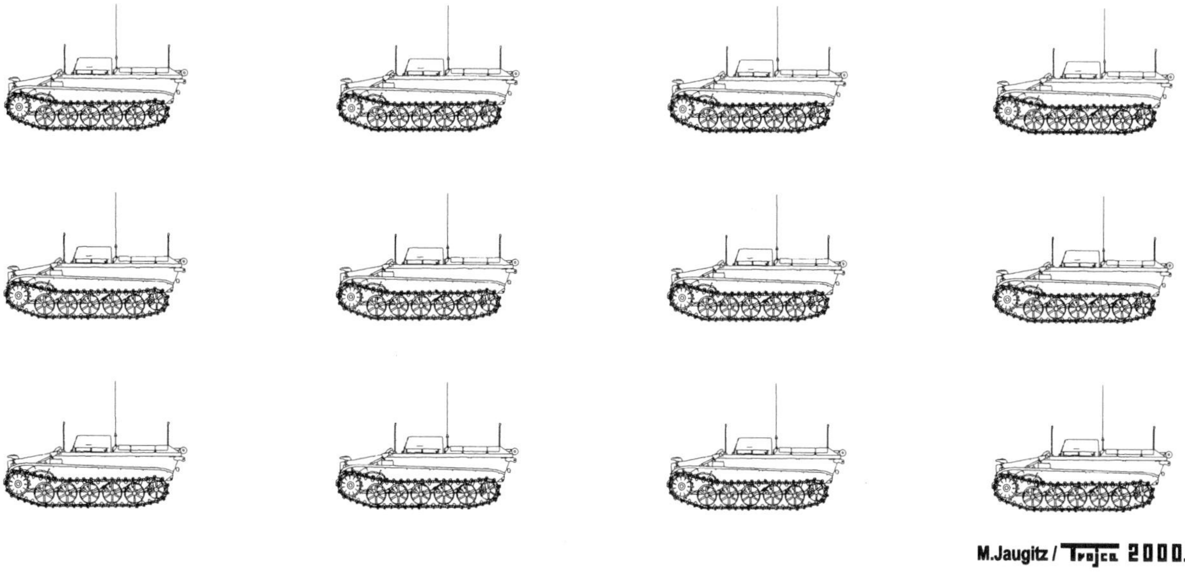

M.Jaugitz / Trojca 2000.

Training of company personnel took place on the *Leitpanzer III, Ausführung J* of the former *2./Panzer-Abteilung 302* or the *2./Panzer-Abteilung 301*. (Geissel)

Panzer-Kompanie (Fkl) 312 received the *Sturmgeschütz III, Ausführung G* in the spring of 1943. (Höland)

The combat elements were billeted at Mühlberg during an exercise at the Ohrdruf Training Area. (Höland)

Additional track links have been mounted on the hull sides in the area of the running gear. (Geissel)

Above and left: Two trains head east on 30 April 1943 carrying *Panzer-Kompanie (Fkl) 312*. The first train was derailed on 3 May 60 kilometers east of Brest-Litowsk after partisans had blown a section of track (Höland)

After its arrival the company moved to the woods at Filosowo on 19 May 1943. (Höland)

A 1-ton prime mover of the maintenance section stands by to recover damaged vehicles. (Höland)

The "Uncle Nolte" insignia of the company can be identified on the rear of this *BIV*. (Hochgreef)

A radio-control platoon sets up in anticipation of a demonstration for generals in June 1943. The left-hand *Sturmgeschütz* has been outfitted with a deep-water exhaust stack. (Höland)

Above and opposite page top: Company personnel enjoy a few days of peace and quiet in the billeting area. (Geissel)

Operation *Zitadelle* starts on 5 July 1943. *Panzer-Kompanie (Fkl) 312* is placed under the operational control of *Panzer-Brigade 21*. (Höland)

Right: The assault gun of *Unteroffizier* Rott. The 2nd Platoon was attached to *schwere Panzer-Abteilung 505* on the first day of the offensive. (Hochgreef)

Above: A *BIV* and its driver wait in a reverse-slope position for their next mission. (Hochgreef)

Right: Command post at Wesch-Tagino. *Oberleutnant* Nolte sits under the tent and studies the map. (Hochgreef)

Left: Ammunition and demolition charges were brought to the front with this 3-ton prime mover. (Geissel)

Above: Operations with demolition-charge carriers are conducted again and again. (Münch)

Left: The bitter fighting continues. The *Sturmgeschütz* crews are at the end of their tether. (Hochgreef)

Above: This 18-ton prime mover of the maintenance section has been continuously employed. (Höland)

Upper right: A tipped-over trailer has to be placed back on its wheels again. (Höland)

Right and opposite page, top left: The commander of *Panzer-Brigade 21*, *Oberst* Burmeister, awards decorations to personnel of *Panzer-Kompanie (Fkl) 312* in the middle of July. (Höland)

The company is loaded at Briansk on 1 August 1943. (Höland)

Right and next page top: Elements of *Panzerjäger-Regiment 656* could also be seen there. That regiment had been employed during Operation *Zitadelle* with the *Ferdinand* heavy tank destroyer. (Höland)

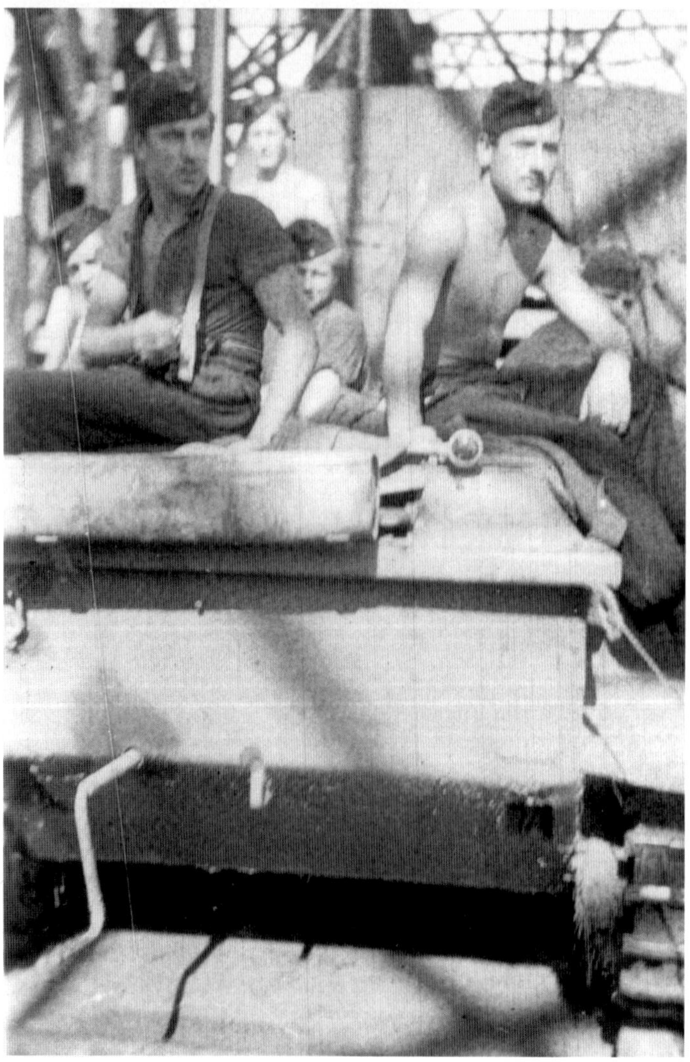

Left: During the rail movement to the Grafenwöhr Training Area. (Schlagloth)

Right and below: Inside a *Sturmgeschütz III, Ausführung G* (Chassis number: 95992). The fighting compartment is very cramped. (Höland)

Above, below and next page top: *Panzer-Kompanie (Fkl) 312* was also issued a *Leitpanzer III, Ausführung M* during its stay at the Oldebroek Training Area. It had belonged to *Panzer-Kompanie (Fkl) 313*. Turret alphanumeric *F 11* is the indicator that this vehicle belonged to *Leutnant* Sigmund, the platoon leader of the 1st Platoon. (Höland)

The crews of the assault guns affixed basic-issue-item tools to their vehicles in a manner each saw fit. All of the armored vehicles were consolidated as the *1./Panzer-Abteilung (Fkl) 301)* in the middle of April 1944. (Höland)

2.2.9: Panzer-Kompanie (Fkl) 313 – 3./schwere Panzer-Abteilung 508

Panzer-Kompanie (Fkl) 313 was created by order of the *AHA I a (721/43 geheim)* on 9 January 1943 by redesignating the *3./Panzer-Abteilung 302* at Eisenach (see Chapter subsection 2.2.5).

The unit continued to be commanded by *Oberleutnant* Fritschken. The 1st Platoon was led by *Leutnant* Sigmund; the 2nd Platoon by *Leutnant* von Rhoden. Radio-control officer was *Leutnant* Starck.

As per *Kriegsstärkennachweisung 1171 f (Ausführung B)*, *Panzer-Kompanie (Fkl) 313* was equipped with ten *Panzerkampfwagen III* command and control tanks in Neuruppin. The company commander and the platoon leaders were each issued a *Panzerkampfwagen III, Ausführung N* armed with the 75 mm *KwK L/24*. The remaining command and control tanks were either the *Panzer III, Ausführung J* or *N*, armed with the 50 mm *KwK 39 L/60*.

All of the command and control tanks had been thoroughly overhauled and modified for their new role, however, the various features of the different variants were still visible. One peculiarity was the identification alphanumeric painted on the turrets of the command and control tanks. A large "*F*" appeared in front of the turret numeral, the first letter of the company commander's name.

The majority of the company personnel were experienced radio-control crews who had served with the *2./Minenräum-Abteilung 1* or the *2./Panzer-Abteilung 300 (FL)*. After several months of training and rest time in Neuruppin, *Panzer-Kompanie (Fkl) 313* received orders to proceed to the Eastern Front.

The unit entrained on 10 June 1943 for transport to the Orel area. Arriving after a journey of several days, the company detrained and was directed into the assembly area of *schweres Panzerjäger-Regiment 656*. As per a telex sent by the Operations Department of the *AHA*, the company was attached to *schwere Panzerjäger-Abteilung 654* (also known as the *II./schwere Panzer-Regiment 656*).

After scouting the area of operations, *Panzer-Kompanie (Fkl) 313* occupied its assigned assembly area near the front in the first days of July. The unit made its final preparations to go into action right beside the Orel — Kursk rail line south of Glasunowka.

Following the start of the offensive on 5 July 1943, the radio-control sections moved out in the early hours of the morning with the *Ferdinands* of *schwere Panzerjäger-Abteilung 654*. The 1st and 2nd Platoons were deployed in separate sectors.

Soon after the start of the attack, several demolition-charge carriers of the 1st Platoon drove into an uncleared friendly minefield. *Oberleutnant* Fritschken contacted radio-control officer Starck, who was in the assembly area with the reserve demolition-charge carriers and a command and control tank of the 2nd Platoon, and instructed him to send replacement vehicles immediately.

A short time later a *BIV* was hit by artillery fire and exploded. Two nearby *BIV's* also went up and the force of the explosion also destroyed *Leitpanzer F23*. Another command and control tank not far away was damaged, as was a *Ferdinand*.

Further radio-control missions were carried out on the day of the attack and there were additional casualties. *Panzer-Kompanie (Fkl) 313* lost 17 killed on 7 July alone. Two more members of the company died of wounds in a field hospital several days later.

After four days of combat the company had suffered such heavy losses in men and equipment that its operational effectiveness was severely reduced and it had to be pulled out of the line. On 7 August 1943 the *AHA* ordered the company home to the training grounds at Grafenwöhr.

The company did not remain in Grafenwöhr for long, however. On 18 October 1943 orders were issued for *Panzer-Kompanie (Fkl) 313* to proceed to the Mailly le Camp training grounds in France. There the unit was to be brought up to strength in equipment and personnel and reorganized as per *Kriegsstärkennachweisung 1171 f.*

On 5 December 1943 a change in equipment was ordered. Beginning in January 1944, the unit was to switch from *Panzerkampfwagen III's* to assault guns.

By mid-December 1943 both the company commander, *Oberleutnant* Fritschken, and the platoon leaders, *Leutnants* Sigmund and von Rhoden, had been transferred to *Panzer-Versuchs-und Ersatz-Abteilung 300 (FL)*.

At the end of December the *AHA* ordered the *Panzer-Kompanie (Fkl) 313* incorporated into *schwere Panzer-Abteilung 508* as the *3./schwere Panzer-Abteilung 508*. *Schwere Panzer-Abteilung 508* had also been transferred to Mailly le Camp. The company was to be organized in accordance with *Kriegsstärkennachweisung 1176 f* of 28 December 1943. All surplus personnel and unneeded *Panzerkampfwagen III's* and *BIV* demolition-charge carriers were to be transferred to *Panzer-Versuchs-und Ersatz-Abteilung 300 (FL)*.

All surplus equipment was to be transferred to the *Ausbildungsstab Mailly* (formerly *Stab Panzer-Regiment 39*). The army field post number 40153 was eliminated, since all former members of *Panzer-Kompanie (Fkl) 313* were assigned the *3./schwere Panzer-Abteilung 508's* field-post number. Once incorporated into the *Tiger* battalion, *Panzer-Kompanie (Fkl) 313* was considered disbanded.

The *3. Kompanie* was equipped with *Panzerkampfwagen VI Tiger* tanks modified to serve as command and control vehicles. Beginning in January 1944 the commanders and gunners of the *3. Kompanie* received intensive training in the use of radio-controlled demolition-charge carriers and this knowledge was put to practice in several exercises.

Following the Allied landing in the Anzio — Nettuno area on 22 January 1944, the German command sent additional armored units to Italy to take part in a planned counterattack (see also Chapter subsection 2.2.6). On 6 February 1944 the unit was transferred to Italy by express train. The route followed by the train was Mailly le Camp — Lunéville — Strassburg — Stuttgart — Innsbruck — Brenner — Verona — Florence — Ficulle. After detraining on 11 February the battalion proceeded overland to Tiburtina near Rome, suffering its first casualties during the march. Traveling the difficult roads with numerous serpentines resulted in engine fires. One *Tiger* was lost when it caught fire and exploded.

From there the tank companies of *schwere Panzer-Abteilung 508* moved into the area near the front. On 16 February the companies took part in the attack on the enemy main line of resistance at Aprilia. During the night of 16-17 February the *Tigers* became involved in a night engagement on the road one kilometer south of Aprilia. Attacks on the Finger Wood and Riserva Campo di Carne followed on 17 February. On 18 February the battalion took possession of the east-west road near Kilometer 25 and advanced toward the Ovita Farm. In the days that followed, the tanks fought defensive engagements for possession of the east-west road. On 24 February the *3./schwere Panzer-Abteilung 508* recorded additional successes. *Unteroffizier* Zinn destroyed 11 American tanks, while *Feldwebel* Hammerschmidt accounted for 6 more.

During these days of fighting *schwere Panzer-Abteilung 508* got a taste of the tremendous firepower of the enemy's artillery and suffered its first combat losses. In the days that followed, the Germans regrouped their forces for a second attack and the battalion was directed via Albano — Arazia — Cenzano into an assembly area near Cisterna.

On 29 February the battalion attacked in the area of Iso Bella / Mussolini Canal. Radio-controlled demolition-charge carriers also took part in this action. Several *Tiger* tanks ran over mines or were put out of action by antitank fire. The unit lost several men killed and many wounded. On 1 March several *Tigers* successfully recovered a number of damaged *Tigers* and several *Ferdinands* of the *1./schwere Panzerjäger-Abteilung 653*.

On 5 March the tank companies moved to Rome for 10 days. During the night of 16 March the companies moved back into assembly areas near the front. In the weeks until the end of May 1944 the *Tigers* were sometimes employed as provisional artillery. A number of radio-controlled demolition-charge carriers were also committed. On 24 May *schwere Panzer-Abteilung 508* received orders to fall back to Rome. There were eight unserviceable *Tigers* in Cori. With the Americans drawing near, the tanks had to be blown up late on the afternoon of 25 May.

At the beginning of June 1944 *schwere Panzer-Abteilung 508* left Rome and moved via Viterbo in the direction of Siena. There followed fighting withdrawals in the direction of Empoli and Florence, where the battalion arrived at the end of June. In July the tank companies took part in defensive fighting near Colle Val d'Elsa, Poggibonsi, Merkatale and Ciaciano. There was a major tank engagement there on 25 July which resulted in further tank losses. This was followed by a further withdrawal through Florence and Pisa to Viareggio.

At the beginning of September *schwere Panzer-Abteilung 508* was moved to Rimini. The commander of the *3./schwere Panzer-Abteilung 508*, *Oberleutnant* Stein, was killed in the fighting there. In the weeks that followed, small numbers of demolition-charge carriers were committed in combat operations in the Imola — Forli — Cesena — Forlimpopoli and Padua areas.

Following the removal of the radio-command transmitters from the command and control tanks, the remaining *BIV* demolition-charge carriers were put aboard trains in Padua and, together with their drivers, were sent to *Panzer-Versuchs- und Ersatz-Abteilung 300 (FL)* in Eisenach on 20 October.

From that point on the *3./schwere Panzer-Abteilung 508* was employed solely as a conventional tank company with no radio-control equipment.

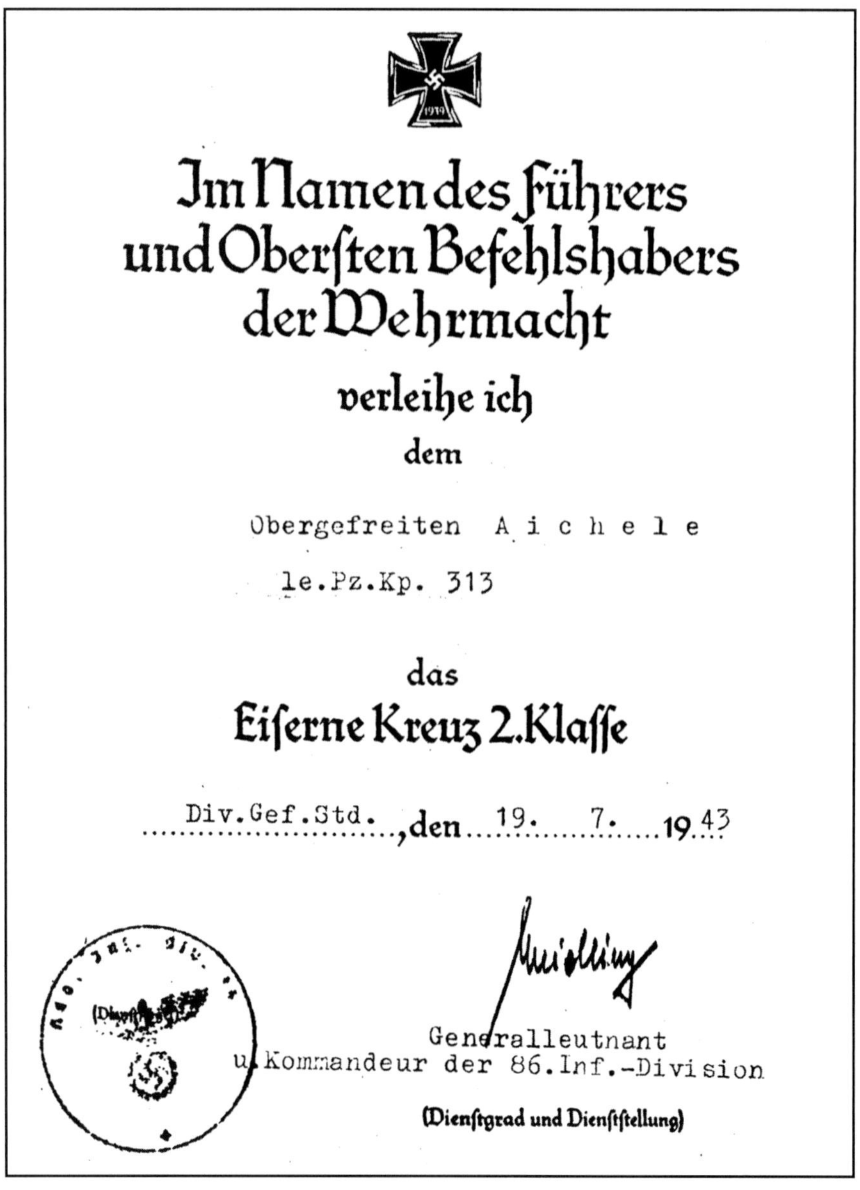

Im Namen des Führers
und Obersten Befehlshabers
der Wehrmacht

verleihe ich

dem

Obergefreiten Aichele

le.Pz.Kp. 313

das

Eiserne Kreuz 2.Klasse

Div.Gef.Std., den 19. 7. 19 43

Generalleutnant
u. Kommandeur der 86.Inf.-Division

(Dienstgrad und Dienststellung)

Award document for the Iron Cross, Second Class to *Obergefreiter* Aichele of *leichte Panzer-Kompanie 313.*

Gliederung der Panzerkompanie (Fkl) 313 (Stand 01.07.1943)

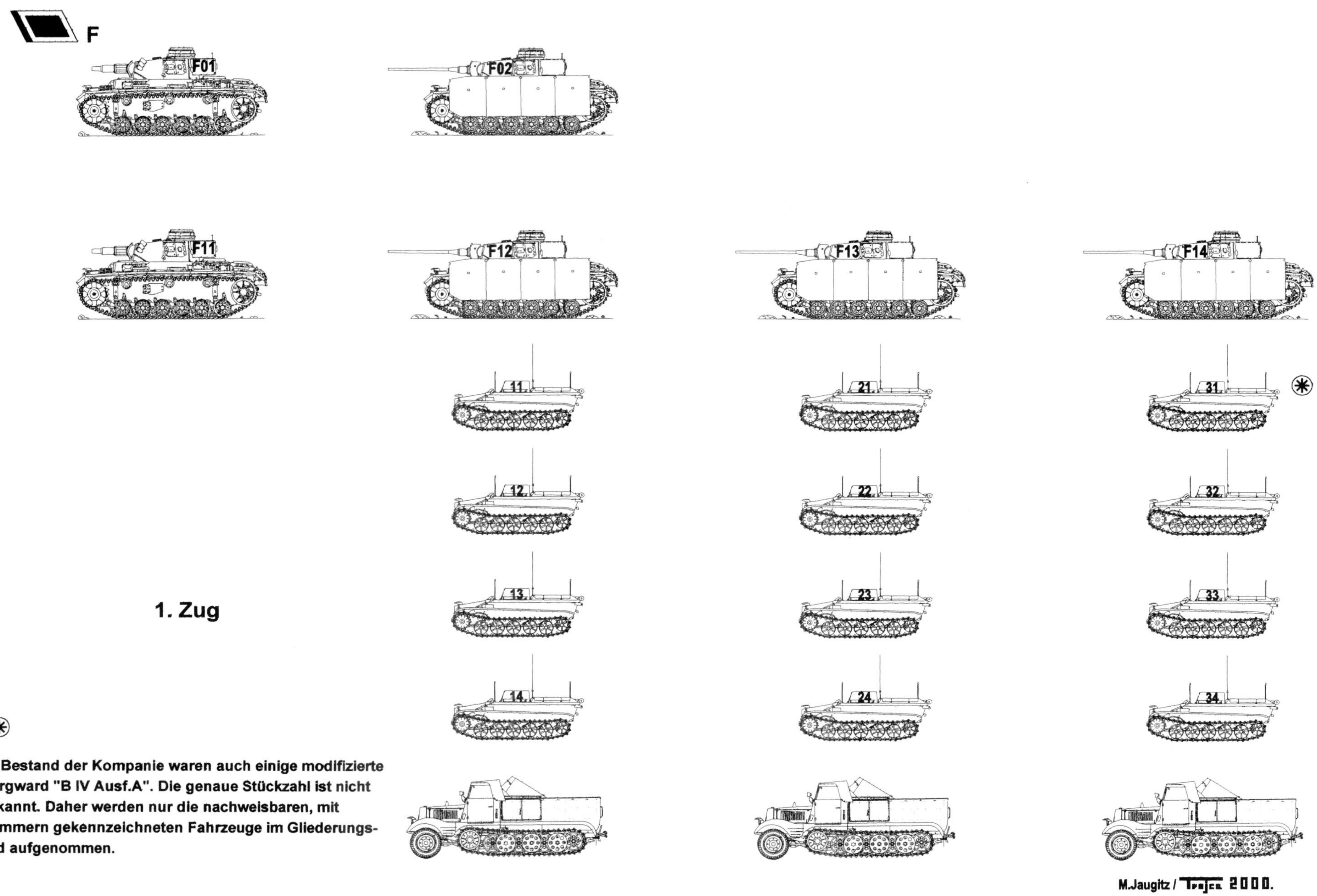

⊛ Im Bestand der Kompanie waren auch einige modifizierte Borgward "B IV Ausf.A". Die genaue Stückzahl ist nicht bekannt. Daher werden nur die nachweisbaren, mit Nummern gekennzeichneten Fahrzeuge im Gliederungsbild aufgenommen.

Gliederung der Panzerkompanie (Fkl) 313 (Stand 01.07.1943)

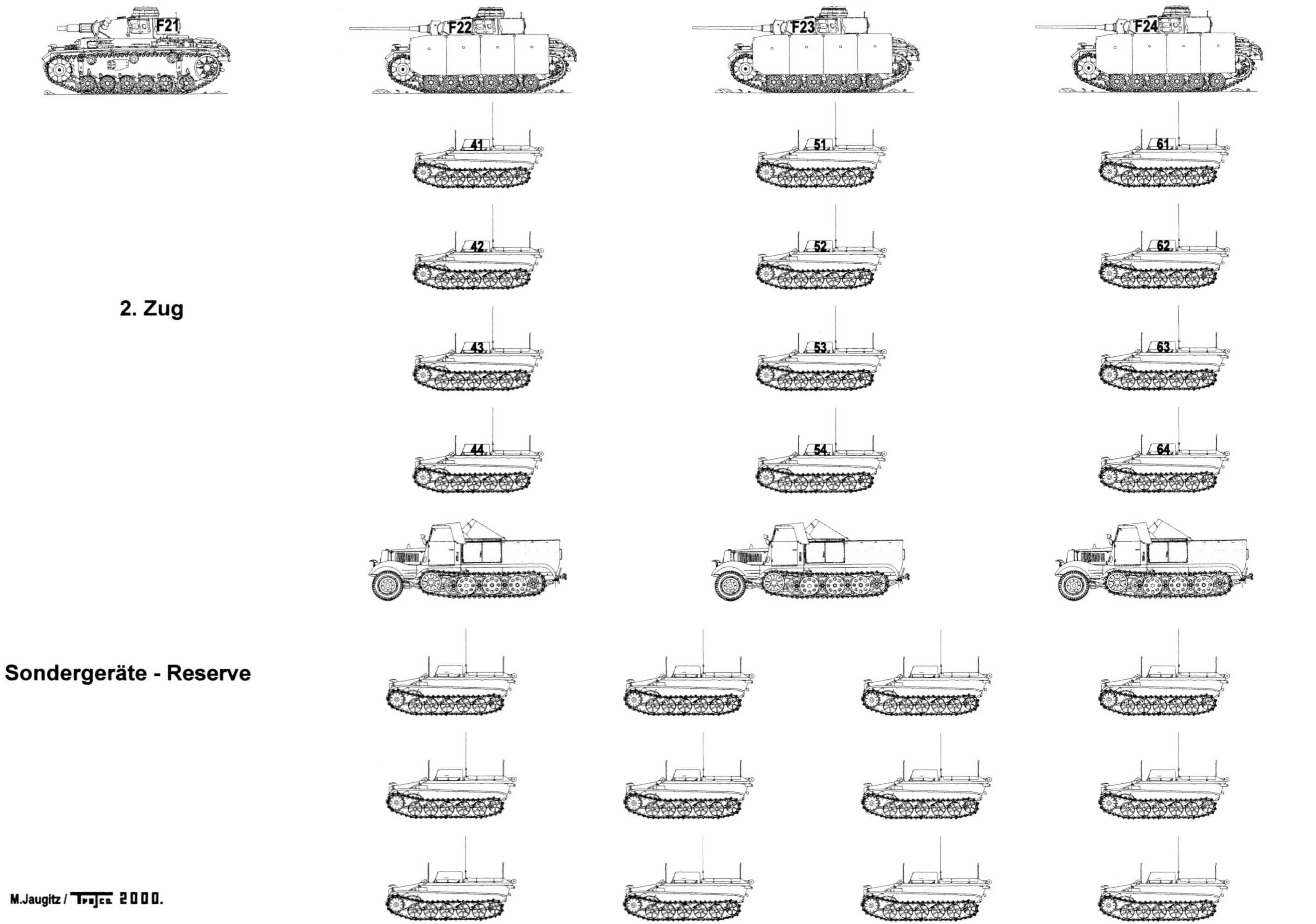

Gliederung der 3. / schweres Panzer - Abteilung 508 (Stand: Januar 1944)

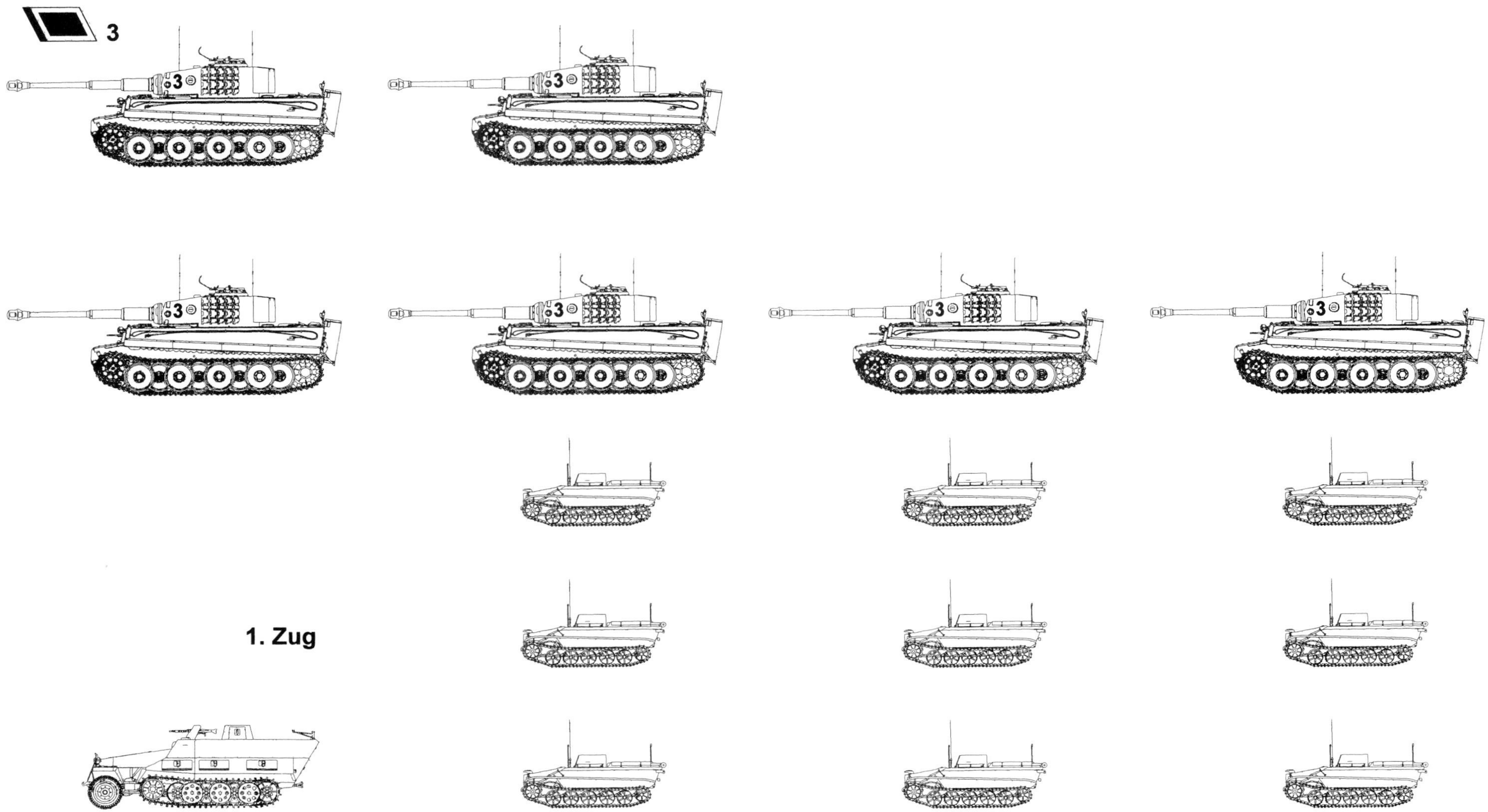

M.Jaugitz / Trojca 2000.

Gliederung der 3. / schweres Panzer - Abteilung 508 (Stand: Januar 1944)

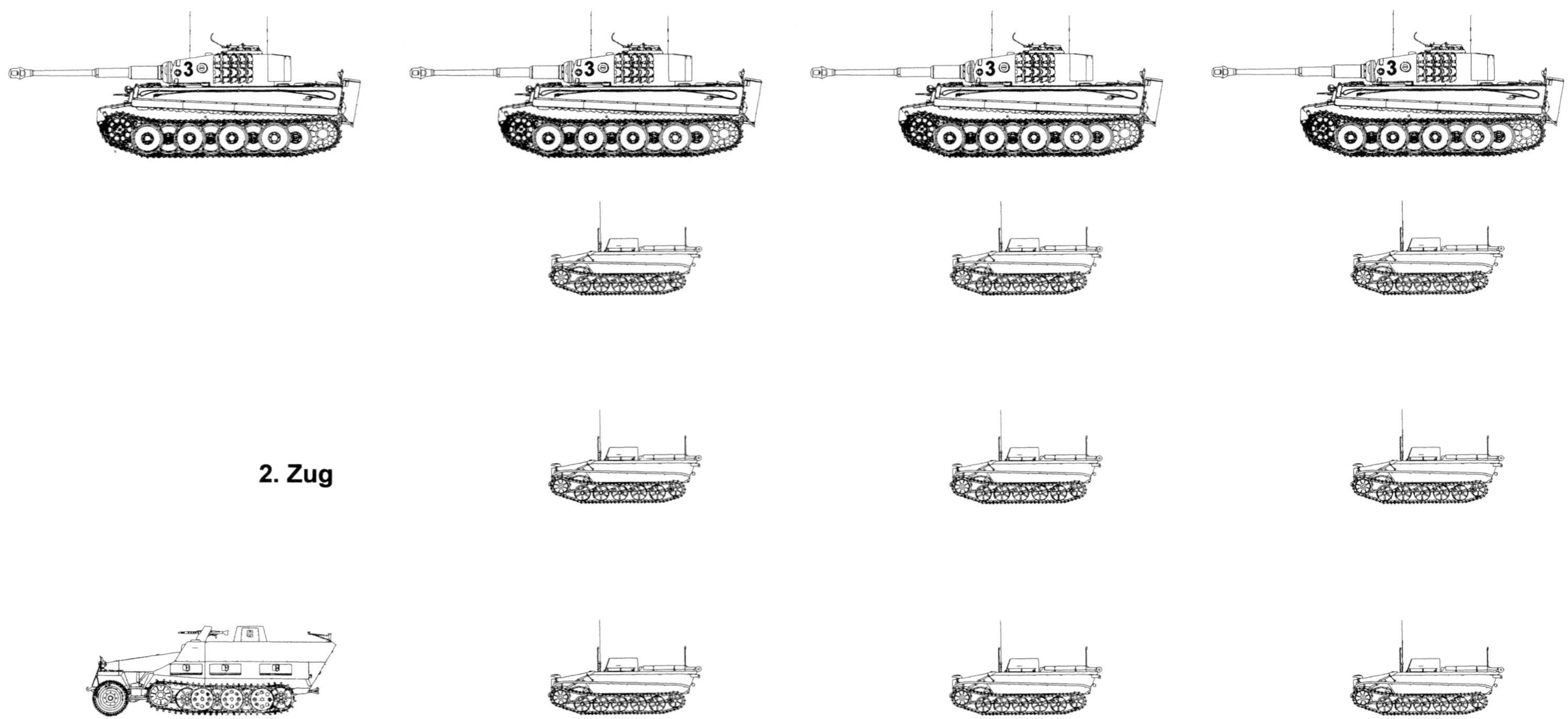

2. Zug

M.Jaugitz / Trojca 2000.

Gliederung der 3. / schweres Panzer - Abteilung 508 (Stand: Januar 1944)

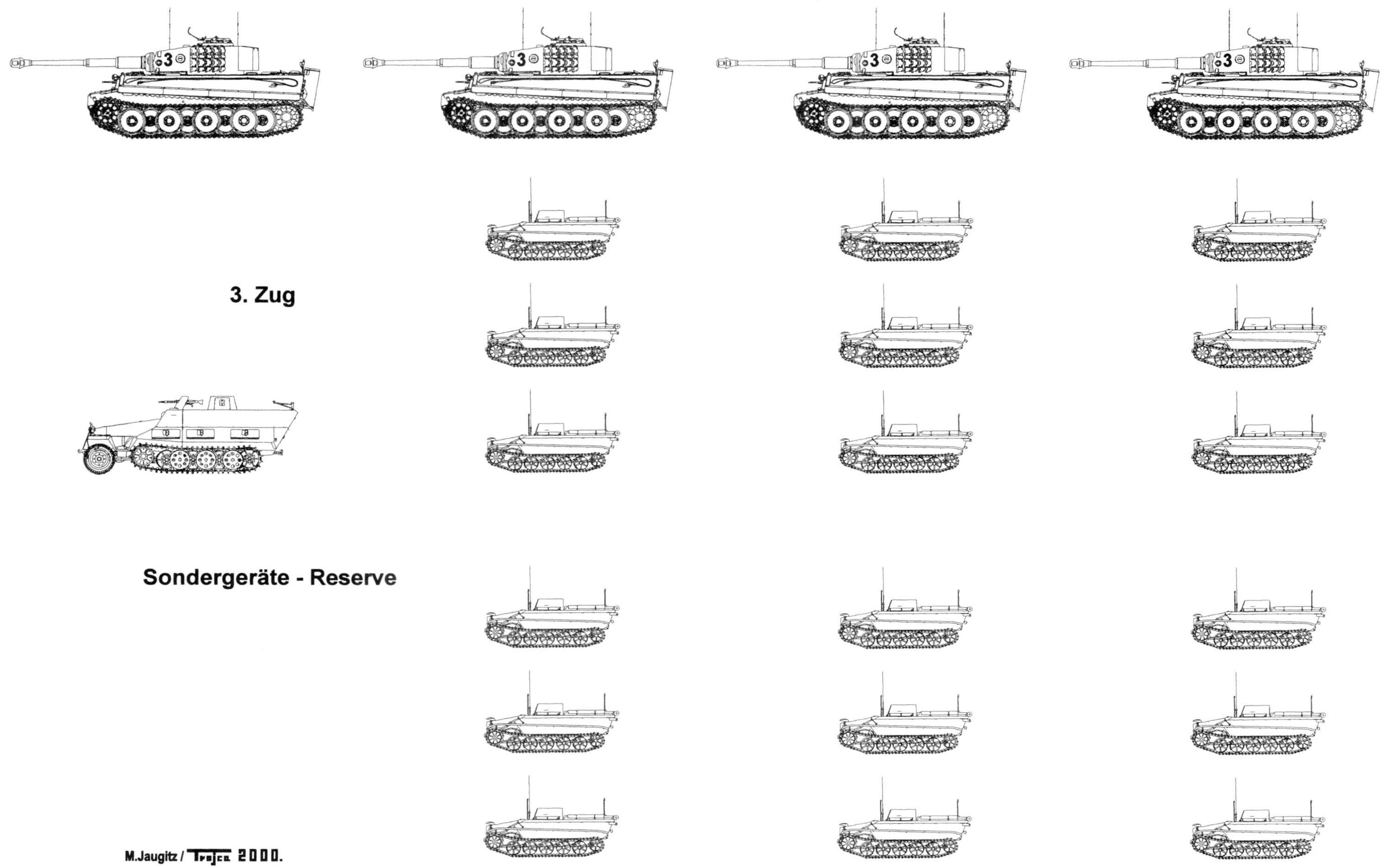

Above: Some of the modified *Panzer III*, *Ausführung J* command and control tanks have engine-compartment covers with the earlier engine-access hatches and ventilation covers. (Hoffmann)

Top left: The newly formed *Panzer-Kompanie (Fkl) 313* received three *Panzer III*, *Ausführung N* and seven *Panzer III*, *Ausführung J/L* command and control tanks. This *Panzer III*, *Ausführung L* has the alphanumeric *F 02*. (Ling)

Left: On the way to the railhead at Neuruppin. A few children enjoy the opportunity to ride along. (Fritschken)

Above left and above: The demolition-charge carriers are prepared for transport. The last *BIV* with the numeral *31* is a modified *Ausführung A*. (Hoffmann)

Left: Several trucks and command and control tanks (*F 12* and *F 13*) have already been loaded on the rail cars. (Hoffmann)

Top row and below: A *BIV*, *Ausführung A* demolition-charge carrier in the assembly area of *schweres Panzer-Regiment 656* at the end of June 1943. This *BIV* was produced in the spring of 1943. (Baisch)

Above: A member of a propaganda company visits *Panzer-Kompanie (Fkl) 313* at the beginning of July prior to the beginning of Operation *Zitadelle*. (Fritschken)

Top left and above: Taking a break in the billeting area. Most of the soldiers wear the light mouse-gray *Panzer* work uniform. (Mohrhäuser)

Left: Members of the maintenance section prepare the side skirts for mounting. (Stolz)

The demolition-charge carriers move forward to the attack positions on 4 July 1943. A *Ferdinand* heavy tank destroyer can be seen in the background. (Münch)

Another modified *BIV*, *Ausführung A* moves forward. (Münch)

The offensive begins! A *Ferdinand* advances to a rail line. (Mohrhäuser)

Clouds of smoke on the horizon identify the first knocked-out vehicles. (Mohrhäuser)

This page and top of next page: The attack is executed with unbelievable intensity. *Schwere Panzer-Regiment 656* suffers its first casualties. (Stolz)

Above and top next page: On the morning of the offensive there was a deadly accident. A *BIV* received a hit from artillery and exploded. Command and control tank *F 23*, which was nearby, was also destroyed by the explosion. (Mohrhäuser)

The running gear of *Ferdinand 623* has been damaged; the left-hand track guard has been torn away. (Mohrhäuser)

The company commander's tank — *F 01* — moves forward in the attack as part of the regiment. (Mohrhäuser)

Command and control tank *F 13* oversees the operation of two demolition-charge carriers. (Mohrhäuser)

Oberleutnant Fritschken and *Leutnant* Sigmund in front of *Ferdinand 722*. (Fritschken)

Obergefreiter Stolz, the radio operator in *F 01*, the company commander's tank. (Stolz)

Ferdinand 702 of *schwere Panzerjäger-Abteilung 654* (*II./schwere Panzerjäger-Regiment 656*) rolls past. (Fritschken)

Oberleutnant Fritschken signals additional vehicles to advance. A *Sturmpanzer IV "Brummbär"* can be seen to the right in the picture. (Sigmund)

Two *Brummbären* of *Sturmpanzer-Abteilung 216* (*III./schwere Panzerjäger-Regiment 656*) support the attack waves. (Scherer)

A *Leitpanzer III, Ausführung L*. The tactical insignia of *Panzer-Kompanie (Fkl) 313* can be seen to the right next to the driver's vision slot. (Sigmund)

The evening of 5 July 1943. *Oberleutnant* Fritschken with the crew of *F 22*. The company had 17 soldiers killed in action by that point. (Hoffmann)

The attack continued on the morning of 6 July 1943. Crew members in front of the company commander's tank and rolling out. *Oberleutnant* Fritschken is in the commander's hatch. (Stolz)

This *Sturmpanzer IV "Brummbär"* is camouflaged in an assembly area. (Mohrhäuser)

Ferdinand II 03 of the battalion staff of *schwere Panzerjäger-Abteilung 654* covers the battlefield. (Mohrhäuser)

Panzer-Kompanie (Fkl) 313 suffered casualties on the second day of the attack as well. A demolition-charge carrier is recovered. (Wild)

F 01, the company commander's tank, looks like it has received its share of damage. (Fritschken)

Both of the remote-control companies attached to *schweres Panzer-Regiment 656* were inspected by the Inspector-General of the Armored Force, *Generaloberst* Guderian, during a fact-finding trip to the front from 10 to 15 July 1943. *Generaloberst* Guderian is in the middle. Next to *Oberleutnant* Fritschken is *Leutnant* Keitel of *Panzer-Kompanie (Fkl) 314*. (Fritschken)

The company is pulled from the front in August 1943 after having suffered considerable personnel and equipment losses. It is transferred back to Germany. The staff car of the company commander — a *Mercedes Benz L 1500 A* — is being towed through the morass. (Stolz)

Above and top left next page: On the way to the railhead. (Stolz)

Above: A *Leitpanzer III, Ausführung N* — probably *F 11* — has just been tied down. *F 24* is likewise loaded. This command and control tank also has the older engine-compartment hatches and ventilation covers. (Stolz)

Left: Saying farewell to nurses at a military hospital. Members of the company were treated there after being wounded. (Scherer)

Above left: At the Mailly le Camp Training Area in France. (Dürr)

Left: The fender in the area of the driver's emergency hatch is coupled with the hatch and can be lowered. (Dürr)

Above: These *BIV's* were manufactured in the spring of 1943. The demolition-charge carriers still have the cross-shaped drive sprocket. (Dürr)

Panzer-Kompanie (Fkl) 313 was incorporated into *schwere Panzer-Abteilung 508* in December 1943 as its *3. Kompanie*. The *Tigers* have already had the radio-control antenna mounting brackets affixed to the right-hand side of the turret. (Herwig)

The officers of the battalion are instructed on the properties of the demolition-charge carriers as well as how to employ them. (Herwig)

Top left and right: The *BIV's* were parked not only in the billeting area but also in the wooded areas of Mailly le Camp Training Area. (Dürr)

Left: Live fire with the 88! The *Tiger I* has been converted to burn propane. Note the gas bottle on the engine compartment. (Dürr)

Basic-issue items have been removed and are checked out. (Dürr)

Leitpanzer Tigers and *BIV's* during an exercise. (Dürr)

Winter fighting exercises are also conducted in the snowy terrain. (Herwig)

Top row: The battalion was transported quickly to the Italian theater after the Allied landings at Anzio/Nettuno. The enlisted personnel have made themselves at home in the boxcars during the trip. (Stolz)

Winter camouflage clothing has been issued to the crews. (Stolz)

Transportation tracks have been put on this *Tiger*. (Stolz)

After detraining at Ficulle at the beginning of February 1944, the battalion continued on to the front. (Stolz)

Above: *Oberleutnant* Stein, the company commander of the *3./schwere Panzer-Abteilung 508*, in the commander's hatch of his *Tiger*. (Stolz)

Above right: A *Tiger* in an ambush position. (Stolz)

A new set of track is placed on the *Tiger* with the help of a *Sd.Kfz. 251*. (Wild)

A 3-ton prime mover of the *3./Panzer-Abteilung 508* at Aprilia. One can see the tactical insignia on the left. (Wild)

Two demolition-charge carriers and a *Tiger* are being prepared to conduct operations. (Schneider)

The *BIV's* carry additional jerry cans. The lead demolition-charge carrier has been outfitted with a smoke-grenade launcher. (Osterried)

A *Tiger* on the East-West Road. (Osterried)

A *BIV* awaits its commitment between the buildings. Cecchina-Albano, 18 March 1944.

Opposite page top: *Schwere Panzer-Abteilung 508* moved back to the Italian capital on 24 May 1944. (Wild)

Opposite page bottom: The battalion trains and a few demolition-charge carriers stayed there for a short period of time. (Stolz)

A break for chow! Replacement demolition charges for the *BIV's* have already been loaded on this 3-ton prime mover. (Dürr)

Two demolition-charge carriers and a 3-ton prime mover on a rural Italian road. (Ling)

Some of the *Tigers* near the front were dug in. (Stolz)

Additional *Tigers* road march through destroyed areas near the front. (Stolz)

Above and top of next page: A *Tiger* of the *3./schwere Panzer-Abteilung 508*. The tank still has the older-style running gear. The bow machine gun has been dismounted. (Stolz)

This *Tiger* has been damaged during the fighting. Note the hits in the area around the bow machine gun and on the gun mantlet. (Stolz)

Above and opposite page top: The battalion reached Siena during its withdrawal at the beginning of June 1944. The demolition-charge carriers and the tanks take a short break. (Wolfer)

The *Tigers* were continuously involved in the fighting during the withdrawal. Radio Operator Stolz in front of the company commander's tank. (Stolz)

This tank received innumerable hits on 25 July 1944. A portion of the fender has been torn off. The *Zimmerit*, used to protect the tank against sticky bombs, has also broken off. (Stolz)

Several hits have also penetrated the engine-compartment hatches. (Stolz)

Schwere Panzer-Abteilung 508 was moved to the Adriatic Coast at Rimini at the beginning of September 1944. (Stolz)

The *Tigers* were loaded on to the special six-axle *SSYMS* rail cars. (Wild)

Opposite page: After the death of *Oberleutnant* Stein, *Oberleutnant* Herwig took over the *3./schwere Panzer-Abteilung 508*. A camouflaged *Tiger* covers the entrance to the assembly area while repairs are conducted on another tank. (Herwig)

The skies are constantly watched during the movement. (Stolz)

Many tanks were knocked out or had to be destroyed in the following weeks and months. Here: Allied soldiers take the opportunity to check out a less-than-lethal *Tiger*. (Leon)

2.2.10: Panzer-Kompanie (Fkl) 314 – 3./schwere Panzer-Abteilung 504

At the beginning of August 1942 *Panzer-Ersatz-Abteilung 25 (neue Abteilung)* (new battalion) was formed in Erlangen from the personnel of *Panzer-Ersatz-Abteilung 25*.

The *2./Panzer-Ersatz-Abteilung (neue Abteilung)* received the designation *2. leichte Panzer-Kompanie "f"* and the military field post number 34368.

The company still had no vehicles and, in case of need, had to fall back on the motor pool of *Panzer-Ersatz-Abteilung 25*.

On 30 September 1942 the *2. leichte Panzer-Kompanie "f"* was transported by rail to Neuruppin, where it was attached to *Panzer-Abteilung 302*, then being formed.

In Neuruppin the company was issued *Panzerkampfwagen III's* and a number of *BIV* demolition-charge carriers. After training on the tanks and technical instruction on the demolition-charge carriers, training of company personnel in the basics of radio control was begun in December 1942.

Of this period Günther Hopf wrote (then a *Gefreiter* and gunner in the 1st Platoon):

Neuruppin, December 1942

Finally, we gunners also began training in the techniques of remote control. We began classes at 0800. Suddenly *Leutnant* Kempe entered the room and at the *Unteroffizier's* command we sprang to our feet. The young armor officer gratefully received his report. With a harsh look his pale eyes swept the faces of his men, who looked at him. Touching his cap briefly, he greeted us with a raspy "Morning men!" His curt order: "Stop your lesson, put on your coats and assemble outside." We swirled out of the room and very soon the section was assembled in front of the company building ready to go. The truck was waiting. Everyone got in, the motors roared and we rushed through the city in the direction of the Genzrode training grounds.

We had driven this route often, but it never became boring. There was always something new to see when we drove past the airfield that lay to the left of the road. Aircraft of various types parked in front of the hangars were cause for a tanker to look curiously. I liked the *Ju 87*.

The tarpaulin on the back, our favorite lookout spot, had been lowered because of the cold, and it wasn't long before the air became so thick as to be almost unbearable. My comrades blathered on endlessly. The bad air came to an end when the brakes squealed and we scrambled out of the vehicle.

Before us lay part of the training grounds — open terrain. And there, on the road to the left, sat long rows of squat, black *BIV's*. The drivers stood near them in groups, stamping their feet and swinging their arms to keep warm. The remote-control transmitting site had been set over to the right near the air-defense vehicle. Not far away two *BIV's* were being readied for remote control.

We marched over to the transmitter at quick time. This got the blood circulating and warmed up our cold feet. Our breath hung in clouds in the cold morning air as we crowded around the apparatus.

Feldwebel Krebs had rechecked the system and now he began explaining the practical side of wireless remote-control to us gunners and tyros. We already knew the theoretical side; in our circle of comrades we had often repeated what we had been taught in class. We were proud of this new weapon and eager to take it into action. Intensely we listened to the instruction, asking questions when something was unclear, and soon we had absorbed the essentials. Now, let's get to work!

One of us gunners put on the control box, a small gray box with various knobs and levers on the top and sides. Gingerly he pressed the starter button. The vehicle's engine sprang to life. He eased the pressure on the large black lever and touched the command "faster". Engine racing, the vehicle jerked, its tracks rattled and the *BIV* moved off. Various commands were issued in quick succession and the vehicle carried them out well, turning its broad snout left, then right, making humorous leaps over the uneven terrain, its long, thin antenna waving back and forth. Following a zigzag course, it drew ever nearer to its goal, a large cardboard target.

The driver, sitting concealed in the vehicle, must have been shaken up when the vehicle passed over ruts and holes at full speed. We were constantly amazed by the *BIV's* tremendous ability to handle the terrain.

Was the small gray fellow at the target? It required practiced eyes to identify it clearly without the aid of binoculars. But our eyes had to become used to estimating distances such as these. We had practiced it often in training: What is the interval between vehicle and target at ranges between 100 and 400 meters?

Everyone stared intensely at the target. The target was still somewhat to the right of the vehicle. The controller tried his best to direct the vehicle in such a way as to overrun the target. But he waited too long. The turn to the right was too great, and the *BIV* raced away like a mischievous pixie with its broadside to the target. Precious seconds passed before the vehicle answered the command and turned left again. It passed just behind the target — bad luck!

The vehicle was brought back in a wide arc and the controller was given another try. This time the *BIV* covered the last 100 meters to the target in a straight line and knocked it over. About face!

By order of the *AHA I a* (*No. 721/43* dated 9 January), the unit was renamed *Panzer-Kompanie (Fkl) 314* on 25 January 1943. *Hauptmann* Braam remained as company commander; *Leutnant* Keitel led the 1st Platoon and *Leutnant* Kempe the 2nd. The remote-control officer was *Leutnant* Hellmann.

The unit was supposed to be immediately reorganized in accordance with *Kriegsstärkennachweisung 1171* (*Ausführung A*) of 1 January 1943. The company took delivery of the necessary assault guns in mid-March 1943.

Because of the reorganization of the radio-control force, for administrative and training purposes *Panzer-Kompanie (Fkl) 314* was attached to *Panzer-Abteilung (Fkl) 301* in Neuruppin.

For the next six weeks the training schedule was filled with remote-control exercises, live firing and radio exercises.

At the beginning of May *1943 Panzer-Kompanie (Fkl) 314* was loaded aboard trains and transported to Bruck an der Leitha in Austria. Also stationed there was *schwere Panzerjäger-Abteilung 653*, equipped with the new *Ferdinand* tank-destroyer.

Generaloberst Guderian inspected the units on 24 and 25 May 1943 and observed a combined exercise by the *Ferdinands* and demolition-charge carriers. The *3./Panzerjäger-Abteilung 653* and *Panzer-Kompanie (Fkl) 314* gave a demonstration on breaching a minefield.

In Bruck an der Leitha the assault guns were retrofitted with side skirts.

On 8 June 1943 orders were received for *Panzer-Kompanie (Fkl) 314* to transfer to *Heeresgruppe Mitte* immediately. On its arrival, the company was to be attached to *schweres Panzerjäger-Regiment 656*, which consisted of *schwere Panzerjäger-Abteilung 653 (I./656)*, *schwere Panzerjäger-Abteilung 654 (II./656)* and *Sturmpanzer-Abteilung 216 (III./656)*.

On 10 June 1943 the company left Bruck an der Leitha and moved to Vienna-Schwechat to entrain. The train departed on the morning of 11 June and headed east. After a five-day trip the company detrained in Smijewka, on the afternoon of 16 June. After a road march the company occupied new quarters in a bivouac area in Davidowo, 35 kilometers south of Orel. *Schweres Panzerjäger-Regiment 656* was located in the surrounding area.

The company remained there until the end of June 1943. Then, beginning on 1 July, it moved up in small sections into the assembly areas near the front. There the vehicles were dug in and camouflaged.

On the evening of 4 July 1943, after a meeting with the regimental commander, *Oberstleutnant Freiherr* von Jungenfeld, *Leutnant* Keitel and several noncommissioned officers conducted a reconnaissance of the assigned sector to see for themselves the lanes the engineers had cleared through the minefields.

The offensive began at 0300 hours on 5 July 1943. Following a thirty-minute artillery barrage and attacks by several waves of bombers, the demolition-charge carriers and command and control tanks rolled forward in front of the *Ferdinands* of *schwere Panzerjäger-Abteilung 653*.

Several *BIV's* ran over mines and exploded. But the mines claimed other victims as well. Several command and control tanks and *Ferdinands* were disabled by mines. The reason for this was the Soviets' new wooden-case mines, which could not be found by the mine detectors used by the German engineers. Furthermore, the Russian minefields had been laid with no discernable pattern. The mines had been laid randomly and were scattered over an area kilometers deep.

On the evening of 5 July 1943 *Panzer-Kompanie (Fkl) 314* lost five men killed and several wounded. Among those killed was the remote-control officer, *Leutnant* Hellmann, who was shot in the head by a Soviet sniper.

Panzer-Kompanie (Fkl) 314 carried out further attacks in the period 6 to 8 July. After the wounding of *Hauptmann* Braam on 7 July, *Leutnant Dr.* Keitel was given temporary command of the company.

After a brief rest from 9 to 16 July, *Panzer-Kompanie (Fkl) 314* was committed in several operations associated with the ongoing withdrawal. On 29 July the company reached Briansk where, after entraining, it was transported to the west.

On 2 August 1943 the new company commander, *Oberleutnant* Diehl, joined the company together with the new remote-control officer, *Leutnant* Hecker, and other replacement personnel.

The company entrained again in Mogilew on 4 August. After a rail trip lasting five days, it arrived at Camp Senne near Paderborn on 9 August.

In his writings, Günther Hopf described the company's transfer to the Eastern Front and its operations in the Kursk — Orel area in July 1943:

10 June 1943: Broke camp at Bruck an der Leitha. Road march to Vienna-Schwechat for rail loading.

11 June 1943: Rose at 0500 hours, formation at 0600, then weapons cleaning. At 0900 the transport departed for Vienna. Traveled in comfort. The radio people had installed radio loudspeakers in most of the cars. Rolled through Brünn, Zwittau, Mährisch-Trübau, Glaz/Silesia.

12 June 1943: Train arrived in Breslau at about 1100 hours. The company commander talked a bit. Topic: "Fire-fighting". 0200 hours Modlin. A huge quantity of bailed hay was loaded aboard the train. Trains proceeding to the front have to be used to the utmost.

13 June 1943 (Pentecost Sunday): The field kitchen offered real coffee. 1100 hours Warsaw. Traces of the war visible. Children came up to the train begging for bread and cigarettes. 2400 hours Brest-Litovsk. Unrest on the train because of the partisan threat. Minsk. On an open stretch of rail a mine was removed from the rail bed. Field kitchen also gave the Russians something to eat.

16 June 1943: 1000 hours Orel. Artillery could be heard firing from the front. Detrained in Smiyewka in the afternoon. Road march to Davidowo, 25 kilometers south of Orel.

16 June 1943 to 30 June 1943: Bivouac at Davidovo.

1 July 1943: At 0530 hours the combat elements departed for Sorochi Kusty via Gostinowo and Smijewka. The attack troops of the *9. Armee*, of which we are a part, began filtering into assembly areas behind the front in small detachments. At the end of the day each crew had to dig a position for its vehicle.

2 July 1943: 0530 hours departure from Sorochi Kusty for Nowopolewo, 15 kilometers away. The vehicles were dug in.

3 July 1943: At 2000 hours departure for the assembly area between Ladyrewo and the Orel — Kursk rail line. Made camp at about 2400 hours, vehicles camouflaged, tents dug in by 0200 hours.

4 July 1943: In the assembly area. In the evening the regimental commander addressed the assembled troops.

5 July 1943: 0330 hours attack against the enemy positions outside of Maslowo left of the Orel — Kursk railway as part of *schweres Panzerjäger-Regiment 656 (Ferdinands)* in the direction of Ponyri. Company suffered heavy losses in equipment through mines. Killed were *Leutnant* Hellmann, *Schütze* Fritz Mehnert, *Panzerschütze* Hinneburg, *Schütze* Oswald, *Schütze* Wittkuhn. Wounded: *Hauptmann* Braam, *Oberschütze* Hans Haberkorn, *Schütze* Arnold Bayerlein, *Schütze* Deutschmann (list incomplete).

6 July 1943: *Gefreiter* Heinz Vith killed.

7 July 1943: *Leutnant Dr.* Keitel acting commander.

8 July 1943: Combat elements in ready position at the railway embankment behind Kleninski. Bombing attack while moving up. *Unteroffizier* Barthel wounded.

9 to 16 July 1943: In the old bivouac area at Ladyrewo as alert company. *Unteroffizier* Jakob slightly wounded in surprise artillery barrage.

17 July 1943: Motor pool pulled back to Starapolewo except for the nine assault guns. The guns positioned themselves outside of Masolowo to serve as combat outposts. Two wounded by artillery fire.

19 July 1943: Combat elements withdrew to Starapolewo and road marched via Lasunowka and Smijewka to Domnini (east of Orel). Guarded the airfield on the near side of the Optuscha and

several security assignments in the Domnino area.

25 July 1943: Elements transferred to Kofanowo (south of Orel). Five assault guns under the command of *Leutnant Dr.* Keitel took part in a counterattack in the Chotetowo area (30 kilometers south of Orel).

26 July 1943: Assault guns of *Feldwebel* Beck and Birner knocked out. Gunner Horst Schmucker and loader Engelbert killed. Assault gun of *Leutnant Dr.* Keitel disabled by enemy fire. Combat train in Stanowoy-Kolodes, company train 20 kilometers west of Orel.

28 July 1943: Three assault guns blown up.

29 July 1943: Road march through Karachew to Briansk. Entrained.

1 August 1943: (Company's first birthday) Lengthy stop in Kritchew. Tracks repeatedly damaged by partisans.

2 August 1943: Arrived in Mogilew, moved into camp outside the town. Company received 20 replacements. New company commander: *Oberleutnant* Diehl; *Leutnant* Hecker the new remote-control officer.

4 to 9 August 1943: Entrained in Mogilew. Return trip via Orscha, Minsk, Brest-Litowsk, Warsaw, Posen, Halle, Paderborn to Camp Senne.

On 18 October 1943 it was ordered that *Panzer-Kompanie (Fkl) 314* was to be moved without changes in personnel or equipment to the Mailly le Camp training grounds in France under the command of the *Oberbefehlshaber West*.

There, in December 1943, the unit began receiving replacement personnel. As part of the refurbishment of its equipment the company was supposed to receive the *Panzerkampfwagen III's* of *Panzer-Kompanie (Fkl) 313*, however, this did not happen. During its stay at Mailly le Camp, which lasted until January 1944, the company took part in exercises with larger armored formations of the army and *Waffen-SS*.

At the beginning of February 1944 *Panzer-Kompanie (Fkl) 314* left Mailly le Camp and moved to new billeting on the Oldebroek/Wezep training grounds in Zwolle, Holland. On 10 March 1944, by order of the *AHA Stab I (1)*, the company was incorporated into *schwere Panzer-Abteilung 504* as its *3. Kompanie*.

After most of the unit's personnel had been transferred to the new unit, *Panzer-Kompanie (Fkl) 314* was considered disbanded as of 15 March 1944.

The *3./schwere Panzer-Abteilung 504* was formed as a *schwere Panzer-Kompanie "Tiger" (Fkl)* in accordance with *Kriegsstärkennachweisung 1176 f* of 1 February 1944. The former assault gun crews were retrained on the *Tiger* in new-equipment training courses given in Paderborn by *Panzer-Ersatz-Abteilung 500*.

The *3./schwere Panzer-Abteilung 504* was commanded by *Oberleutnant* Wriedt. *Leutnant Dr.* Keitel and *Leutnant* Hecker became platoon leaders in the *3. Kompanie*. *Leutnant* Kempe was transferred to the battalion headquarters as its signals officer.

After a three-month stay in Holland, *schwere Panzer-Abteilung 504* was transferred to Maille near Poitiers in western France, detraining there on 1 May 1944. In the weeks that followed, the battalion took part in joint exercises with the *17. SS Panzer-Grenadier-Division "Götz von Berlichingen"* in the Niort — La Rochelle area.

At the beginning of June 1944 *schwere Panzer-Abteilung 504* received orders to proceed to Italy. The battalion entrained in Parthenay on 4 and 5 June. Several transport trains carried the unit to Italy via France, southern Germany and Austria. There it detrained at Sarzana east of La Spezia. The battalion then proceeded via road march to Val di Castello, east of Pietrasanta.

There were numerous mechanical problems on the road. In the second half of June the battalion assembled near San Vicenzo north of Piombino. The *1./schwere Panzer-Abteilung 504* saw its first action near Grosseto on 20-21 June 1944.

The first operation of the *3./schwere Panzer-Abteilung 504*, albeit without demolition-charge carriers, took place on 22-23 June 1944 from Follonica in the direction of Massa Marittima. During the night of 25 June several demolition-charge carriers were moved from the battalion's trains at Saline di Volterra south to Larderello, where the *Tigers* were engaged in a defensive struggle against American forces together with *Panzer-Aufklärungs-Abteilung 26* of the *26. Panzer-Division*. When the Americans moved in, the demolition-charge carriers were detonated on a bridge, destroying it.

On 29 and 30 June 1944 *schwere Panzer-Abteilung 504* carried out further withdrawal movements in the direction of Cecina — Rosignano, south of Livorno. The *3./schwere Panzer-Abteilung 504* was committed with units of the *16. SS Panzer-Grenadier-Division "Reichsführer SS"*. On 12 August the battalion received marching orders; it traveled by road to Parma by way of Pisa, Massa Carrara, Aulla, Pontremoli and the Cisa Pass. From there it was transported by rail via Voghera to Forlimpopoli south of Ravena, detraining there on 10 September. Travelling by night, the battalion moved to its new area of operations on the Rimini front.

In autumn 1944 the demolition-charge carriers and carrier drivers were transferred by rail back to the replacement battalion in Eisenach. There had only been a few remote-control missions during the withdrawals of the previous three months, although the remote control weapon's positive effect on morale had been demonstrated on occasion.

After the autumn of 1944 the *3./schwere Panzer-Abteilung 504* served as a pure tank company and the suffix *Fkl* was dropped from its designation.

Besitzzeugnis

Dem Gefreiter

(Dienstgrad)

Günter H o p f

(Vor- und Zuname)

Pz. Komp. (FKL) 314

(Truppenteil)

wurde das

Panzerkampfabzeichen

— Silber —

verliehen.

Rgts.Gef.St.,den 18.10.1943

(Ort und Datum)

(Unterschrift)

Oberstleutn.u.Rgts.-Kdr.

(Dienstgrad und Dienststellung)

Award document for the *Panzerkampfabzeichen* in Silver for *Gefreiter* Günter Hopf of *Panzer-Kompanie (Funklenk) 314*.

Gliederung der Panzer - Kompanie (Fkl) 314 (Stand 01.07.1943)

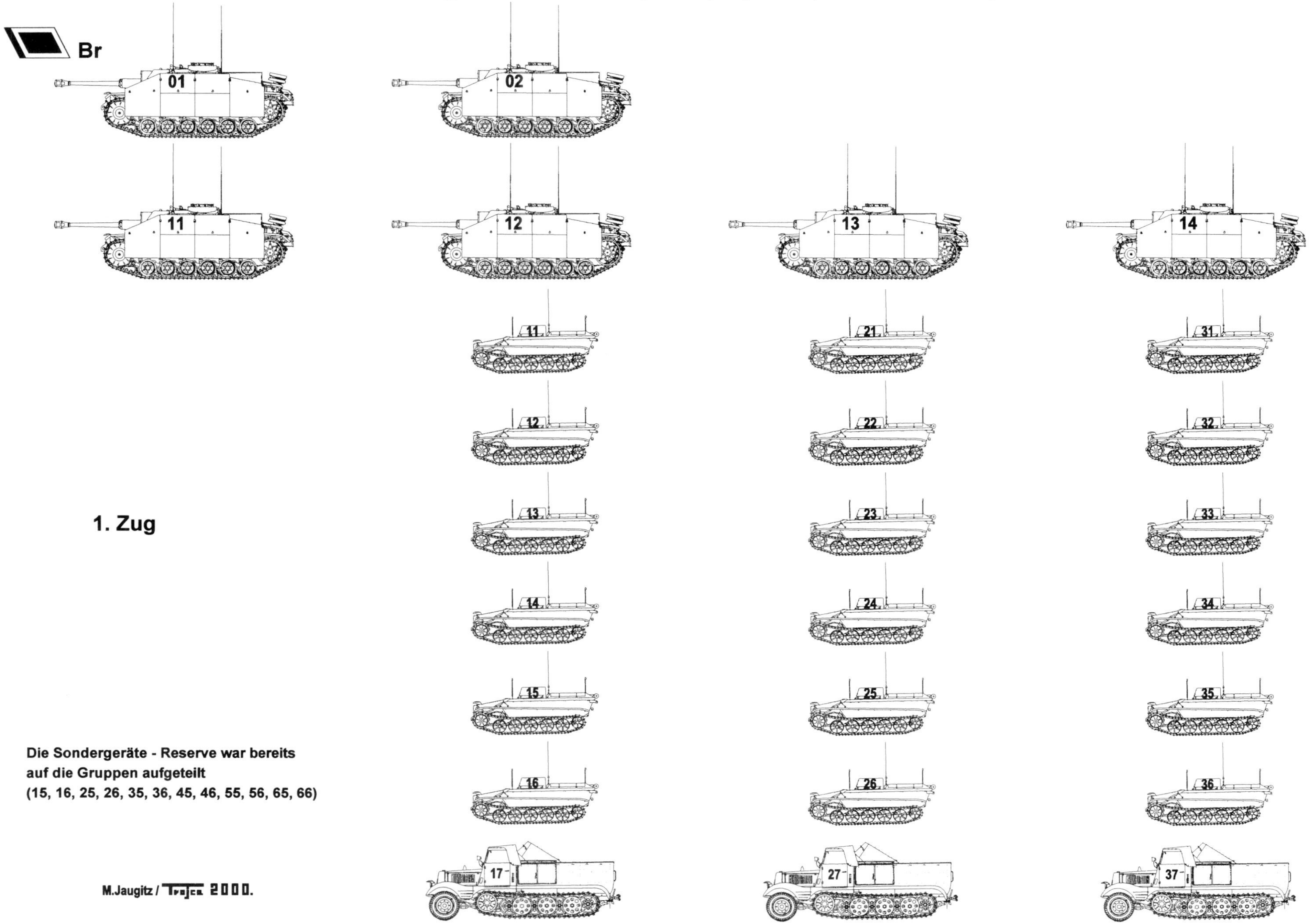

Die Sondergeräte - Reserve war bereits auf die Gruppen aufgeteilt (15, 16, 25, 26, 35, 36, 45, 46, 55, 56, 65, 66)

M.Jaugitz / Trojca 2000.

Gliederung der Panzer - Kompanie (Fkl) 314 (Stand 01.07.1943)

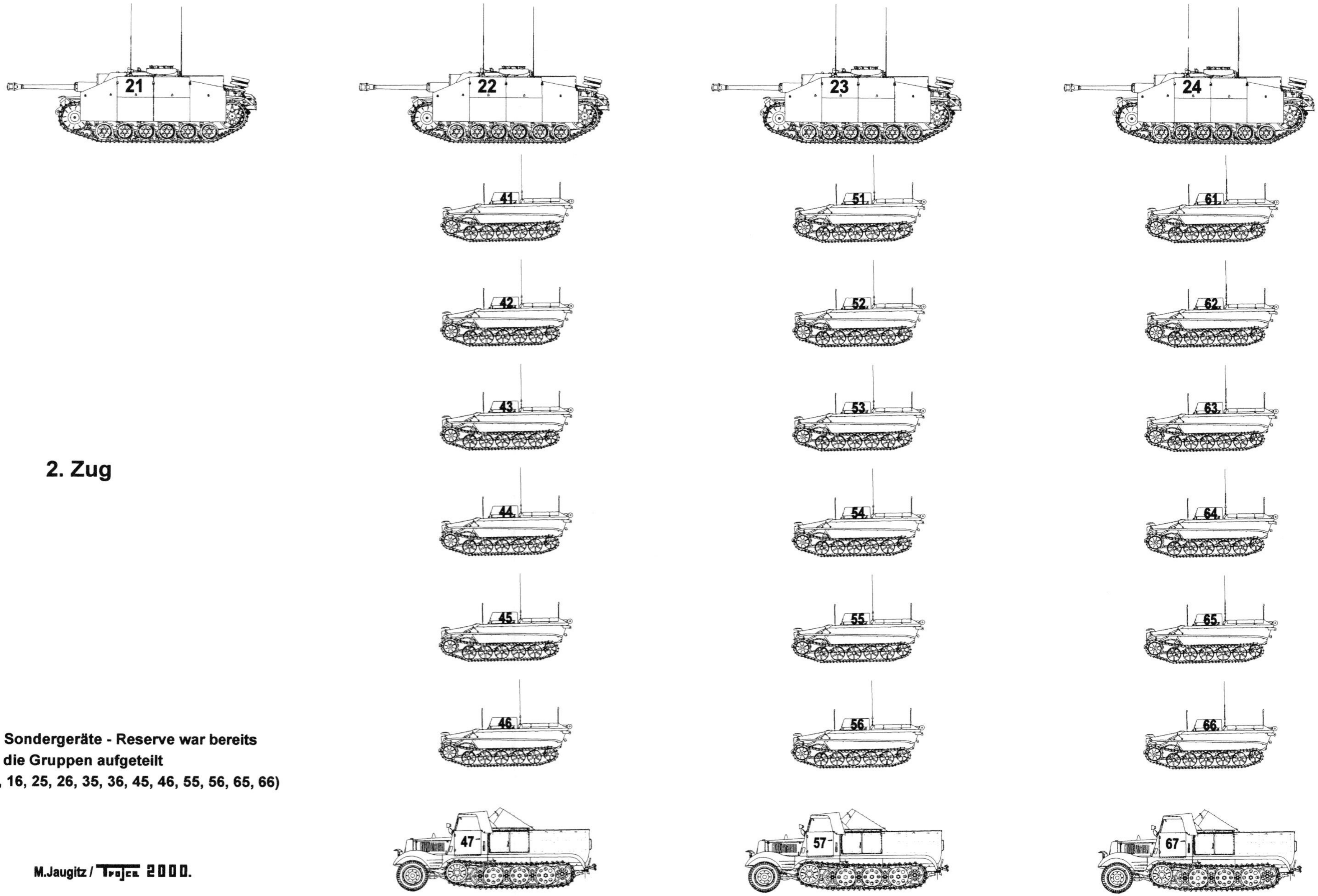

2. Zug

Die Sondergeräte - Reserve war bereits auf die Gruppen aufgeteilt (15, 16, 25, 26, 35, 36, 45, 46, 55, 56, 65, 66)

M.Jaugitz / Trojca 2000.

Gliederung der 3./schweren Panzer - Abteilung 504 (Stand: Mai 1944)

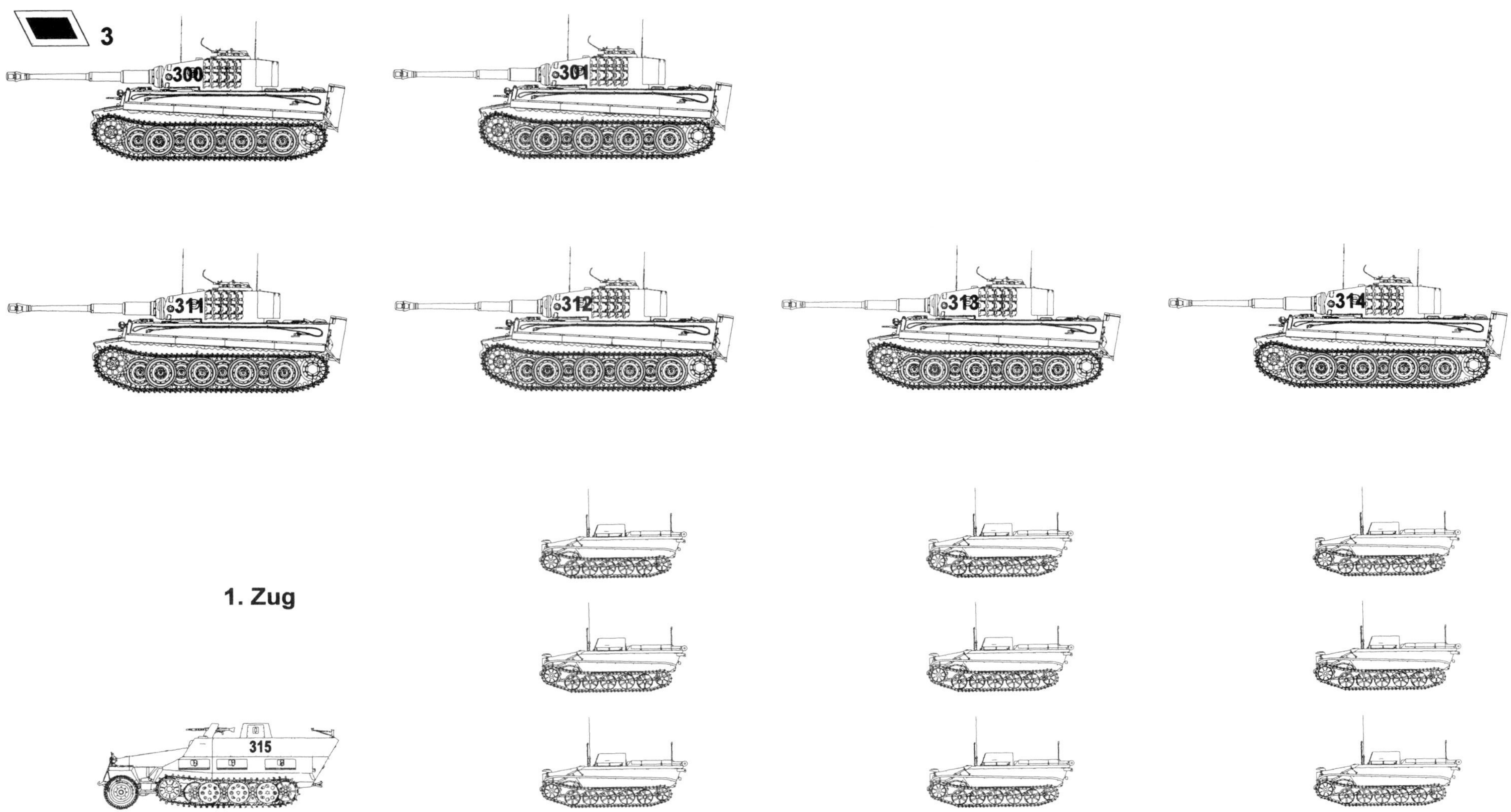

M.Jaugitz / Trojca 2000.

Gliederung der 3./schweren Panzer - Abteilung 504 (Stand: Mai 1944)

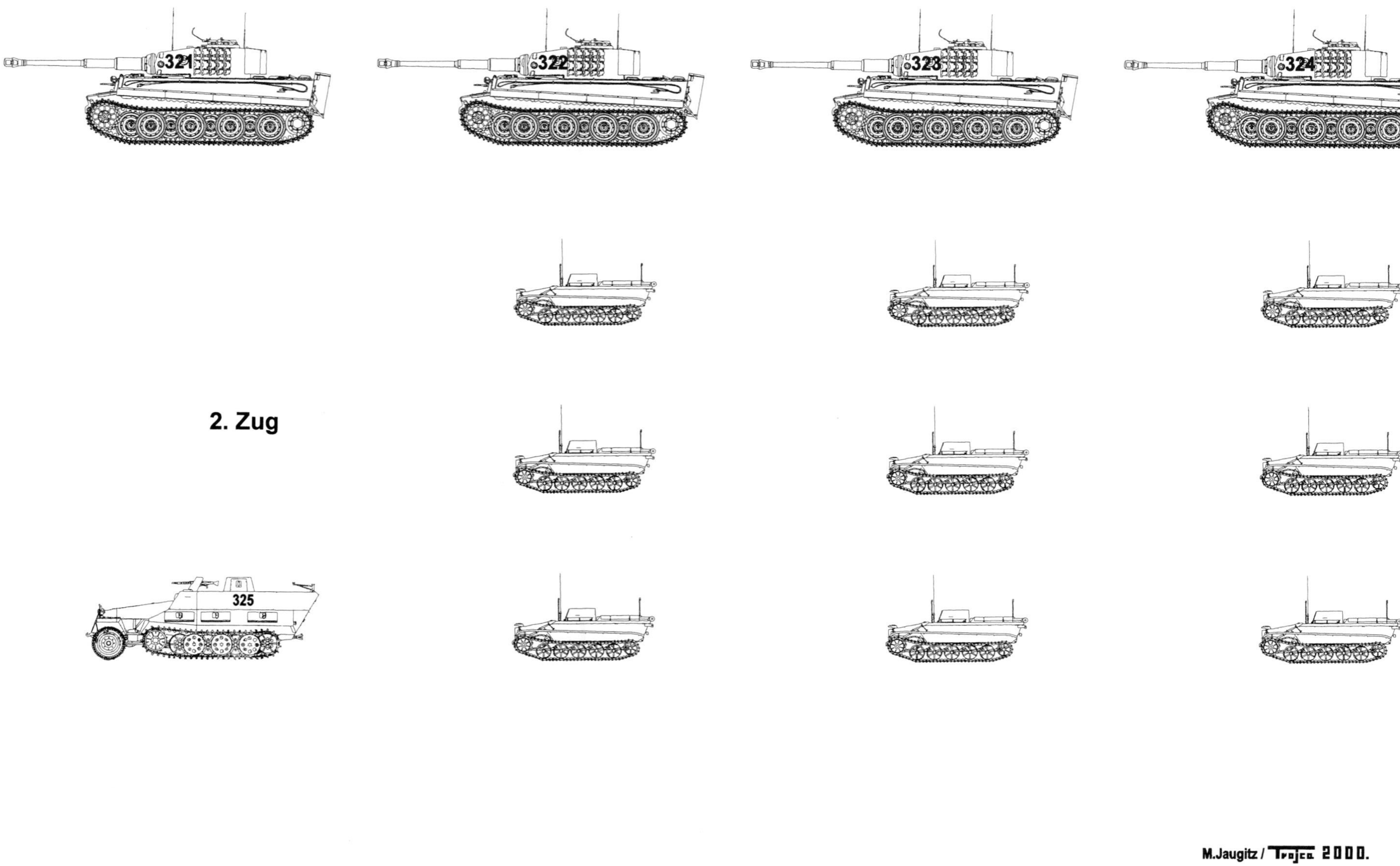

Gliederung der 3./schweren Panzer - Abteilung 504 (Stand: Mai 1944)

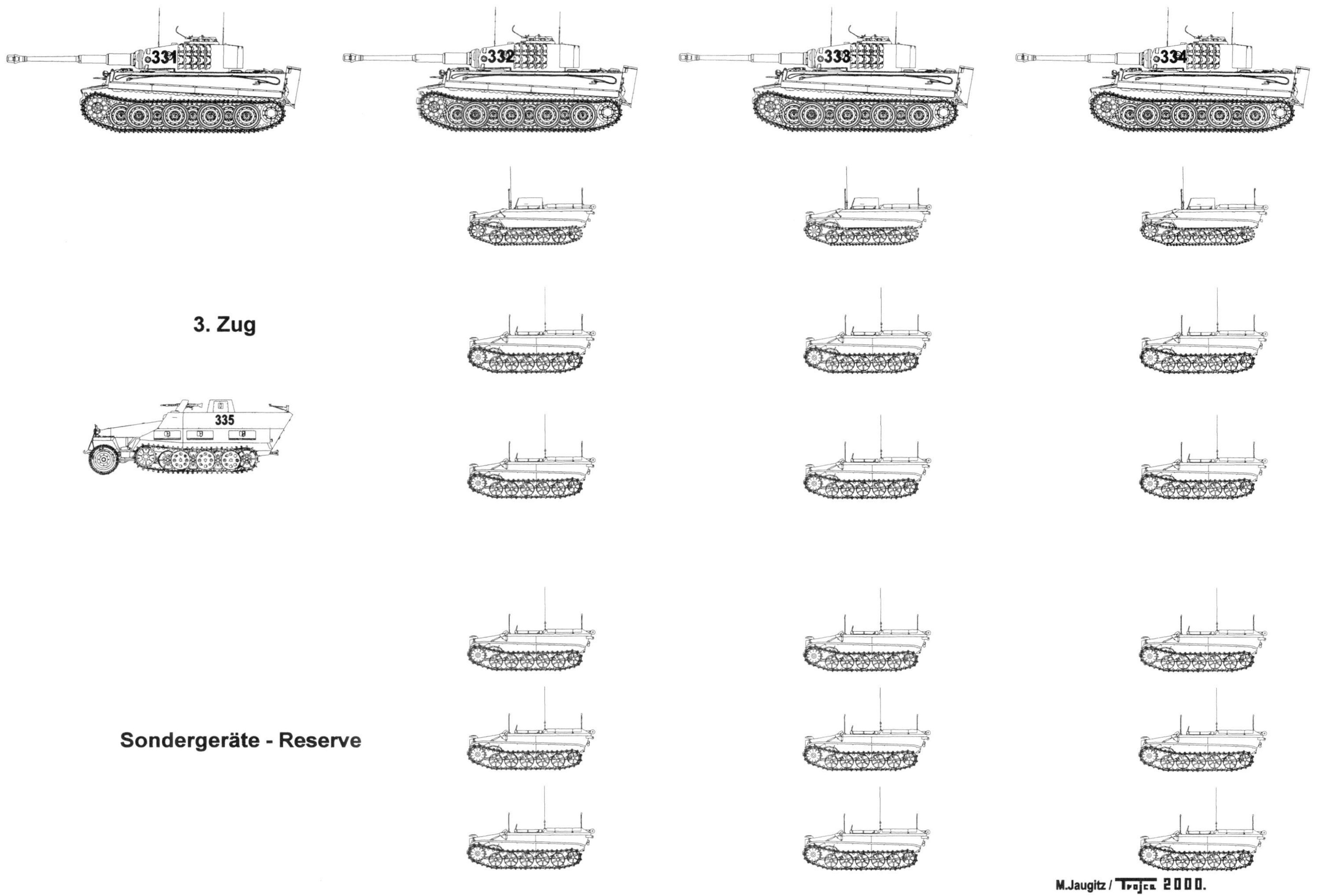

Training personnel of the *leichte Panzer-Kompanie "f"* on the *Panzer III* and the *BIV* demolition-charge carrier at Neuruppin. (Hopf)

The noncommissioned officer with the visor cap has put on the command transmitter. The command transmitter and associated equipment have been stored provisionally in the staff car. (Hopf)

The carrier drivers are instructed on how to operate the *BIV* at the Genzrode Training Area. (Keitel)

Above and 2 photos next page: This *BIV* slid into a ditch during a radio-control exercise. The damaged vehicle is recovered with the aid of a 3-ton prime mover. (Keitel)

19

Ten new *Sturmgeschütz III's* were delivered to the unit in March 1943. In the meantime, it had been redesignated as *Panzer-Kompanie (Fkl) 314*. (Hopf)

A radio-control section has set up for a demonstration. The *BIV's* are still in their *panzergrau* paint scheme. (Hopf)

A single *BIV* is guided by radio. There were still early production models of the *BIV* being used in the spring of 1943. (Hopf)

A few of the *Sturmgeschütze* have welded supplementary armor while others have the bolted-on type. (Held)

The company commander for *Panzer-Kompanie (Fkl) 314*, *Hauptmann* Braam, with the platoon leader of the 1st Platoon, *Leutnant* Keitel. (Keitel)

Leutnant Keitel and the platoon leader of the 2nd Platoon, *Leutnant* Kempe. (Keitel)

Above: In April 1943 the sand-yellow *Sturmgeschütze* received a green and brown camouflage pattern. (Held)

Left and Opposite page: *Panzer-Kompanie (Fkl) 314* was moved to the Bruck Training Area in Austria in May 1943, where it conducted exercises with *schwere Panzerjäger-Abteilung 653* which was also there at the time. (Keitel)

Generaloberst Guderian inspected the units on 24 and 25 May 1943. (Keitel)

This page and two top photos next page: In addition to the Ferdinands on display, demolition-charge carriers deposited and ignited charges during the demonstration. The effectiveness of the charge was clearly demonstrated. (Keitel)

Above: On the rails. Easily identifiable is the insignia of *BIV 14* (First Section, Fourth Vehicle). The meaning of the numeral *2* underneath could not be determined. (Hopf)

Left: On 11 June 1943 *Panzer-Kompanie (Fkl) 314* headed east on trains. This 3-ton prime mover has mounted an antiaircraft machine gun in the crew compartment to defend against aerial attack. (Hopf)

After its arrival the company occupied a bivouac site at Dawidowo. This well-camouflaged *Sturmgeschütz* has been outfitted with two deep-water wading exhaust stacks. (Held)

Below and next page: Starting on 1 July 1943 the armored vehicles moved towards the front to occupy assembly areas. There they were carefully camouflaged. (Keitel)

Next two pages: The offensive started on 5 July 1943. Well-camouflaged *Sturmgeschütze* guide *BIV's* toward the enemy lines while Ferdinands provide support. (Held)

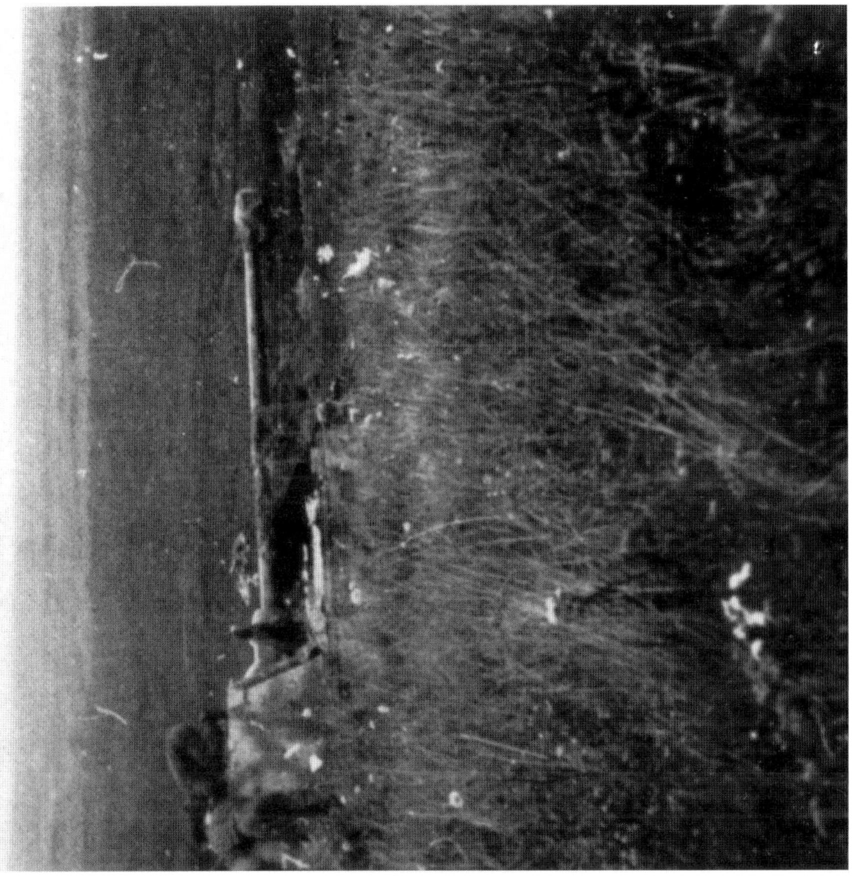

The *Ferdinands* of *schweres Panzer-Regiment 656* follow. (Hopf)

NW

Opposite page top: The company suffers its first casualties. (Hopf)

Opposite page bottom and this page: Several *BIV's* got stuck in ditches and trenches. The *BIV* drivers remain where they are and wait until darkness to return to their own positions. (Keitel)

The combined attacks with the *Ferdinands* were continued on the fourth day of the attack. (Hopf)

The number of disabled *Sturmgeschütze* climbs. *Sturmgeschütz III 02* has driven on to a mine. (Keitel)

The side skirts have been shot away on most of the assault guns. (Keitel)

Above and top next page: The graves of fallen comrades were visited during the short break in the middle of July. (Hopf)

Many company personnel were decorated by *Leutnant* Keitel for their performance. (Keitel)

Opposite page: Five assault guns of the company were employed at Chotetowo on 25 July 1943. (Keitel)

Top two images this page: After returning to Germany, *Panzer-Kompanie (Fkl) 314* stayed at the Senne Training Area near Paderborn for two months. During that time company personnel were instructed intensively in how to operate the radio-control equipment of the *BIV*. (Held)

Opposite page top left: *Panzer-Kompanie (Fkl) 314* was loaded by rail in October 1943 and sent to the Mailly le Camp Training Area in France. (Held)

Opposite page top right and bottom and three photos on following page: Training immediately started for the newly arrived company personnel. (Held)

Left: Vehicle numeral *61* gives an indication of what element this vehicle belonged to (6th Section, 1st Vehicle). (Held)

RX

RZ

RZ

Right: The company was moved from Mailly le Camp to Oldebroek Training Area in Holland at the beginning of February 1944. In March 1944 the company was incorporated into *schwere Panzer-Abteilung 504* as its *3. (Fkl) Kompanie*. (Held)

Below: One of the few *Sturmgeschütze* of the company which survived the operations at Orel. *Unteroffizier* Held, the gun commander, in commander's hatch. Note the replacement road-wheel mounting bracket on the side of the superstructure. (Held)

Late-model *Tiger I's* were delivered as command and control tanks. (Hopf)

Above and top of next page: Because the unit had been previously equipped with *Sturmgeschütze*, it was necessary to provide orientation and training on the *Tigers*. Here: *Tiger 323* of *Oberfeldwebel* Wagner. (Großkopf)

Schwere Panzer-Abteilung 504 was rail loaded at Parthenay and sent to Italy on 4 and 5 May 1944. (Hopf)

The train rolled through France, southern Germany and Austria. (Hecker)

A *BIV* behind a *Sd.Kfz. 251.* The numeral *315* indicates the halftrack is a transport vehicle for the replacement demolition charges of the First Platoon. (Schneider)

Sarzana (near La Spezia) is reached. (Hopf)

Leutnant Keitel was a platoon leader in the *3./schwere Panzer-Abteilung 504* until the summer of 1944. In the background is a *Sd.Kfz. 251* of the *2./schwere Panzer-Abteilung 504*. (Keitel)

The demolition-charge carriers initially remained with the battalion trains. The first *BIV* operation took place on 25 June 1944. (Großkopf)

Top two images: The battalion began to move over the Cisa Pass to Parma on 12 August 1944. (Hecker)

Left and next page: The demolition-charge carriers were sent back to Eisenach in the fall of 1944. The *3./schwere Panzer-Abteilung 504* was employed as a pure tank battalion. The mounting brackets for the radio-control antenna on the turret sides were not removed, however. Many *Tigers* were lost in the next few weeks and months in the defensive fighting. (Leon)

Smashed
BY
D.A.F.

2.2.11: Panzer-Kompanie 315 (Fkl)

Panzer-Kompanie 315 (Fkl) was formed in Eisenach as a separate radio-control company by order of the *AHA* (*I a II No. 25489/43 geheim* dated 6 July 1943). The company's personnel came from *Kompanie Abendroth* and the *1./Panzer-Abteilung (Fkl) 301* (see Chapter subsections 2.2.3 and 2.2.6). The company assumed army field post number 08914, formerly used by the *1./Panzer-Abteilung (Fkl) 301.*

Oberleutnant von Abendroth remained as company commander, while the remote-control officer was *Leutnant* Pieperhoff. The 1st Platoon was led by *Leutnant* Teich, the 2nd Platoon by *Leutnant* Quilitz.

The company received its vehicles and equipment from mid-July to the beginning of August 1943 and, on 15 August 1943, reported it was ready for operations. The company's command and control vehicles, the *Sturmgeschütz III, Ausführung G*, were new vehicles. The *Borgward BIV's*, however, were mostly older vehicles, some with rubber-padded tracks. The demolition-charge carriers were retrofitted in the workshop. Armor plates were welded onto the sides and rear and the folding driver's shield was fixed in place.

The company took part in exercises with the *Panzer-Versuchs- und Ersatz-Abteilung 300 (Fkl)* at the Berka and Ohrdruf training grounds. During this time there were several demonstrations for senior officers and representatives of the *Waffenamt* and *IN 6*.

On 23 September 1943 orders were issued for *Panzer-Kompanie 315 (Fkl)* to be sent to the *25. Panzer-Division* under the command of the *OB West*. The unit was loaded aboard trains and departed for France on 27 September. It arrived two or three days later and detrained near Arras. The company was temporarily quartered at Oeuf.

Tactically, the company was attached to *Panzer-Regiment 9* of the *25. Panzer-Division*. On 21 October the Army General Staff ordered the radio-command transmitters removed from the command and control vehicles. The 10 assault guns were to be handed over to *the 25. Panzer-Division* for use in the formation of the *3. (Sturmgeschütz)/Panzerjäger-Abteilung 87*. This order was soon rescinded, however, as the division was transferred to southern Russia at short notice.

No longer under the operational control of the *25. Panzer-Division, Panzer-Kompanie 315 (Fkl)* was transferred to Clairy-Saulchoix near Amiens in the second half of October. There it was placed under the operational control of the *9. SS Panzer-Grenadier-Division "Hohenstaufen"*. Starting in December the company was administratively attached to *Panzer-Abteilung (Fkl) 301*, which was quartered in the same area. Tactically, the company remained under the operational control of *"Hohenstaufen"*, which had since been reorganized as an armor division.

By order of the *AHA I* (*Nr. 15255*), the unit was attached to the *21. Panzer-Division* on 15 January 1944. On 21 January *Panzer-Kompanie 315 (Fkl)* moved to Vesly near Gisors east of Rouen and soon afterward to St. Marc-le-Blanc. At the end of April 1944 the company again occupied new quarters in the town of Sassy about 35 kilometers southeast of Caen.

In the final weeks before the invasion new security measures were adopted and fortifications were bolstered. In mid-May *Panzer-Kompanie 315 (Fkl)* began installing post obstacles, so-called "Rommel Asparagus", in its billeting area around Sassy. Five assault guns stood guard at both the north and south entrances to the town. The vehicles were dug in and camouflaged. In addition, one assault gun was kept on alert with a full crew at all times.

On 6 June 1944 *Panzer-Kompanie 315 (Fkl)* set out for Caen, however Allied air superiority made movement almost impossible. The unit crossed the Orne headed northwest, but then it was forced to retire over a badly damaged railroad bridge. The company took up position within sight of the Carpiquet airfield. With the situation changing constantly, the company was forced to change position frequently over the next few days, although there was no contact with the enemy.

On 9 June 1944 *Leutnant* Quiltz, platoon leader of the 2nd Platoon, who had been wounded on the first day of the invasion, was killed in a bombing raid while in the military hospital in Alencon.

Panzer-Kompanie 315 (Fkl) made its first contact with the enemy on 18 June 1944 in the Bavent Forest northeast of Caen. The assault guns were supposed to clear the forest with infantry support. In the process, they came under fire from enemy snipers hidden in the trees.

On 23 June 1944 the company carried out its first radio-control operation west of the Bavent Forest near the village of St. Honorine la Chadronnette. *Panzer-Kompanie 315 (Fkl)* attacked advancing British units as part of *Kampfgruppe* von Luck (*21. Panzer-Division*).

This engagement was described by then *Gefreiter* Joachim Brandtner, loader in *Sturmgeschütz 21* of *Panzer-Kompanie 315 (Fkl)*:

> Near St. Honorine on 23 June 1944, the radio-controlled *BIV's* and four assault guns of our 2nd Platoon, together with several *Panzer IV's* of *Panzer-Regiment 22*, launched an attack north through a wheat field against British troops.
>
> We committed the *BIV* special vehicles in an effort to draw the enemy's fire. This effort was unsuccessful, however, because the enemy could scarcely see the squat radio-controlled BIV's in the tall wheat field. Even we had a difficult time following the vehicles' path. The enemy did not open fire.
>
> Our assault guns and tanks then moved off from the assembly area. We soon came under heavy antitank fire. Beside us, two *Panzer IV's* of *Panzer-Regiment 22* were hit and caught fire. Our assault gun was hit on the front slope by an antitank round. The weld seam was burst open. Our attack was halted by the heavy antitank fire.

The English offensive, code-named "Epsom," began west of Caen between Bretteville-sur-Odon and Tilly-sur-Seulles at 0500 on 25 June. Several times *Panzer-Kompanie 315 (Fkl)* was committed with other armored elements of the *21. Panzer-Division* and the *12. SS-Panzer-Division "Hitlerjugend". Panzer-Kompanie 315 (Fkl)* was committed primarily in the area around Mouen, where radio-controlled demolition-charge carriers were used on 26, 27 and 28 June 1944.

At the end of June 1944 *Panzer-Kompanie 315 (Fkl)* had five serviceable *Sturmgeschütz III's* and 19 functioning *Sd.Kfz. 301's*.

Meanwhile an order from the *AHA* (*Stab I (1) No. 27220/44 geheim* dated 22 June 1944) had reached the company. The order stated that *Panzer-Kompanie 315 (Fkl)* had been renamed the *2./Panzer-Abteilung (Fkl) 302* and was to join the battalion then being formed in the Reims area.

Gliederung der Panzerkompanie 315 (Fkl) (Stand 01.06.1944)

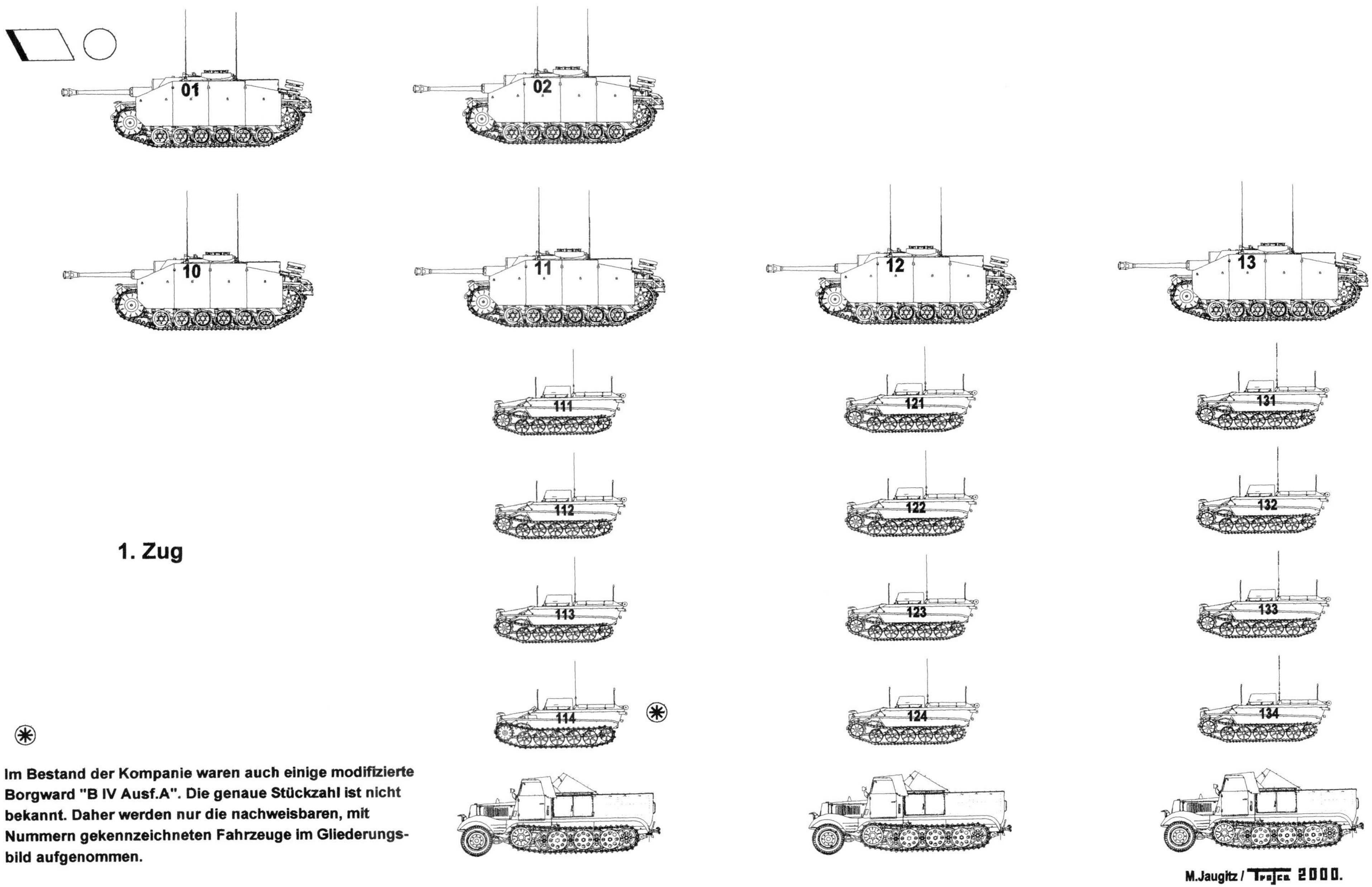

✱ Im Bestand der Kompanie waren auch einige modifizierte Borgward "B IV Ausf.A". Die genaue Stückzahl ist nicht bekannt. Daher werden nur die nachweisbaren, mit Nummern gekennzeichneten Fahrzeuge im Gliederungsbild aufgenommen.

Gliederung der Panzerkompanie 315 (Fkl) (Stand 01.06.1944)

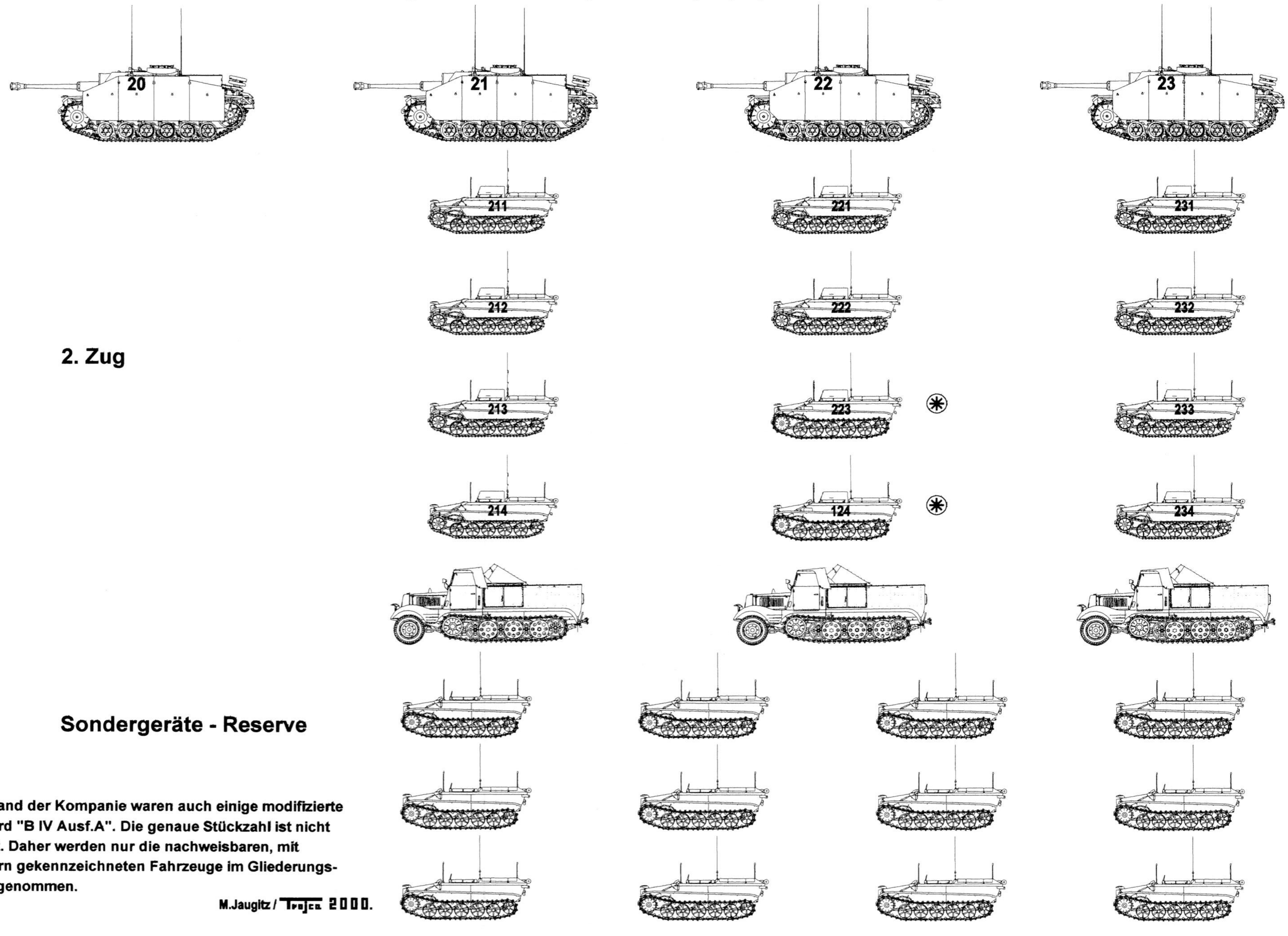

2. Zug

Sondergeräte - Reserve

✻ Im Bestand der Kompanie waren auch einige modifizierte Borgward "B IV Ausf.A". Die genaue Stückzahl ist nicht bekannt. Daher werden nur die nachweisbaren, mit Nummern gekennzeichneten Fahrzeuge im Gliederungsbild aufgenommen.

M.Jaugitz / Trojca 2000.

Exercises of the newly constituted *Panzer-Kompanie (Fkl) 315* at the Berka Training Area in the summer of 1943. Note the tactical insignia on the fender and the mounting bracket for the radio-control antenna. (Susenbeth)

Oberst Mildebrath of *IN 6* and *Oberleutnant* von Abendroth, the company commander, observe the employment of the demolition-charge carriers. *Major* Weicke and *Oberleutnant* Fischer of *Panzer-Versuchs- und Ersatz-Abteilung 300 (Fkl)* stand next to the *Sturmgeschütz*. (Susenbeth)

Above: A demolition-charge carrier follows the assault gun. (Susenbeth)

Right: *Oberleutnant* von Abendroth received the German Cross in Gold on 3 August 1943. (Susenbeth)

A driver moves the demolition-charge carrier into the jump-off position. (Susenbeth)

Another *BIV*, *Ausführung A* waits to be employed. (Susenbeth)

Left: A third *BIV* is brought up. This one is a modified *BIV, Ausführung A*. (Susenbeth)

Left: A few officers of *Panzer-Versuchs- und Ersatz-Abteilung 300* at the exercise after-action review. The tactical insignia of *Panzer-Kompanie 315 (Fkl)* can bee seen on the *Schwimmwagen*.

In the garrison at Eisenach. The tactical insignia has been added to the left-hand fender.

Several *BIV*, *Ausführung A* demolition-charge carriers have been parked in the open.

Departure to the railhead. The numeral *121* has been added to the superstructure of the *Sturmgeschütz*. The numerals were quickly painted over since the incorporation of the company into *Panzer-Abteilung (Fkl) 301*(as the *1. Kompanie*) never took place. (Susenbeth)

The demolition-charge carriers follow. (Susenbeth)

The column moves through Eisenach… (Susenbeth)

…in the direction of the rail station. (Susenbeth)

Top left: The *BIV's* and some of the trucks have already been loaded. All of the vehicles have a base coat of sand-colored paint. (Susenbeth)

Top right:Higher open cars are placed between the flat cars. Antiaircraft machine guns have been set up there. (Susenbeth)

Right: The first train sets out. The destination is the *25. Panzergrenadier-Division* in Arras. (Susenbeth)

Right: The company was outfitted with the *BIV, Ausführung A* demolition-charge carrier at the time. Carriers produced from June 1942 until June 1943 were in the unit. (Poggensee)

The *Sturmgeschütze* were parked at Oeuf after the arrival of the company. At Oeuf all of the vehicles received the usual three-color camouflage scheme. The superstructures were covered with tarpaulins. (Poggensee)

The stay at Oeuf did not last long. The company was loaded on trains yet again in the middle of October 1943. The replacement track links on this assault gun have also been overpainted in the camouflage colors. (Poggensee)

Right: A 1-ton prime mover of the maintenance section moves over a ramp on to the flat car. (Jenckel)

Left: The side skirts on the assault guns have not been dismounted. (Poggensee)

Right: The loading of all of the wheeled and tracked vehicles takes some time. (Poggensee)

Numerous changes to the vehicles were undertaken by the soldiers themselves. A personal-effects container has been mounted to this fender. (Poggensee)

The demolition-charge carriers are loaded onto the train. (Jenckel)

Two versions (early and late) of the *BIV, Ausführung A.* Notice the difference in the running gear. (Poggensee)

The forward *BIV* is a modified *Ausführung A.* The tactical insignia of the company has been added to the right-hand side of the demolition charge. (Poggensee)

After moving to Clairy-Saulchoix in the middle of October 1943, *Panzer-Kompanie 315 (Fkl)* was placed under the operational control of the *9. SS-Panzergrenadier-Division "Hohenstaufen"*. Operational possibilities for the employment of demolition-charge carriers were demonstrated to officers of the army, the *Luftwaffe* and the *Waffen-SS*. (Walther)

Hauptmann von Abendroth in his medium staff car. Note the drawing of "modest Helene" on the passenger door. (Grumptmann)

The assault guns of the company have also been assembled. (Walther)

A hole for inserting the blasting cap can be seen in the front slope of this modified *BIV, Ausführung A.* (Kornmeier)

Unfortunately, the quality of this photo is poor. Despite that, it is interesting how the former company insignia — a setting red sun — has been painted on these vehicles (right front fender). (Poggensee)

Spring of 1944. Training on the gunner optics. The complete sighting system of the 75 mm *Sturmkanone 40 L/48* has been mounted on a tripod. (Jenckel)

Break at Sassy. Two *BIV, Ausführung A* demolition-charge carriers in the background. One of them has the older running gear with the rubber-padded tracks. The other one has the dead track. (Poggensee)

The assault guns were maintained operationally ready at all times. (Knott)

Panzer-Kompanie 315 (Fkl) was moved from Sassy towards Caen on 6 June 1944. Demolition-charge carrier *114* became disabled during the road march there. (Poggensee)

Sturmgeschütz III 10 has been camouflaged with foliage. The crew and some personnel of the platoon headquarters wait to be employed. The platoon leader, *Leutnant* Pieperhoff, can be seen to the right. (Walther)

Top left: *Sturmgeschütz III 21* was damaged by an antitank round to its superstructure during the fighting at St. Honorine on 23 June 1944. (Walther)

Top right: A portion of the side skirts have also been torn off. (Walther)

Opposite page: The air supremacy of the Allies forced the vehicles to remain camouflaged all the time. (Walther)

Left: The necessary repair work was conducted on the battlefield. (Walther)

These two pages: *Panzer-Kompanie 315 (Fkl)* received the order to move to Reims at the end of July 1944. Those *Sturmgeschütze* still operational were loaded on the train. (Walther)

Zw Br. Unt Gd 25.4.44

2.2.12: Panzer-Kompanie (Fkl) 316

In April 1943 *Panzer-Versuchs- und Ersatz-Abteilung 300 (Fkl)* formed the *1. leichte Panzer-Kompanie* from extra personnel. On 1 June 1943 the unit was temporarily given the unofficial designation *6./Panzer-Versuchs- und Ersatz-Abteilung 300 (Fkl)* and received the army field post number 56041.

When the *OKH/GenStdH/Org.Abteilung I* issued an order (*No. 6039/43 geheim* dated 13/7/1943) for the formation of *Panzer-Kompanie (Fkl) 316*, it turned to the trained personnel of *the 6./Panzer-Versuchs- und Ersatz-Abteilung 300 (Fkl).*

Implementing instructions followed on 4 August 1943. In accordance with *Kriegsstärkennachweisung 1171 f* of 1 January 1943, the company would be equipped with 36 *Borgward BIV* demolition-charge carriers and ten assault guns.

The company commander was *Oberleutnant* Meinhardt. Platoon leaders were *Leutnants* H. Fischer and F. Schneider, while the remote-control officer was *Leutnant* Stölten.

After the issuance of equipment was completed by mid-September 1943, on 21 September 1943 *Panzer-Kompanie (Fkl) 316* was transferred from Eisenach to the Fallingbostel training grounds and attached to the *Panzer-Lehr-Regiment*. There the company would be tested in operations with a regiment-size formation.

On 30 September 1943 the *Panzer-Lehr-Regiment* received three *Panzerkampfwagen VI Tigers*, followed by another seven on 8 October. These were issued to *Panzer-Kompanie (Fkl) 316*, which was to test the *Tiger* in the command and control tank role.

In mid-January 1944 it was ordered that *Panzer-Kompanie (Fkl) 316* was to be equipped with *Panzerkampfwagen VI Tiger II* command and control tanks prior to reorganization as an experimental remote-control company. It subsequently received the designation *1. schwere Panzer-Kompanie (Fkl).* At that time the *Panzer-Lehr-Division* was in northern France. *Panzer-Lehr-Regiment 130* was stationed in Verdun.

At the end of February 1944 ten tank crews plus technical personnel went to *Panzer-Ersatz-Abteilung 500* in Paderborn for training on the *Tiger II* heavy tank.

Since *Panzerjäger-Lehr-Abteilung 130* had not yet equipped with the *Panzerjäger IV* and was thus not operational, the remote-control company was ordered to hand its 10 *Sturmgeschütz III's* over to the *3./Panzerjäger-Lehr-Abteilung.*

On 6 March 1944 the *Panzer-Lehr-Division* received orders to transfer to the Vienna area. It was feared that Hungary was about to withdraw from its alliance with Germany. The German command therefore wanted sufficient German units in the border area in the event that entry into Hungary became necessary. The cover designation for this operation was "Margarethe".

The division was broken down into four march serials. March Group B included *Panzer-Lehr-Regiment 130*, other division elements and the *1. schwere Panzer-Kompanie (Fkl).* All units of March Group B were billeted on the Kaisersteinbruch training grounds on Lake Neusiedler.

The first five *Panzerkampfwagen VI Tiger II's* to be built arrived there on 14 March 1944. On 22 March the personnel sent to Paderborn for *Tiger* instruction returned to their unit. By this time the number of *Tiger I's* in the remote-control company was down to three.

The company's exact vehicle complement and its state of readiness is contained in the following report. (Source: Bundesarchiv / Militärarchiv Freiburg)

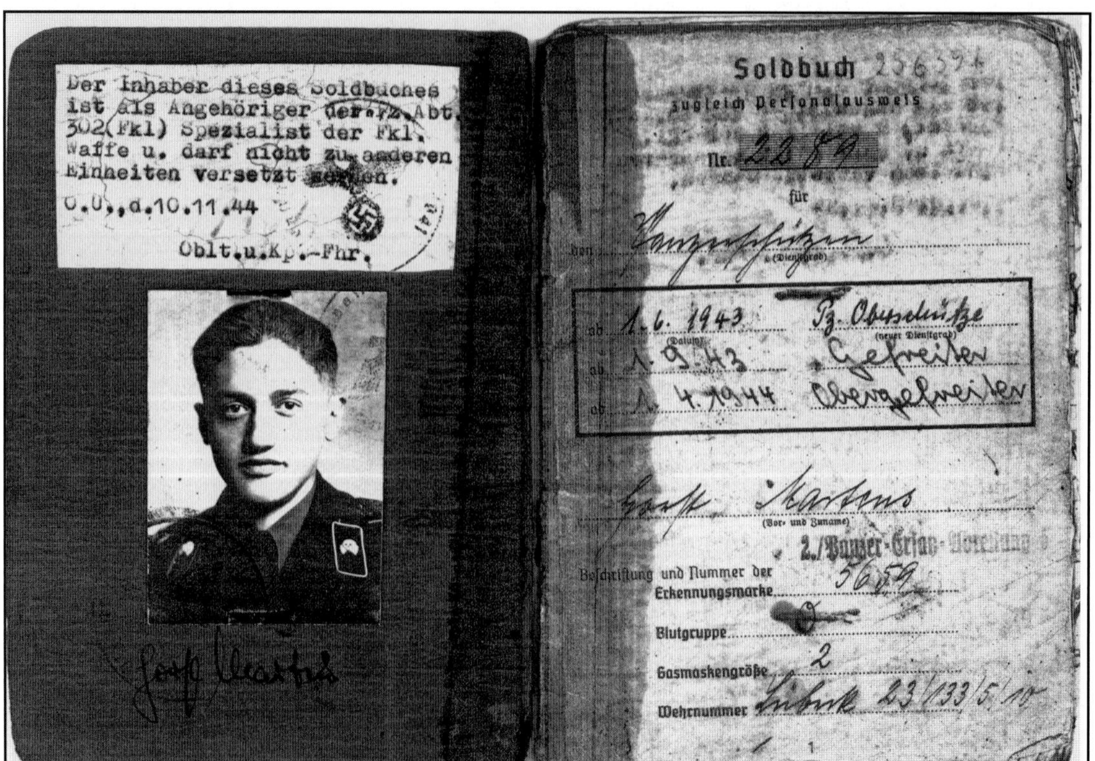

Left: A soldier's paybook identifying him as a member of a special unit (in this case *Panzer-Abteilung 302 (Fkl)*) and cannot be transferred to other units. This entry often proved to be extra "life insurance" for members of the *Funklenkwaffe*.

1. schwere Panzer-Kompanie (Fkl) location classified, 2 April 1944
To the *Panzer-Lehr-Division*
SUBJECT: Company Operational Readiness Report
Reference: Request from *Panzer-Lehr-Division* Rear Detachment

The company reports:

Vehicles available in the company:

Vehicle Type	Number	Operational	Non-Operational	Operational (by 10 April)	kdt.	Remarks
Tiger II's	5	2	3	4	-	One tank requires a new power plant
Tiger I's	3	3	-	3	-	
Demolitions-Charge Carrier *Sd.Kfz. 301*	51	45	6	51	-	
3-Ton Prime Movers	4	3	-	3(4)	1	With the Supply Company of the *II./Panzer-Lehr-Regiment*
1-Ton Prime Movers	1	1	-	1		
Heavy Trucks	9	7	1	7(8)	3	With the *Stab* of the *Panzer-Lehr-Regiment*
Light Tracked Trucks	3	3		3(3)	3	1 with the Stab of the *Panzer-Lehr-Regiment*; 2 with the divisional staff (field kitchens)
Medium Trucks (Calibration)	1	1	-	1	-	
Light Staff Cars	3	3	-	1(3)	2	Regimental staff
Motorcycle/Sidecars	6	3	3	6	-	
Light Motorcycles	2	2	-	-	-	

Vehicle shortfalls in accordance with *KStN 1176 f:*

Tiger II's 9
Light Staff Cars 3
Mediums *SPW's* 3
Kettenkräder 1

Shortfalls in personnel: 4 *Tiger* crews, supposedly to be provided by *Panzer-Ersatz-Abteilung 500* (already requested through the *Panzer-Lehr-Regiment*).

For supply, urgently needed are: 13 heavy trucks with which to supply 14 *Panzerkampfwagen VI Tiger II's* with fuel and ammunition. Additionally, there are no spare parts for the *Tiger II's*. The company urgently requests that they be requisitioned.

In closing, the company reports that it is in position to be operational by 10 April 1944 with 4 *Panzerkampfwagen VI Tiger II's*, 3 *Panzerkampfwagen VI Tiger I's* and a corresponding number of demolition-charge carriers.

In this regard, however, the company urgently requests the return of the detached vehicles, since these are required to make the company operational in the field (field kitchens!).

As the company has had serious engine and transmission difficulties with the *Panzerkampfwagen VI Tiger II*, it has contacted Henschel, and has learned that these are prototypes which the company never declared releasable for operational use. In fact, the previously reported problems have mostly been overcome through modifications.

Use of these 5 vehicles in the field can only be considered if the company is assured it has the necessary recovery vehicles. This requirement can be met, as the regiment has 13 prime movers (18-ton) on hand.

The company requests more detailed information on when it can expect to move or be committed, so that it might prepare a precise schedule for further technical work on the *Tiger II's*. Should the expected move be no more than a local one, the company requests that it be left in its present quartering area until its formation activities are complete, since our proximity to the Vienna Arsenal offers the best possibilities for technical support.

signed Meinhardt, *Oberleutnant* and Company Commander

At the end of April 1944 the *Panzer-Lehr-Division* received orders to proceed to France to act as the operational reserve of the *OB West* in case of an Allied landing. By mid-May all division units were quartered in the Chatres — Le Mans — Orléans area. The remote-control company was quartered in St. Denis near Chateaudun.

On 18 May 1944, as a result of the serious technical problems affecting the *Tiger II* tanks, the division headquarters ordered the remote-control company reequipped at once. The *3./Panzerjäger-Lehr-Abteilung* was ordered to return the assault guns detached to it from the company. (Source: Bundesarchiv/Militärarchiv Freiburg)

Telex

To: General Inspector of the Armored Forces

From: Operations Officer, *Panzer-Lehr-Division*

Based on statements by Henschel and experience to date, the five *Tiger II's* (chassis numbers 280001 – 280005) are not yet ready for front-line service.

Should they be used in combat there exists the danger that they might fall into enemy hands as a result of mechanical failure. The division therefore requests that, for operational security reasons, these 5 *Tiger II's* be sent to a garrison in Germany for further service trials.

Since the delivery of operational *Tiger II's* is unlikely at present, the division has — as a temporary measure —assigned 9 assault guns to the remote-control company from the *3./Pz.Jäg.-Lehr-Abteilung 130* to supplement its 3 *Tiger I's*. Company is therefore deemed conditionally operational.

Request your approval.

Panzer-Lehr-Division Operations
No. 1436/44 geheim, 19 May 1944

The Allies landed in Normandy on 6 June 1944. The *Panzer-Lehr-Division* received its operational orders at 1430 hours the same day. The division's columns were attacked by fighter-bombers during its approach march. The remote-control company suffered its first killed. The *Tiger II's* were left behind at Chateaudun.

After all elements of the division had arrived, the *Panzer-Lehr-Division* was supposed to launch its initial counterattack into the Bretteville — St. Croix area. On 9 June the division was ready to attack toward Bayeux, however, the attack was soon called off.

On the same day British units occupied St. Pierre, a suburb of Tilly-sur-Seulles. On 10 June the town was recaptured from the enemy.

On 11 June the British launched a heavy attack into the area of operations of the *Panzer-Lehr-Division*. The remote-control company played a successful role in the subsequent counterattacks. The commander of the *II./Panzer-Lehr-Regiment* and several officers were killed in the fierce fighting.

In the days that followed, the assault guns of the remote-control company were used to support *Panzer-Grenadier-Lehr-Regimenter 901* and *902*. Only a few remote-control missions were carried out. On 19 June the remote-control company ran into an ambush by about 40 enemy tanks. Most of the company's assault guns were damaged in the engagement, with several completely destroyed. *Oberleutnant* Meinhardt, the company commander, was killed on this day, as was the leader of the 2nd Platoon, *Leutnant* Fischer.

After this engagement the badly decimated remote-control company was scarcely fit for operations and from this point on it was, for the most part, committed with the *Panzer-Grenadier-Lehr-Regimenter*.

By order of the *AHA* (*Stab I (1) No. 27220/44 geheim*, dated 22/6/1944) the company was redesignated as the *1./Panzer-Abteilung (Fkl) 302*. After being detached from the division, the company was sent to Reims to join *Panzer-Abteilung (Fkl) 302*.

IM NAMEN DES FÜHRERS

VERLEIHE ICH

DEM

Obergefreiten

Horst Martens

Pz.Kp.(Fkl.)Pz.Lehr-Rgt.130

DAS

EISERNE KREUZ

2. KLASSE

Div.Gef.St., den 12. Juli 1944

Generalleutnant und Kommandeur

(DIENSTGRAD UND DIENSTSTELLUNG)

Iron Cross, Second Class award document to *Obergefreiter* Horst Martens of the *Panzer-Kompanie (Funklenk)/Panzer-Lehr-Regiment 130*.

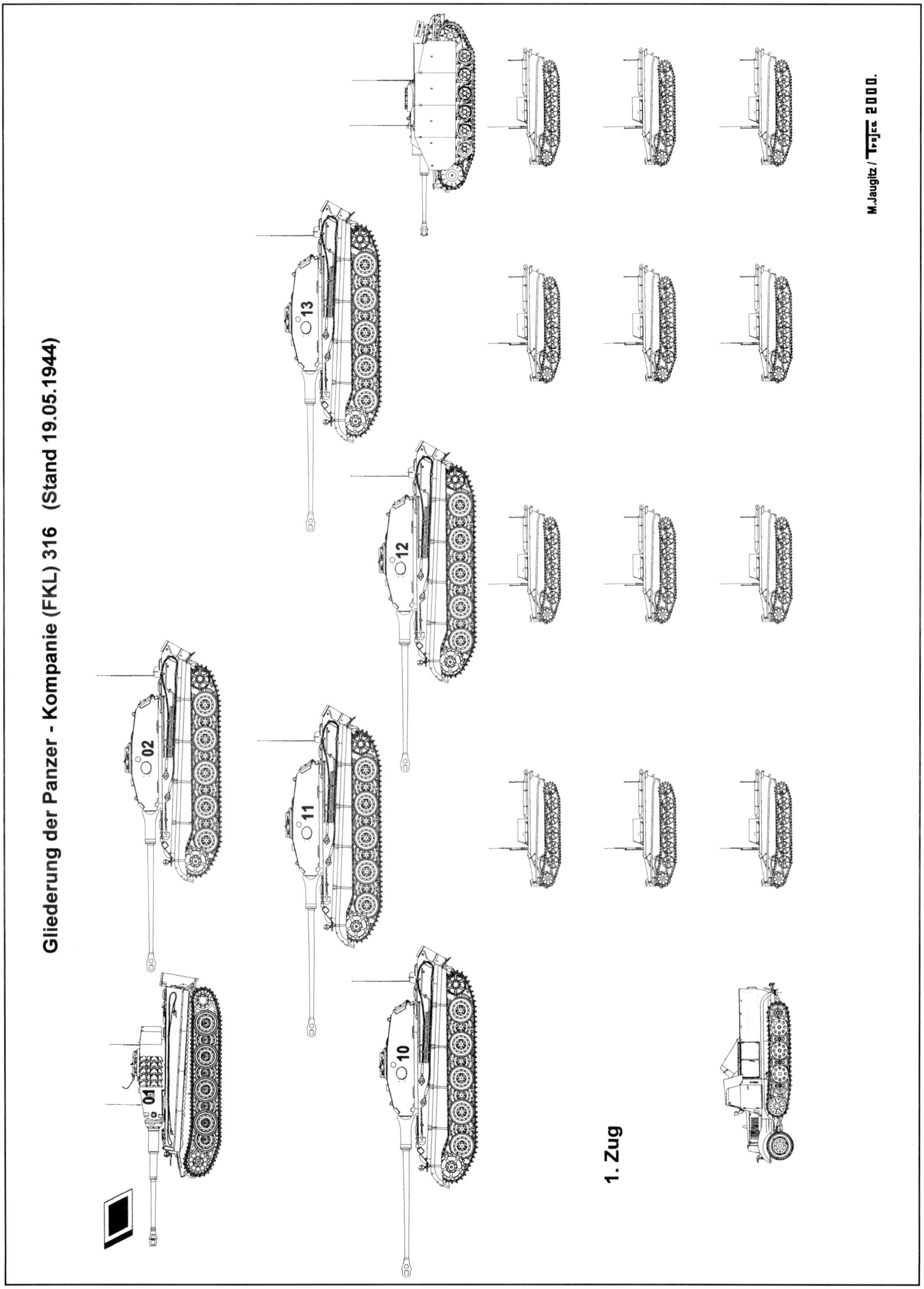
Gliederung der Panzer - Kompanie (FKL) 316 (Stand 19.05.1944)
01
02
10
11
12
13
1. Zug
M.Jaugitz / 2000

Gliederung der Panzer - Kompanie (Fkl) 316 (Stand 19.05.1944)

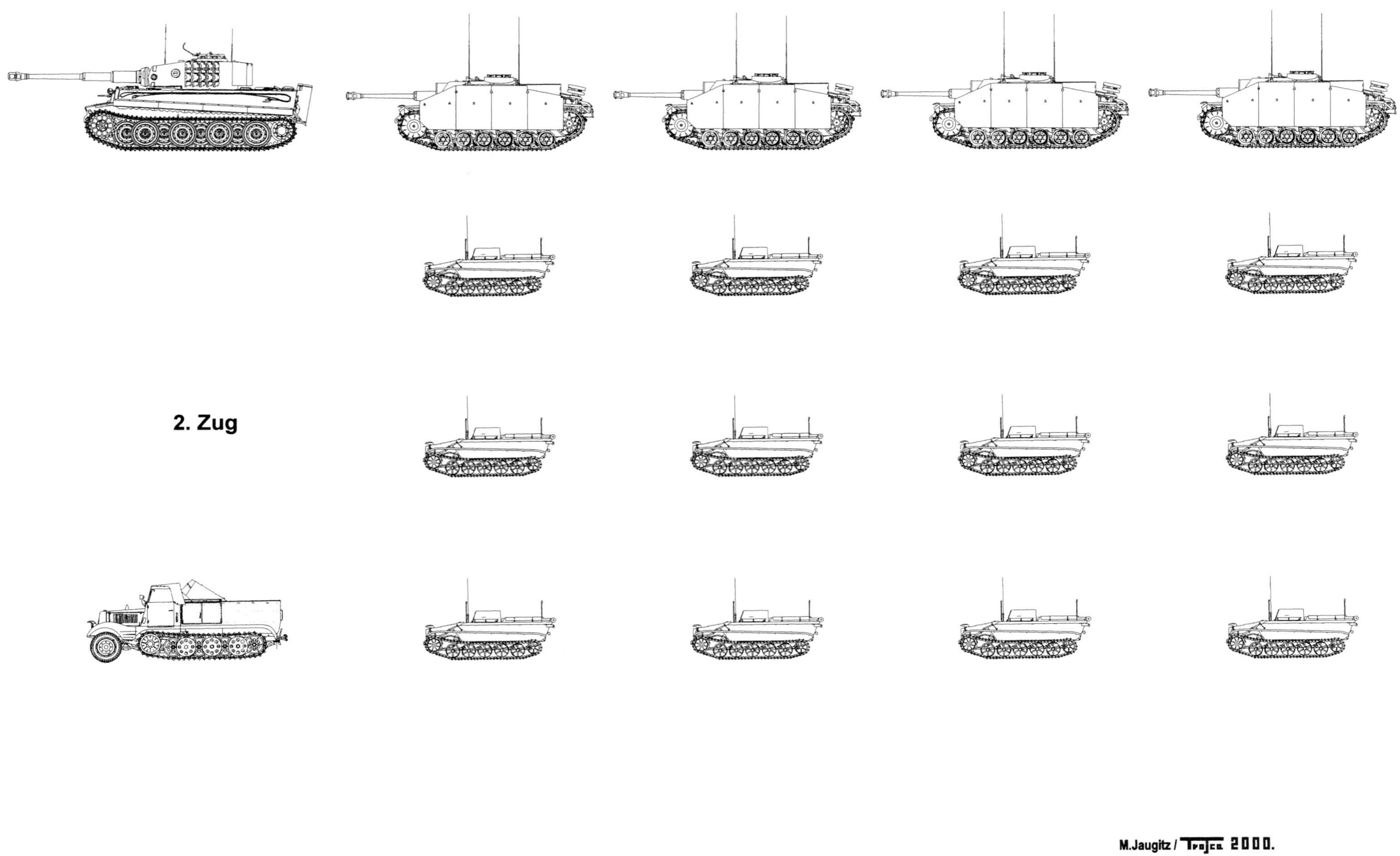

Gliederung der Panzer - Kompanie (Fkl) 316 (Stand 19.05.1944)

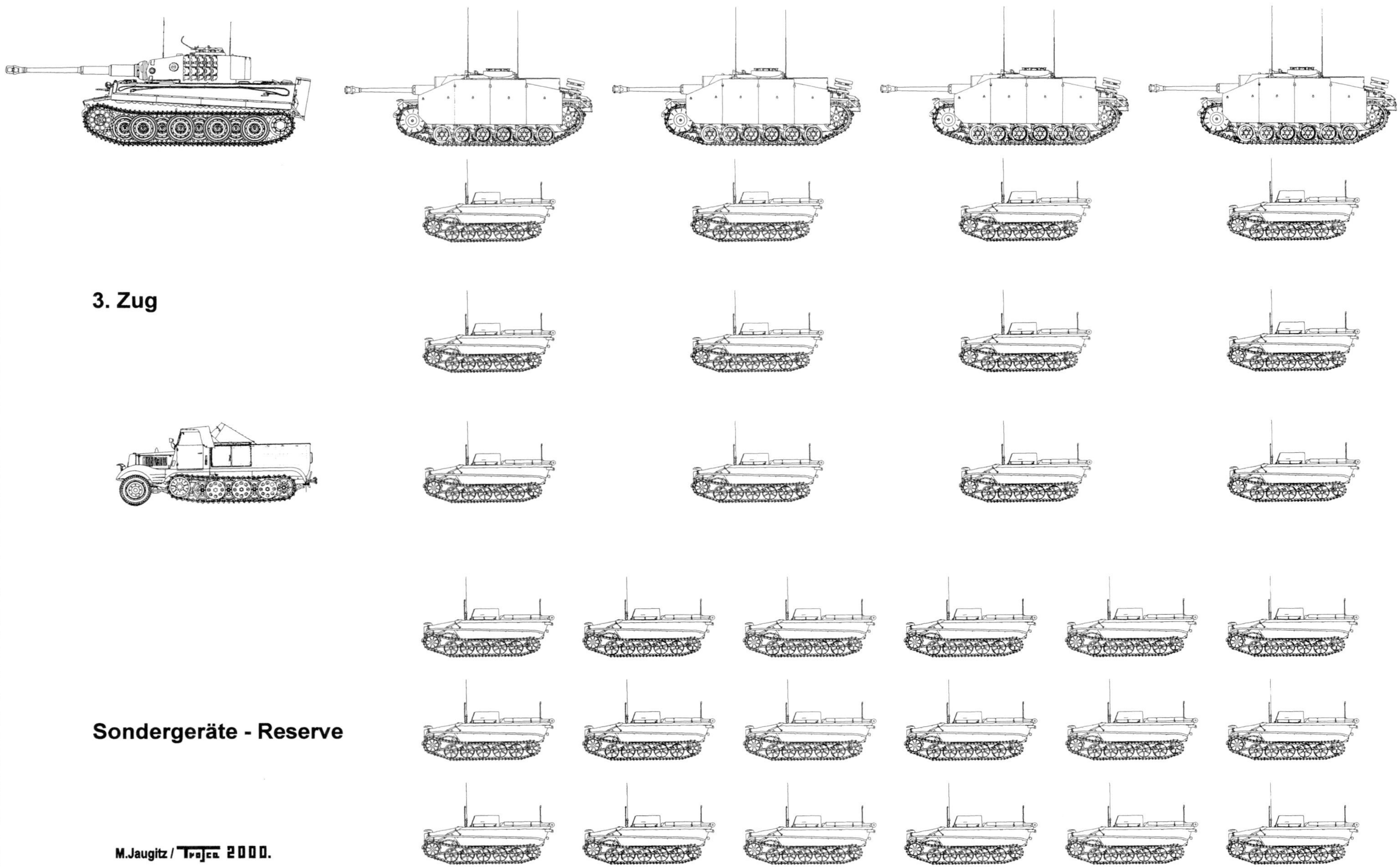

Panzer-Kompanie (Fkl) 316 received its tracked and wheeled vehicles in the late summer of 1943. (Lock)

Top left and above: The unit remained in Eisenach for only a short period of time. By September 1943 it had been moved to the Fallingbostel Training Area. (Lock)

Left and next page top: This assault gun was christened *"Erna"*. (Lock)

Combined exercises were held with the *Panzer-Lehr-Regiment* at Fallingbostel. (Lock)

Assault gun commanders of *Panzer-Kompanie (Fkl) 316*. (Lock)

Above and next page top left: *Unteroffizier* Lock, an assault gun commander. Notice the mounting for the radio-control antenna on the front part of the superstructure roof. (Lock)

Top right: The unit received the first five *Tiger II's* produced at the Kaisersteinbruch Training Area on 14 March 1944. By then, the company had been redesignated as the *1. schwere Panzer-Kompanie (Fkl)*. (Schatz)

Left: The large, two-digit numerals on the turrets of the five tanks are very noticeable. (Schatz)

By the middle of May 1944 all of the elements of the *Panzer-Lehr-Division* were moved to northern France. This included the *1. schwere Panzer-Kompanie (Fkl)* which was attached to the division. (Martens)

2.2.13: Panzer-Kompanie (Fkl) 317

On 15 August 1943 the *leichte Panzer-Kompanie (Fkl)* (cadre) was formed in Eisenach from elements of the disbanded *Versuchs-Kommando (F.L.) Tropen* and members of *the 3./Panzer-Versuchs- und Ersatz-Abteilung 300 (Fkl).*

Oberleutnant Faßbeck assumed command of the company. The 1st Platoon was led by *Leutnant* Knoblauch, while the 2nd Platoon was led by *Leutnant Graf* Kinsky.

As a cadre unit, the company had no tanks or vehicles of its own and was thus had to rely on the support of *Panzer-Versuchs- und Ersatz-Abteilung 300 (Fkl)*.

In an order dated 4 November 1943 (*Org.Nr. 1780/43 g.Kdos.*) the Inspector-General of Armored Forces first made mention of plans to assign ten new production *Sturmgeschütz III's* to *Panzer-Kompanie (Fkl) 317.*

At that time no activation order had yet been issued for a *Panzer-Kompanie (Fkl) 317*. Nevertheless, pay books and identity discs reveal that this designation was used by the cadre unit.

On 6 April 1944 the unit was transferred from Eisenach to the training area at Fallingbostel, where it remained about eight weeks. Soon after the Allied invasion on 6 June the unit was put aboard trains and transported into the Reims area in northern France. There it was quartered in Vouziers.

On 16 June 1944 instructions were issued by the chain of command for the formation of *Panzer-Kompanie (Fkl) 317*, a separate unit. Several days later, however, the *AHA (I (1) 27220/44 geh.)* ordered that the company was to be incorporated into the newly-formed *Panzer-Abteilung (Fkl) 302* and renamed the *3./Panzer-Abteilung (Fkl) 302* (see Chapter subsection 2.2.14).

Leichte Panzer-Kompanie (Fkl) (Personaleinheit) conducted a few exercises at the Berka Training Area. (Lehmann)

Oberleutnant Faßbeck (right), the company commander, in conversation with officers of *Panzer-Versuchs- und Ersatz-Abteilung 300 (Fkl)*. (Lehmann)

Company personnel take a break.

After moving to the Fallingbostel Training Area in April 1944 the company received a few *BIV's* and *Leitpanzer III's* for training purposes. (Lehmann)

The company commander of *Panzer-Kompanie (Fkl) 317)*, *Oberleutnant* Faßbeck. (Lehmann)

The platoon leader of the 2nd Platoon of *Panzer-Kompanie (Fkl) 317*, *Leutnant Graf* Kinsky. (Lehmann)

2.2.14: Panzer-Abteilung (Fkl) 302

The *AHA* issued an order for the formation of another large remote-control formation — *Panzer-Abteilung (Fkl) 302* — on 22 June 1944 *(Stab I (1) Nr. 27220/44 geh.)*.

The necessary personnel for the headquarters, headquarters company (without scout and engineer platoons) and the maintenance platoon were to be provided from *Wehrkreis IX*, after which they would be sent to the *OB West*. The scout and engineer platoons were to be provided from *Wehrkreis IV*.

Existing separate units were used for the three remote-control panzer companies. *Panzer-Kompanie (Fkl) 316* became the *1./Panzer-Abteilung (Fkl) 302*; *Panzer-Kompanie 315 (Fkl)* became the *2./Panzer-Abteilung (Fkl) 302*; and, *Panzer-Kompanie (Fkl) 317* formed *the 3./Panzer-Abteilung (Fkl) 302*.

The unit was established in the Reims area of northern France. To this end it was attached to the *Arbeits-Stab Reims*, which was part of *Panzer-Brigade 10*. The commanding officer of *Panzer-Abteilung (Fkl) 302* was *Major* Reinel, who had previously commanded *Panzer-Abteilung (Fkl) 301* until relieved by *Hauptmann* von Abendroth at the end of May 1944.

While the headquarters, headquarters company, the maintenance platoon and the *3./Panzer-Abteilung (Fkl) 302* were at almost at full strength in terms of personnel and equipment, the *1.* and *2./Panzer-Abteilung (Fkl) 302* still had to make good the losses which the units had taken on the invasion front in June.

According to a status report dated 1 August 1944, the battalion had a personnel complement of 22 officers, 2 civilian officials, 185 noncommissioned officers and 423 enlisted men. Available armored vehicles included 24 *Sturmgeschütz III's*, 108 *BIV* demolition charge carriers and 10 *SPW's*. The remaining six *Sturmgeschütz III's* and the unit's three armored command vehicles (*Befehlspanzer IV's*) were en route.

Most of the demolition-charge carriers were the new *C* version of the *BIV*. Since the personnel were as yet unfamiliar with this variant and because the line companies required practice in operating as part of a battalion formation, the battalion commander requested *Panzer-Abteilung (Fkl) 302* not be committed until mid-August 1944.

On 22 June 1944 the Soviets launched a major offensive which resulted in the collapse of *Heeresgruppe Mitte*. Reinforcements were rushed to the Eastern Front, however, it was not until mid-July 1944 that the German forces were able to regroup and stabilize the front line in Poland and the Ukraine.

Panzer-Abteilung (Fkl) 302 was one of the units affected. On 5 August 1944 it received orders to proceed to Krakow to join *Heeresgruppe Nordukraine*. This order was amended on 8 August and the battalion was directed to proceed to Warsaw.

In Warsaw the Polish Home Army (Armia Krajowa), commanded by General Count Bor-Komorowski, had launched a rebellion at 1700 hours on 1 August 1944. The German authorities in Warsaw had been expecting an uprising for some time, and the rebels failed to achieve the hoped-for element of surprise.

On the first day of the rebellion the Polish Home Army occupied large parts of the old city and the inner city as well as extensive areas in the Wola, Ochota, Zoliborz, Mokotow, Okecie and Czerniakow districts and Praga on the right bank of the Vistula. Weeks earlier the Germans had reinforced their guard and security forces and key positions remained in their hands.

Serious Polish losses, about 2,000 men, created the impression among the Armia Krajowa that the uprising had failed. Some of the rebels subsequently pulled back into the forests around the city and the first suburbs were lost again on 2 August 1944.

The organized dispatch of police and *Waffen-SS* units began on 4 August. Most of the personnel of these units were foreign volunteers in German service. Since these troops lacked heavy weapons, the fighting with the rebels had the character of an infantry battle initially. *Einsatzgruppe Reinefahrt*, which had been assigned to quell the rebellion, called for the accelerated issuance of tanks, combat engineers, flamethrowers and assault guns.

In northern France *Panzer-Abteilung (Fkl) 302* began entraining in the first weeks of August. The trains were subjected to attacks by enemy aircraft and the line companies sustained their first losses in wheeled vehicles and demolition-charge carriers while en route.

The first to arrive, the *2.* and *3./ Panzer-Abteilung (Fkl) 302*, rolled straight from the offloading ramps into action. By 21 August 1944 all five of the transport trains carrying the remote-control battalion had arrived from northern France.

Additionally, the *AHA* ordered the formerly separate *Panzer-Kompanie (Fkl) 311* consolidated with *Panzer-Abteilung (Fkl) 302* as its *4. Kompanie* at the end of July 1944. This remote-control company was undergoing rest and refit in Germany. After the company personnel, who were spread out among various garrisons, had been alerted, the *4./ Panzer-Abteilung (Fkl) 302* was supplied with the necessary command and control vehicles, demolition-charge carriers and wheeled vehicles within two days and transported to Warsaw.

Furthermore, *Sturmpanzer-Kompanie z.b.V 218*, which had been formed on 16 August 1944 solely for this operation, was tactically attached to *Panzer-Abteilung (Fkl) 302*. The special-purpose assault-tank company was equipped with ten *Sturmpanzer IV Brummbär* ("Grizzly Bear") armed with 150-mm *Sturmhaubitze* (assault howitzers).

After their arrival, all elements of the battalion occupied billets in the suburb of Wlochy, west of Warsaw.

The battalion's complement of officer in mid-August 1944 was as follows:

Battalion Commander: *Major* Reinel
Headquarters and Headquarters Company Commander: *Leutnant* Schiller
Battalion Surgeon: *Oberarzt Dr.* Krauskopf
1./Panzer-Abteilung (Fkl) 302: *Oberleutnant* Dettmann
2./Panzer-Abteilung (Fkl) 302: *Leutnant* Weichard
3./Panzer-Abteilung (Fkl) 302: *Oberleutnant* Fassbeck
4./Panzer-Abteilung (Fkl) 302: *Oberleutnant* Bachmann
Maintenance platoon: *Leutnant* Harbort

The main German attack began on 12 August 1944 with the objective of establishing a secure east-west corridor through the city.

The combat elements of *Panzer-Abteilung (Fkl) 302* were directly attached to the German attack force, since renamed *Kampfgruppe Reinefahrt*, and saw their first operations with demolition-charge carriers in the old city on 14 August. In the days that followed, there were further remote-control missions in the old city and the inner city.

It soon became apparent that committing assault tanks, assault guns and demolition-charge carriers piecemeal was ineffective. Therefore it was necessary to change tactics and concentrate all available heavy weapons, including dive-bombers, on the objective. Following the concentrated bombardment, infantry would move in and occupy the battered objective. Assault guns and assault tanks accompanied the riflemen and kept the enemy's head down with sustained fire.

In keeping with this tactic, clearing parties cleared the streets for the demolition-charge carriers, which were guided to barricades and detonated. The riflemen advanced under cover of the cloud of smoke produced by the explosion into the adjacent houses and occupied them to the top floor.

On 1 September 1944 the battalion reported that it was 6 command and control vehicles and 65 *BIV* demolition-charge carriers short of its authorized strength of 40 command and control vehicles and 144 demolition-charge carriers.

Resistance in the old city ceased at the end of August 1944, after which the combat elements were used in the inner city. In mid-September they were moved to the area of the suburb of Praga-Grochow.

Following the arrival of the *19. Panzer-Division*, *Panzer-Abteilung (Fkl) 302* began joint operations with it to subdue the Polish rebels surrounded in the suburbs of Mokotow (24-27 September 1944) and Zoliborz (23-30 September 1944). The Warsaw rebels surrendered on 2 October 1944.

On 24 September 1944 *Major* Reinel handed command of the battalion over to *Hauptmann* Nolte.

By order of the *AHA*, the *4./ Panzer-Abteilung (Fkl) 302* left the battalion on 7 October and was transferred back to its home garrison at Eisenach.

It was planned to reorganize the unit at a location near the front, however, the increasingly threatening situation in East Prussia would not permit this. As a result, following a brief rest, the unit, renamed *Panzer-Abteilung 302 (Fkl)*, was transported to the northeast beginning on 10 October. The first elements detrained in Tilsit on 12 October.

On 16 October 1944 the 3rd White-Russian Front launched an offensive between Suwalki and the Memel into the East Prussian border region. Facing the outnumbered German divisions on a front approximately 100 kilometers long was a huge Soviet army of 40 rifle divisions and numerous armored units. *Panzer-Abteilung 302 (Fkl)* was immediately moved into the area between Eydtkau and Ebenrode which was in the area of operations of the *XXVI. Armee-Korps*.

Between 18 and 25 October 1944 the assault guns of *Panzer-Abteilung 302 (Fkl)* took part in several fierce tank engagements in the Jocken, Nickelsfelde, Parkhof, Föhrenhorst, Kleinlucken and Kornfelde areas. Demolition-charge carriers were also employed on several days. During the fighting in that area the battalion commander, *Hauptmann* Nolte, was killed on 21 October 1944.

The front in the East Prussian border region was stabilized again by the end of October 1944. Losses had been very heavy on both sides. For example, in a period of ten days the 11th Russian Guards Army had lost 616 tanks in the Goldap — Ebenrode — Schloßberg areas.

At the beginning of November 1944 *Panzer-Abteilung 302 (Fkl)* was pulled out of the line and transferred into the Korschen area northwest of Rastenburg for rest and reorganization in accordance with the new tables of organization. Under the command of the new battalion commander, *Major* Sahmel, the battalion carried out joint exercises with *Panzer-Regiment "Großdeutschland"* in November and December.

Beginning on 25 December 1944, *Panzer-Abteilung 302 (Fkl)* moved to the training grounds at Mielau in northern Poland.

After completing its buildup at the beginning of January 1945, the Red Army launched a major offensive against the *XXVI. Armee-Korps'* sector between Gumbinnen and Ebenrode on the morning of 13 January. On 14 January the combat elements of *Panzer-Abteilung 302 (Fkl)* took up position southeast of Praschnitz with units of *Panzer-Grenadier-Division "Großdeutschland"*.

On 15 January 1945 the Germans launched a counterattack into the Praschnitz-Ciechanow area north of the Narew. By 16 January, however, the Soviet units had regained the upper hand. Shortages of fuel and ammunition severely limited the freedom of movement of the German armored units.

In the days that followed, the battalion's assault guns regularly engaged the enemy in a defensive role. Demolition-charge carriers were also used. *Panzer-Abteilung 302 (Fkl)* had to keep pace with the withdrawal movements of *Panzer-Grenadier-Division "Großdeutschland"*, however, and was forced north through Willenberg, Ortelsburg, Heilsberg and Landsberg. The number of assault guns on strength with the combat elements was reduced very quickly by a succession of defensive battles and counterattacks. At the end of January, the fuel shortage made it necessary to blow up damaged armored vehicles and superfluous equipment prior to the planned withdrawal.

According to a status report by *Panzer-Abteilung 302 (Fkl)* dated 1 March 1945, eight officers and 306 noncommissioned officers and enlisted men had been released for use as infantry.

The last three serviceable assault guns were incorporated into the *gemischte Panzer-Abteilung "Großdeutschland"* as an assault gun platoon and, in the second half of March, they were committed southwest of Königsberg between Ludwigsort, Pörschken and Maulen.

The rest of the battalion stood ready for evacuation by sea to northern Germany. After their arrival in various Baltic ports, many members of the battalion made their way west and were taken prisoner by the British.

The following reports by *Panzer-Regiment "Großdeutschland"* and *Panzer-Abteilung 302 (Fkl)* describes the latter unit's actions from January 1945 on (Source: Bundesarchiv / Militärarchiv Freiburg):

Copy
Panzer-Regiment Command Post,
Großdeutschland 5 February 1945
Operations
SUBJECT: Operations by *Panzer-Abteilung 302 (Fkl)*
To:

Panzer-Grenadier-Division "Großdeutschland"

Panzer-Abteilung 302 (Fkl) did not prove itself as a remote-control battalion, instead it was only used in operations as a normal tank battalion. The reasons for this are as follows:

(1) The (in places) heavy snowdrifts and the severe cold usually prevented the *BIV* from driving cross-country and caused serious disruptions in the radio-control equipment.

(2) As a result of the constant withdrawals and the associated marches, the equipment was generally not combat ready and required further checks, which took several hours. By the time these were completed the situation had usually changed, ruling out their use in action for purely tactical reasons.

(3) Usually the regiment faced the decision of whether to use the controlling assault guns in less than promising operations, or to have them available as conventional assault guns in a much shorter time. The constantly changing situation and constant crises always made the assault gun role appear more appropriate, especially since our engagements were usually not against fortified targets but against widely separated attacking infantry and enemy tanks.

(4) In the regiment's view the remote-control weapon's equipment is still too complicated and therefore cannot be made combat ready quickly enough to be employed with success at the right moment in mobile warfare, especially during withdrawal movements. Even if the use of remote-control tanks against towns or woods unquestionably promises a degree of success, it still requires thorough and time-consuming technical and tactical preparations, for which the time was never available during the operations of the past few weeks.

An additional factor was that the fuel situation — always completely unsatisfactory — made it impossible to provide the majority of the explosives carriers with sufficient fuel. Fuel was barely sufficient for the other elements of the regiment and for the most necessary repair work.

signed von Wietersheim
Major

—

Copy

Panzer-Abteilung 302 (Fkl.) Battalion command post,
11 February 1945

After-Action Report on the Employment of the Remote-Control Force during the Fighting in January 1945

A. Preparation and Employment of *Panzer-Abteilung 302 (Fkl.)*

Working closely with *Panzer-Regiment "Großdeutschland"* during the quiet period in East Prussia in November and December 1944, numerous map exercises and exercises were conducted before many of the division's officers; the purpose being to show the nature and effect of the remote-control force. Almost without exception, full understanding was shown by every side.

Beginning on 25 December 1944, the battalion transferred to the Mielau Troop Training Area. Because of lack of shelter and the threat of air attack, the vehicles had to be widely dispersed in the forests and all were exposed to the negative effects of weather. The supply of electricity was spotty and vehicle checks were only conditionally possible because — in spite of repeated requests — no fuel was allocated for operational checks on the remote-control vehicles.

On 14 January 1945 the battalion moved into ready positions southeast of Praschnitz with *Panzerregiment "Großdeutschland"* (according to plan). On the evening of 13 January 1945 the 2./Pz.Abt. 302 (Fkl) was attached to the heavy tank battalion of *"Großdeutschland"* so as to be in position when the battalion attacked. On orders from the *Tiger* battalion at about midnight it occupied security positions in the main line of resistance east of the bridge at Podos Stary. One platoon was deployed with no infantry at all.

During the attack towards Gansewo, which was planned for daybreak, the battalion's *1.* and *3. Kompanien* were placed under the command of *Panzerregiment "Großdeutschland."* Shortly before the start of the attack, on order of the division, the *3. Kompanie* was pulled out of the line and placed under the command of the *Grenadierregiment*. It subsequently attacked to the south. Because of a lack of time, the company was not able to establish contact with the grenadiers until the attack was already under way. This company, which was unable to commit any *BIV's* on account of terrain difficulties, achieved good success. Three tanks, seven antitank guns, a large number of antitank rifles and heavy infantry weapons were destroyed in a short time.

The *1. Kompanie*, which was deployed to secure the left flank, remained with the battalion. There it was possible to commit *BIV's* by remote control. One *BIV* was detonated during the attack by the fusiliers on Jozefofo and helped them take the village without loss.

In the second half of the attack the employment of *BIV's* failed on account of numerous steep-sided, icy ditches. The vehicles were recovered undamaged and the attack was carried out with assault guns with outstanding success.

Damage inflicted on the *BIV's* by very heavy artillery, antitank and mortar fire was considerable. In two cases B IV's were detonated by direct artillery hits, causing damage to nearby vehicles.

On the following day a combined attack with tank support into the deep flank of the enemy advance made it possible to almost completely destroy the enemy.

On the same evening the battalion was split up by the division. In order to ensure logistical resupply of the battalion, it was ordered that each unit should take with it just four to six *BIV's*. The remaining *BIV's* were to be assembled at the forward support base and checked over thoroughly.

In the following days of fighting a serious fuel shortage made itself felt. This situation made it impossible to recover unserviceable *BIV's*, which had to be demolished in place.

With the situation becoming more acute from day to day on account of being outflanked by the enemy, and with the will to fight of the infantry — especially of security units, *Volkssturm* etc., weakening day by day — the dwindling number of *Tigers* soon made it necessary for the assault guns to be used in counterattacks. In most cases, they covered the division's withdrawals almost single-handedly. In crisis situations the division even ordered the commitment of small armored groups of two or three assault guns. Although the assault guns were poorly suited to the task because of their lack of a revolving turret, in many cases the battalion was employed as rear guard and on its own initiative brought about the withdrawal of abandoned security units and *Volkssturm*.

Later all of the remaining assault guns were placed under the command of *Kampfgruppe Bock* to be used in a purely armored role along with the remaining tanks of *Panzerregiment "Großdeutschland."*

Previously the division had made its withdrawal movements in relatively small steps for armor units but now a major move was to be made to the north. In order to assure that the necessary fuel was available, the division ordered on 25 January 1945 that all unserviceable tanks, recovery vehicles and prime movers, plus excess baggage and maintenance facilities — except for one maintenance section per battalion — were to be blown up.

After carrying out this order the battalion's vehicle complement was reduced to the following level, including vehicles lost to enemy action:

31 wheeled vehicles
5 prime movers
1 heavy flatbed trailer
4 *SPW's*
22 tanks
65 *BIV's*

As per a regimental order of 27 January 1945, three personnel units were formed from *Panzer-Abteilung 302 (Fkl)* plus elements of the *I.* and *III./Panzerregiment "Großdeutschland."* Each of these was to form two or three tank-killing companies. Transport was provided by emptied trucks and vehicles with special superstructures.

As per a division order on 5 and 7 February 1945, 250 men were released to the *Grenadier-Division*. These 250 men, who came from every part of the battalion but did not include specialists such as radio technicians, signals mechanics, etc., were transferred to the division's antitank, grenadier and fusilier units.

In order to retain as many specialists as possible, the battalion formed a motor transport company of four platoons with a total tonnage of 76.5 tons. Most specialists, such as *BIV* drivers, signals mechanics and so on were assigned to this company. The column was seconded to the commander of the logistical resupply services of *"Großdeutschland."*

B. Supply

In general supply was assured if it was in the hands of a capable and cautious supply company commander. In spite of all the supply difficulties, the only complaints about supply occurred in the early days, when enemy action prevented the field kitchens from keeping up with the tanks. It is advisable that a remote-control unit's own *SPW's* be used to supply it with rations, ammunition and fuel, and this later became standard procedure.

Starting in November there was a shortage of winter clothing and spare parts, which was inexplicable given the repeated requests for these items. Not until the dumps were emptied, opened by the civilian population or blown up during the withdrawals was the battalion supplied with them. (The same situation as in the winter of 1941-42!) Unfortunately the spare parts obtained in this way came too late, for the division's order to blow up unserviceable armored vehicles meant that they were total losses.

C. Summary

Even under enemy fire, the employment of explosives carriers must be coordinated with other arms to the smallest detail.

Employment is pointless if vehicle checks must be eliminated or are only partially carried out.

Mass employment is often only possible with a very high loss rate in remote-control vehicles, since under heavy enemy fire from artillery, antitank guns and dug-in tanks even veteran tank regiments proceed much too hesitantly.

The cross-country capability of the *BIV b* and *c* with standard tracks is inadequate, especially on ice. Powerful grips are an absolute necessity. (Welding on of cleats.)

If the use of such a costly weapon as the remote-control vehicle is planned, command must take much more consideration of its special characteristics.

If the fighting characteristics of the remote-control force are to be clearly shown to the infantry, then fuel must not be handed out liter by liter.

In spite of all the advantages which the remote-control force can potentially place in the hands of command, it must be said that if the above mentioned points are not taken into consideration, the remote-control force, given the present technical state of the *BIV b* and *c*, must be rejected as too costly!

signed Sahmel
Major and Battalion Commander

—

Panzer-Abteilung 302 (Fkl)
Battalion command post, 10 February 1945

Secret

After-Action Report on Operations by the
Supply Company of *Panzer-Abteilung 302 (Fkl)*

Experience with the supply company in the most recent operations was satisfactory and it was clearly demonstrated that the consolidation of supply services within the battalion is appropriate. The personnel composition of the supply company in accordance with the table of organization was adequate. The only shortcoming was the absence of a maintenance platoon in the supply company, which delayed major repairs to the assault guns and *BIV's* as they first had to return to the maintenance facility. It would be advisable to provide two more dispatch riders for the operations section.

A. Standard Operating Procedures of the Supply Company

The supply company was organized into a forward section and the main body.

The forward section included:

3 Maintenance Sections
3 Field Kitchens
Some of the ammunition and fuel vehicles
The radio-control repair section
Vehicles for the company commander, *Spieß*, communications noncommissioned officer, and armorer
Elements of the engineer and scout platoon
Elements of the headquarters company

The forward section was always three to six kilometers away from the battalion command post and was led by the supply company's special-staff officer. To reduce the burden on the leader of the forward section, all repair services were concentrated under the command of the motor pool officer, whose only concerns were the necessary repairs to the arriving assault guns and *BIV's* and their dispatch to the maintenance facility and back to the combat elements.

The maintenance services also included the recovery platoon, which accompanied the forward section with its two 18-ton prime movers and the recovery tank. Two *Feldwebel* were made responsible for the recovery platoon, one to maintain contact between the combat elements and the forward section and the other to carry out recoveries.

B. Changes

Experience has shown that the following points must be changed or given more consideration:

1.) The forward section is a large grouping of vehicles which is difficult to move. The leader of the forward section has difficulty keeping track of this huge assembly of vehicles and personnel. It is therefore advisable that the repair services and the remaining supply vehicles be set up separately, in different areas.

2.) It is necessary that the leader of the forward section establish a point where all of the battalion's vehicles and personnel can report their arrival and departure. In this way the leader of the section has an exact picture of vehicle and personnel strength and also receives accurate information on fuel and ammunition. By combining efforts it avoids unnecessary supply runs.

3.) A reliable radio station must be available for communications with the actual supply section, remaining in contact with the battalion and the maintenance facility simultaneously. During the fighting withdrawal the situation often became such that range from the battalion to the maintenance facility became too great (for the existing radios). An 80-Watt set should therefore be requested.

4.) Communications are vital for a smooth-functioning supply system. Contact must be maintained between the combat elements and the forward section. Throughout the day the First Sergeants of the companies must determine from the forward section, by way of the battalion command post, exactly how they will supply their companies during the coming night.

5.) As part of the evening resupply runs, a precise ammunition and fuel report must be received from the companies. These will be used by the forward section to determine its daily supply needs and prevent being burdened with excess ammunition which cannot be transported.

6.) The release of transport to superior units (regiment, division) often results in a shortage of trucks and it is therefore vital that the forward section send back all excess vehicles by the shortest route (strict checks on excess cargo!).

7.) To facilitate rapid movement by the forward section, it is necessary that it scout in detail arrival and departure routes immediately after its arrival in a new area. Furthermore, a well-marked route must be prepared from the battalion command post via the forward section to the actual supply section, so that every dispatch rider and every vehicle coming from up front can find its way without maps.

8.) Efforts must be made to locate the forward section, the actual supply section and the maintenance facility on a continuous route, in order to avoid lengthy searching. Furthermore, it is important that every vehicle proceeding to or from the front contact each of these positions for possible combining of tasks.

9.) To enable faster recovery of knocked-out tanks, the recovery tank must be equipped with a Fu 5 and follow some distance behind the tank attack. In this way it can be summoned quickly at any time for a recovery.

C. Supply Company Area of Operations and Communications

The actual supply section is often about 10 – 15, sometimes even 20 kilometers from the forward section. This distance was selected in order to allow the supply company, especially the specialists, to work without interruption during withdrawals. Close cooperation with the division *Ib* and the regiment *Ib* and other staff sections in the regiment (*Iva, V (K), V (N)*) simplifies the supply effort significantly.

Prior to the supply section being moved contact is made with higher headquarters to ensure the units are as close as possible in the new area. Use is often made of a telephone link with the *Ib* of the regiment. Transporting fuel, ammunition and rations to the front was especially difficult, because the roads were often clogged with refugees and there were also frequent air attacks. For this reason supply runs were usually made at night.

D. Maintenance Facility

1.) Administratively, the workshop was not included in the supply company. It was shown, however, that closer cooperation is desirable, especially when committed, to allow complete daily reports to be made. The maintenance facility was always an extension of the supply section, and every vehicle that drove from the front to the rear or from the rear to the front had to contact the supply section. This saved many dispatch trips.

2.) The steady withdrawal meant that it was especially important for the maintenance facility to complete necessary tasks as they arose. After the first short jumps to the rear, the moves were immediately extended to 30 kilometers, providing at least two to three days continuous working time.

signed Sahmel
Major and battalion commander

—

Telex Message:

To the Inspector of Armored Forces, *Generalmajor* Thomale:

Panzer-Abteilung 302 (Fkl). 3 assault guns still in action.

As ordered, all *BIV's* blown up because of lack of fuel.

Battalion requests the radio-guidance specialists be withdrawn and further orders.

Cadre unit *Panzer-Reghiment "Großdeutschland"* also possible.

Signed Sahmel, *Major* and Battalion Commander of *Panzer-Abteilung 302*

—

Telex message:

Sent: 23 march 2400. Received: 24 march 0130

By radio from *Heeresgruppe Nord*

To Inspector General of Armored Forces (*Arbeitstaffel Abteilung-Organization*)

Remnants of *Panzer-Reghiment "Großdeutschland"* and *Panzer-Abteilung 302* in area north of Pillau with 20 officers, 300 noncommissioned officers, 500 enlisted men and 40 special vehicles.

It is planned to use these specialists as an infantry battalion.

Losses while employed as infantry so far: 900 men.

Regiment can be reformed quickly with existing personnel as soon as replacement tanks are delivered.

Request that existing orders be rescinded and new order issued for *Panzer-Regiment "Großdeutschland"* and *Panzer-Abteilung 302* to proceed with existing personnel and materiel to Germany.

Departure contingent on provision of necessary transport space.

To save transport space possible transfer with *schwere Panzer-Abteilung 511.*

Request you advise by telex via *Heeresgruppe Nord.*

Panzer-Regiment "Großdeutschland"

Signed Gahl, *Oberstleutnant* and Regimental Commander

—

Diary entries by Gottfried Beierlein (†) (*2./Panzer-Abteilung 302 (Fkl)*) give insight into the mood of the troops and the course of the withdrawal:

Saturday, 13 January 1945: I had to stand watch again this evening. Morale is anything but good.

Sunday, 14 January 1945: When will it begin? We are supposed to stow away all baggage which is not absolutely necessary. The order to move out came at 1700 hours. We moved into the assembly area at night. Commanding my vehicle was *Feldwebel* Brenk. He is Platoon Leader of the 2nd Platoon. We warmed ourselves around a stove in a partially shot-up house. Together with "GD", we are going to attack at dawn. Everyone is lost in his thoughts. I have come to think that difficult days lay before us.

Monday, 15 January 1945: We attack with about 100 tanks. But soon the force is so dispersed that every tank is fighting on its own. We succeed in breaking through an antitank belt and I am able to knock out seven. We overrun much equipment and the Ivans run so fast their socks are smoking. The attack moves so fast that our infantry cannot keep up. Visibility deteriorates and the attack is halted. We take up a hedgehog position around a burning village.

Tuesday, 16 January 1945: When it becomes light we move into a new assembly area. None of us knows what is going on. Where is the front? We attack once again. Only a few tanks left. The Ivan has set a trap for us. Infantry to the front, Russians on the right flank, antitank guns! With difficulty we managed to knock out a Josef Stalin, but our gun is finished. The recoil mechanism has failed. About face! Back to the maintenance facility. We reach our barracks in the evening, but the first thing we want is to sleep.

Wednesday, 17 January 1945: Watch out, the maintenance facility is jumping to a new position. Well, we'll just go with it. On the contrary, I am glad to have a few days rest. In the new maintenance facility area they start looking after our "ride" immediately. On the road columns roll to the rear. I cannot shake the feeling: This looks like retreat. We remain in the godforsaken hole for the night.

Thursday, 18 January 1945: The gun was adjusted and zeroed quickly; then off to the supply company to refuel. There I saw Karl Krüger for the last time; he was killed on the 15th. We continued on to the support base. The roads were clogged. We had to be quite ruthless in making our way to the front. In the evening we arrived at the support base and then immediately moved back with it. Our engine sputtered terribly. We slept in a room with 13 civilians and 6 soldiers.

Friday, 19 January 1945: This is what you call bad luck! We had to move across a frozen pond. Crash! And then we were sitting in it. One 18-ton and two 12-ton prime movers finally got us out. Everything was full of water up to the gunner's seat. The repair section worked until late into the night. At 0100 hours we withdrew to the rear.

Saturday, 20 January 1945: Work goes on at a frantic pace. We wanted to reach the front by noon to tow back an assault gun. Back at the support base I went to see the medic again. I had to leave my crew and go to the supply company. At night I went there in an ammunition truck; I checked the route. When we arrived in Wapplitz the supply company had already pulled out. I slept in a *Gasthaus*.

Sunday, 21 January 1945: Meanwhile we were joined by a truck and *Leutnant* Heringer. After driving around for a long time we found the supply company. I picked up the salve in the dispensary, otherwise I did what I liked. I stowed my things in the repair truck. We moved out again in the late afternoon, taking with us some refugees from the village. We spent the night in Groß-Albrechtsdorf.

Monday, 22 January 1945: Early in the morning we set out again in the direction of Seeburg. We looked for quarters in a village. But then came the order to fall back. We were just about to leave when all civilians were ordered out of the vehicles. I will spare myself the scenes that followed. In any case I will never forget the suffering of the refugees. Those poor people! We moved on to Glockstein and spent the night in the meeting room of a *Gasthaus*.

Tuesday, 23 January 1945: Today is my birthday! Thank God there are things bought from the field exchange. We sat down in quarters and brewed some grog. Fritz Hermann arrived at noon to wish me all the best. His gun was damaged and he had to go to the maintenance facility. From *Oberleutnant* Dettmann I received a drink of *schnapps* and we celebrated my birthday merrily. In spite of everything we did not forget to listen to the news, for it can only be a matter of days before we are surrounded here. I wrote a letter to my parents. Will they ever receive it?

Wednesday, 24 January 1945: Bear it with humor! That is our motto. We can't change things. I have an uncomfortable feeling, but I can't believe that we are finished. My inner feelings have never betrayed me. I looked for new billets. A nice lady took us in.

Thursday, 25 January 1945: I slept in a real bed! In the morning I had to go to the doctor, perhaps then I would be able to return to the company. Fritz Hermann is still here, because they have no fuel. The company commander is now in my vehicle. Am I now to become the company commander's gunner?

Friday, 26 January 1945: Our battalion has made an attack on Allenstein. Judging by the rumbling from the front, it is not over yet. In fact it can't go on much longer. The battalion is being pulled out of the line and a *"Springer"* battalion is being formed.

Saturday, 27 January 1945: The 27th was a black day for the *2./Panzer-Abteilung 302 (Fkl)*. Almost all of the gasoline vehicles had to be destroyed because of the fuel shortage. Three armored vehicles were left. The supply company was disbanded. An infantry platoon is being formed.

Sunday, 28 January 1945: Near Gallingen I again witnessed the misery of the refugees. Nevertheless we were fed well. A pleasant evening with our hosts.

Monday, 29 January 1945: We are to be ready to move out by noon. We met some members of the *Volkssturm*. I was shaken. We drove to Banditten. The place was so full that it was impossible to find quarters. We literally camped in the street. I can barely look the civilians in the eye any more.

Tuesday, 30 January 1945: I wandered around Banditten all day. We are supposed to move out this evening. I succeeded in procuring 120 liters of fuel for the company. Morale is sinking slowly but surely.

Wednesday, 31 January 1945: We are supposed to depart at 2330 hours today. Hurray, there is fuel again. But it comes too late. Our tanks are here. We moved to the train station in two trucks. The streets are frightfully clogged. It takes two hours to move two kilometers. The move back is no different. The company moved out at 0400. It left without me, however. I had to dole out the fuel to the individual companies. The thaw is the greatest hindrance. My fur-lined boots have had it.

Thursday, 1 February 1945: I am still in charge of the fuel. No rations. Late in the evening we set out to the northeast, in the direction of Königsberg. We didn't get far; soon it was dark. We spent the night in a train station. My feet are wet all the time now.

Friday, 2 February 1945: We are on the road again. We even drive on the *Autobahn*. Our company camped in a nice housing development. At battalion we handed out the fuel. All of our infantry was there. New rumor: Our battalion is to be disbanded and we are joining the infantry. God save us from that. My vehicle was knocked out yesterday. Siegfried Neff was badly wounded. I could easily have been in it.

Saturday, 3 February 1945: *Panzer-Abteilung 302 (Fkl)* is disbanded. General transfer to the infantry.

Sunday, 4 February 1945: Farewell by the battalion commander.

Monday, 5 February 1945: I have landed in the *5. Kompanie* of a *Panzer-Sturmpionier-Bataillon*. [Editor's Note: A euphemism for an infantry company.]

Tuesday, 6 February 1945: I have been assigned to a heavy machine-gun section as a gun commander. I am responsible for the good or ill of six men.

Wednesday, 7 February 1945: Into battle!

Monday, 12 February 1945: I am one of the few *Sturmpioniere* who has not yet bit the dust.

Wednesday, 14 March 1945: I have become a truck driver — Ford V 8.

Wednesday, 18 April 1945: Shipped out on ferries from Pillau to Hela.

Thursday, 19 April 1945: Arrived in Hela.

Friday, 20 April 1945: Still in Hela.

Saturday, 21 April 1945: Boarded the *"Adler Traber"*. 1700 hours submarine attack north of Hela.

Monday, 23 April 1945: Arrived in Swinemünde.

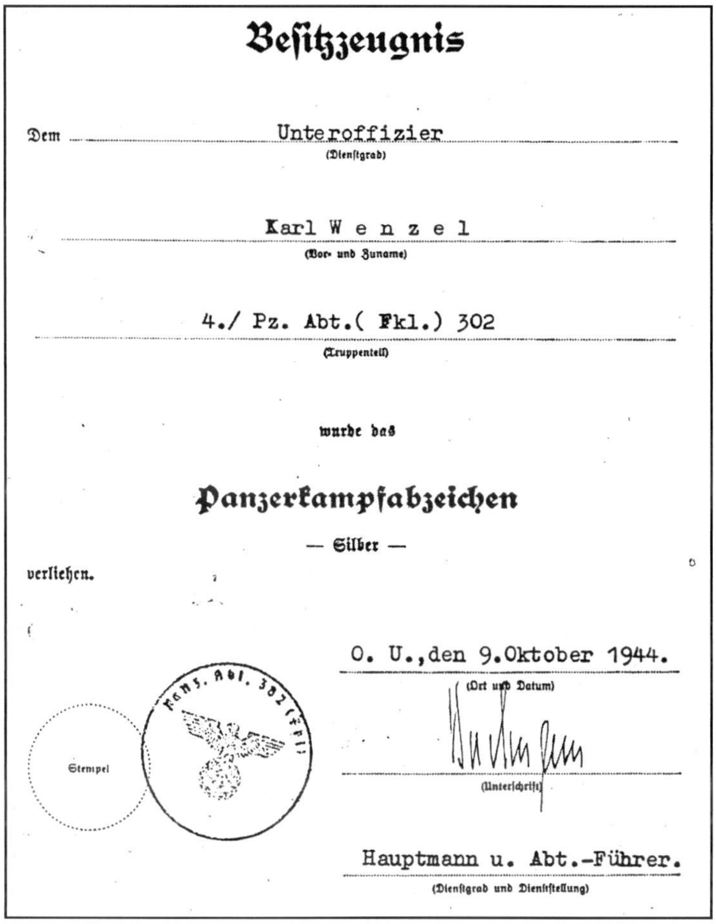

Besitzzeugnis

Dem Unteroffizier
(Dienstgrad)

Karl W e n z e l
(Vor- und Zuname)

4./ Pz. Abt.(Fkl.) 302
(Truppenteil)

wurde das

Panzerkampfabzeichen

— Silber —

verliehen.

O. U.,den 9.Oktober 1944.
(Ort und Datum)

Stempel

(Unterschrift)

Hauptmann u. Abt.-Führer.
(Dienstgrad und Dienststellung)

Award document for the *Panzerkampfabzeichen* in Silver to *Unteroffizier* Karl Wenzel of the *4./Panzer-Abteilung (Funklenk) 302*.

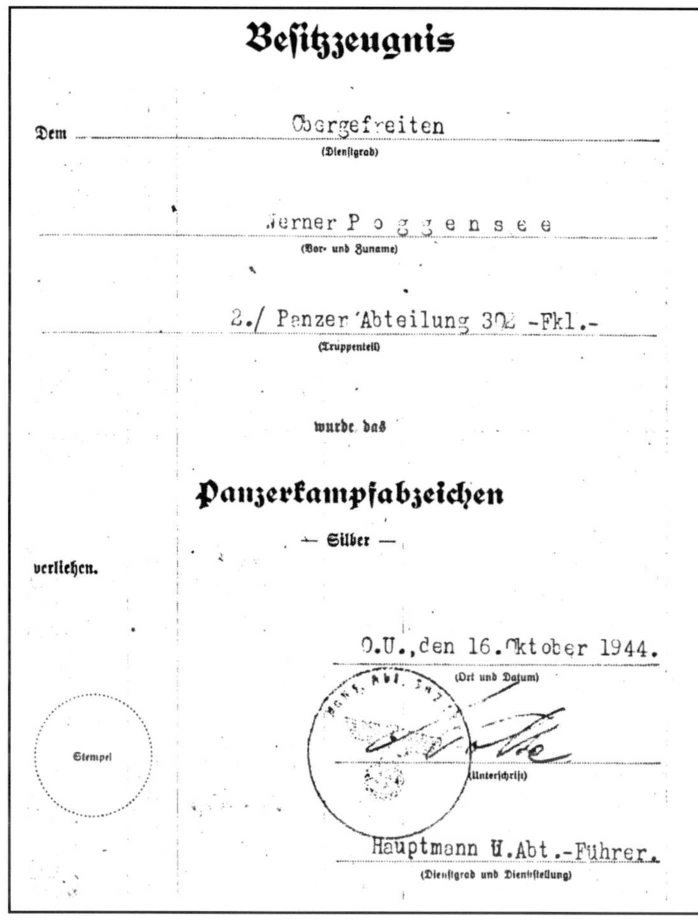

Besitzzeugnis

Dem Obergefreiten
(Dienstgrad)

Werner P o g g e n s e e
(Vor- und Zuname)

2./ Panzer Abteilung 302 -Fkl.-
(Truppenteil)

wurde das

Panzerkampfabzeichen

— Silber —

verliehen.

O.U.,den 16.Oktober 1944.
(Ort und Datum)

Stempel

(Unterschrift)

Hauptmann u.Abt.-Führer.
(Dienstgrad und Dienststellung)

Award document for the *Panzerkampfabzeichen* in Silver to Unteroffizier Werner Poggensee of the *2./Panzer-Abteilung (Funklenk) 302*.

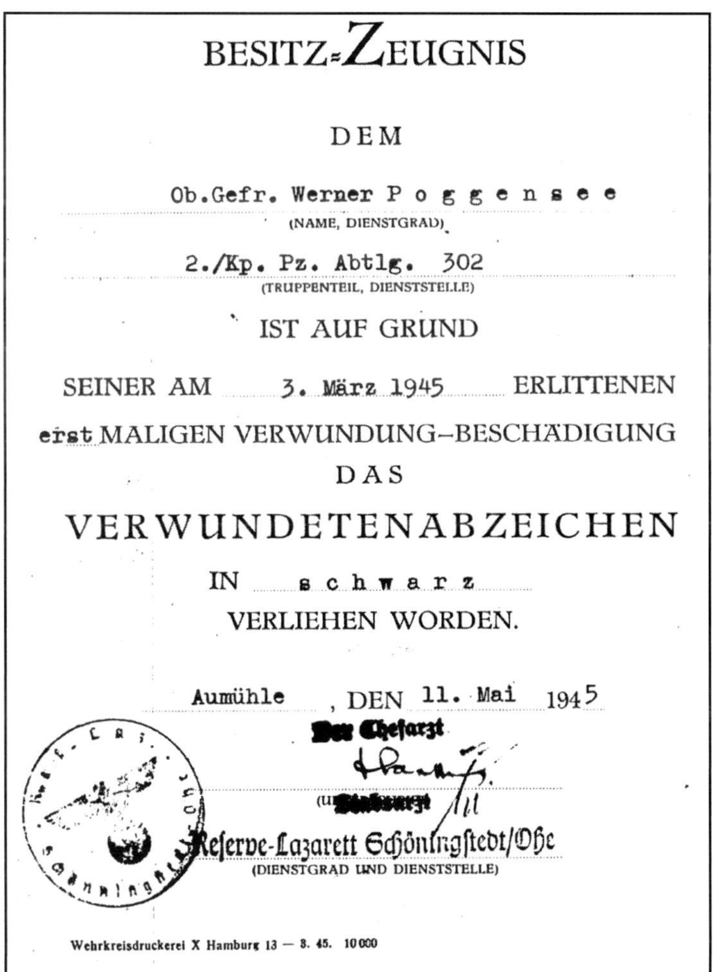

BESITZ-ZEUGNIS

DEM

Ob.Gefr. Werner P o g g e n s e e
(NAME, DIENSTGRAD)

2./Kp. Pz. Abtlg. 302
(TRUPPENTEIL, DIENSTSTELLE)

IST AUF GRUND

SEINER AM 3. März 1945 ERLITTENEN

erst MALIGEN VERWUNDUNG–BESCHÄDIGUNG

DAS

VERWUNDETENABZEICHEN

IN s c h w a r z

VERLIEHEN WORDEN.

Aumühle , DEN 11. Mai 1945

Der Chefarzt

Stabsarzt

Reserve-Lazarett Schöningstedt/Ohe
(DIENSTGRAD UND DIENSTSTELLE)

Wehrkreisdruckerei X Hamburg 13 — 3. 45. 10000

Award document for the Wound Badge in Black (first-time wound) to *Obergefreiter* Werner Poggensee of the *2./Panzer-Abteilung (Funklenk) 302*.

Panzer - Abteilung (Fkl) 302 (Stand 15.08.1944)

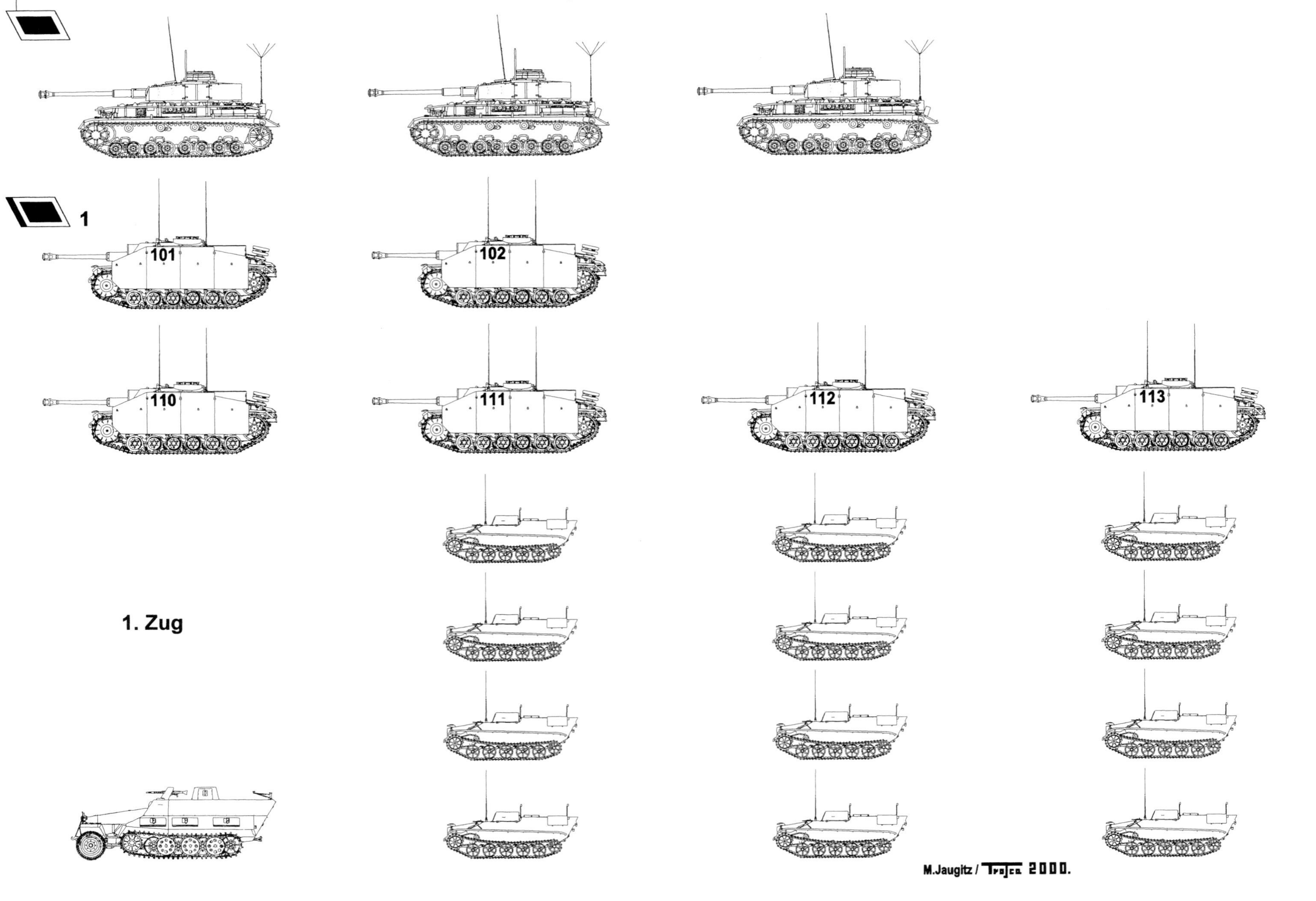

Panzer - Abteilung (Fkl) 302 (Stand 15.08.1944)

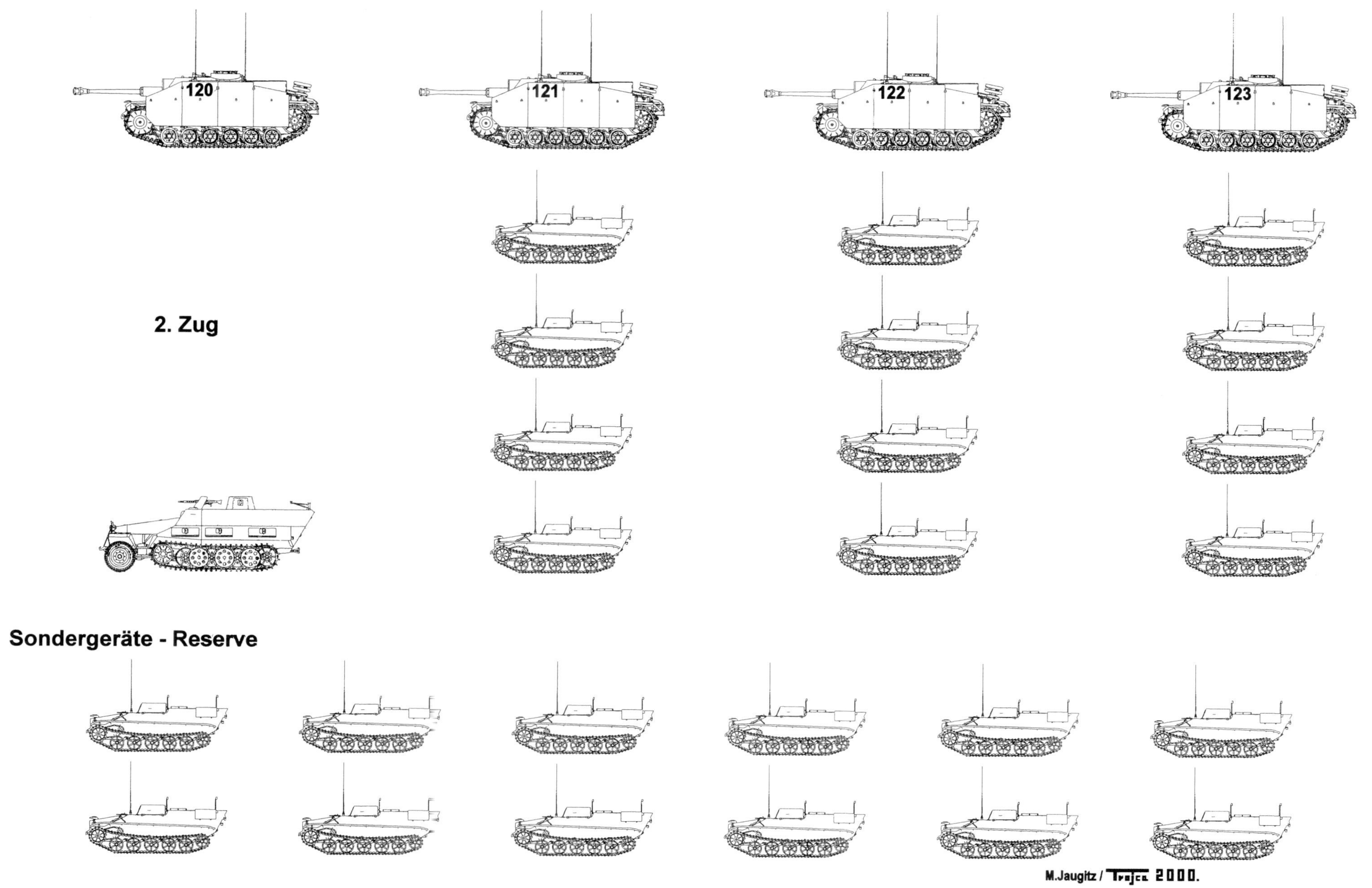

Panzer - Abteilung (Fkl) 302 (Stand 15.08.1944)

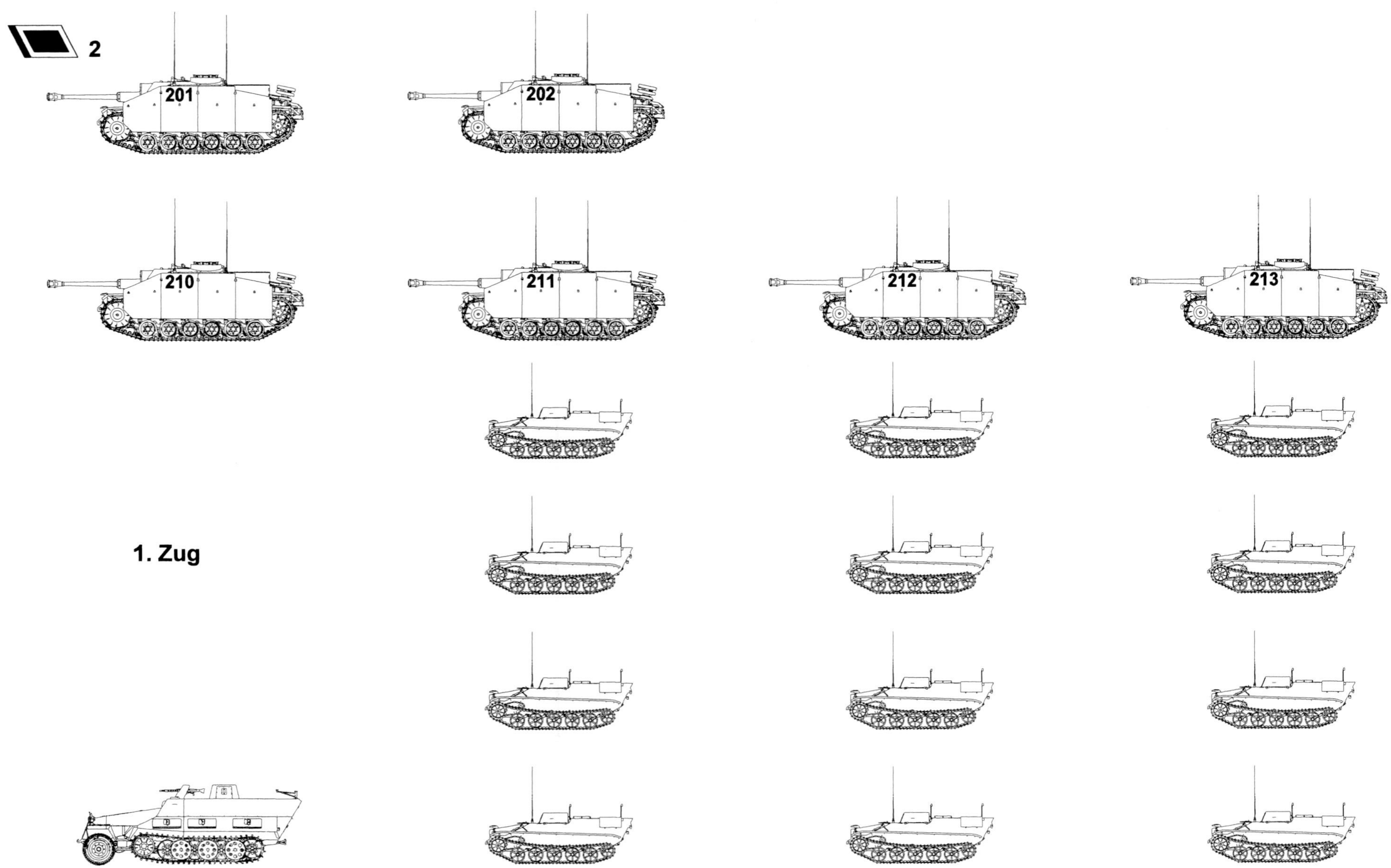

M.Jaugitz / Trojca 2000.

Panzer - Abteilung (Fkl) 302 (Stand 15.08.1944)

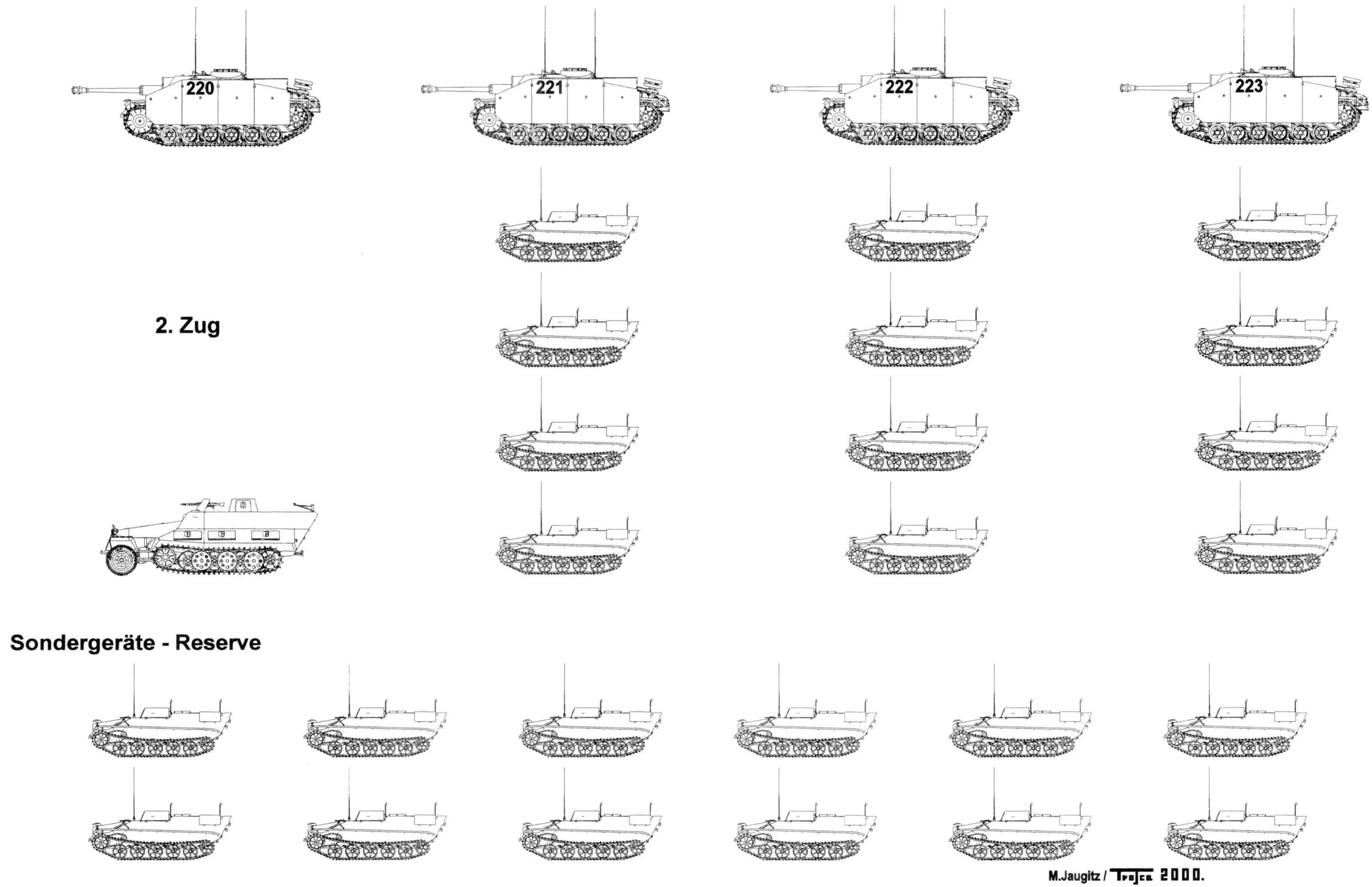

Panzer - Abteilung (Fkl) 302 (Stand 15.08.1944)

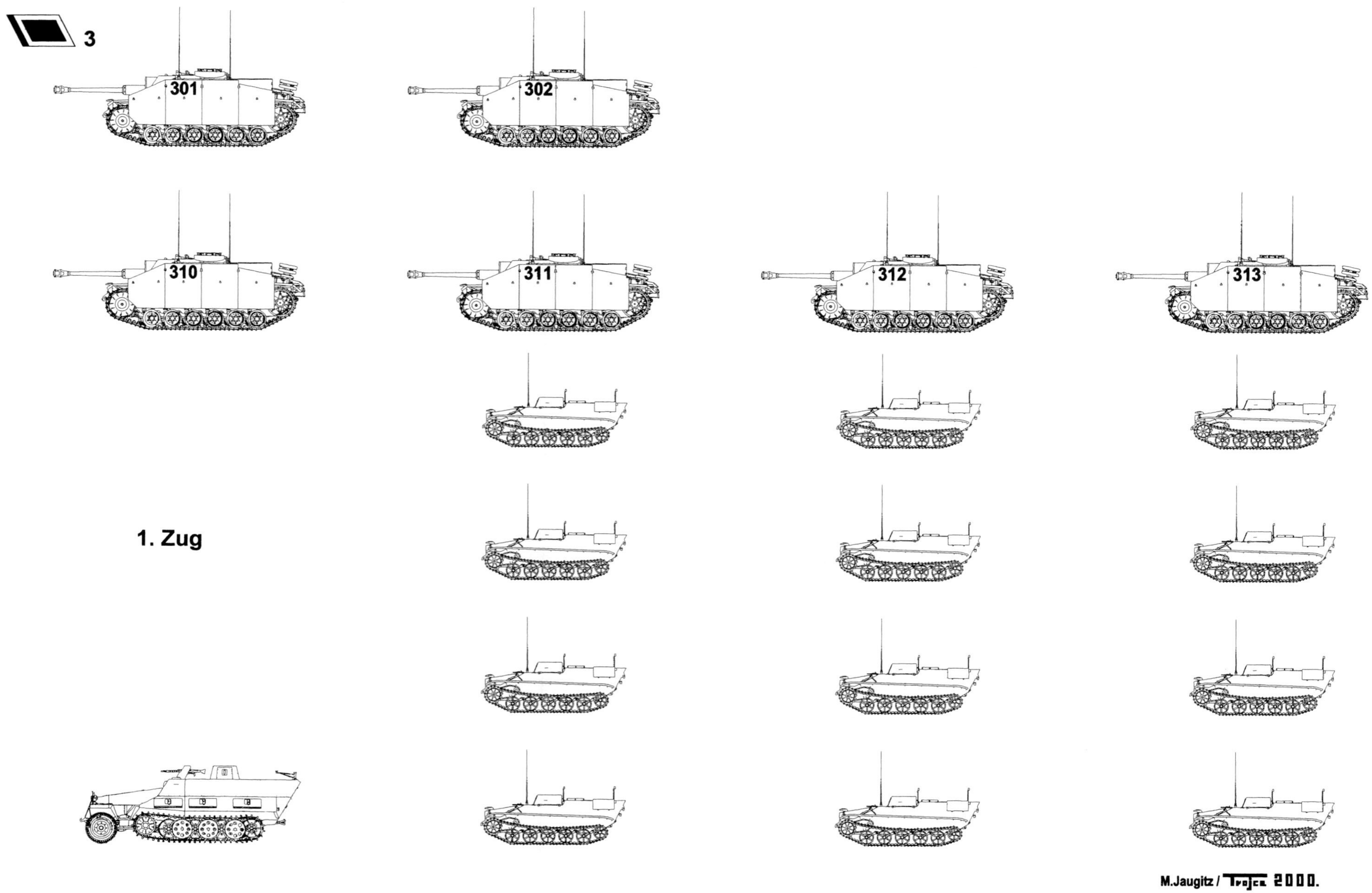

Panzer - Abteilung (Fkl) 302 (Stand 15.08.1944)

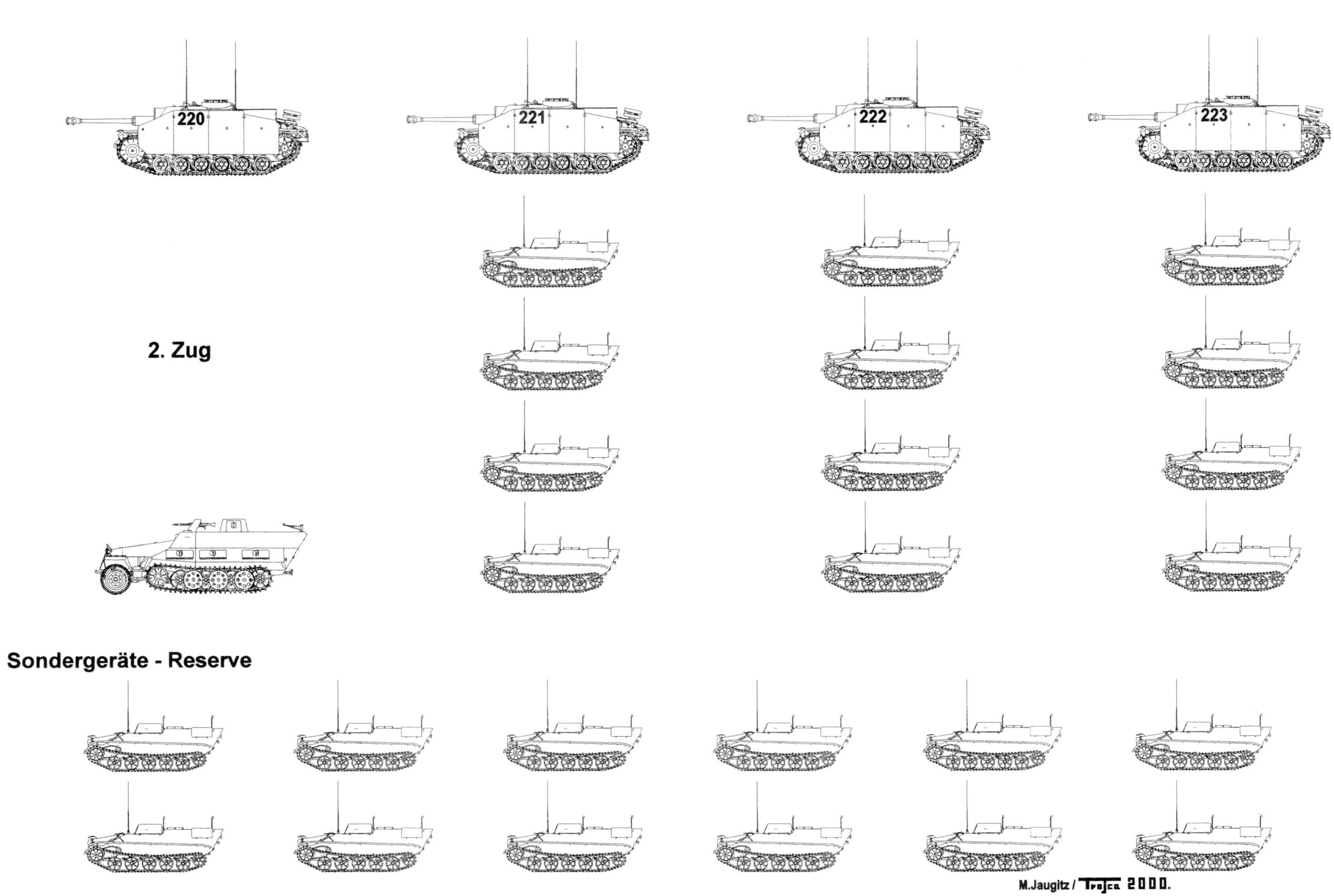

Just a few weeks after it was established, *Panzer-Abteilung (Funklenk) 302* was dispatched to the Eastern Front at the beginning of August 1944 where it was sent to assist in putting down the Warsaw Uprising. The assembly area of the battalion outside of Warsaw was well camouflaged. (Harbort)

A 3-ton *Opel Blitz* truck of the battalion. This type of truck with a built-up rear cab was frequently used by the signals mechanics as a calibration vehicle. (Tamine)

The mood among the *BIV* drivers is good. The battalion has been equipped with the new *Borgward BIV, Ausführung C.* (Müntz)

This page and next: The available time is used to maintain and repair vehicles. (Martens)

The members of the maintenance section are busy repairing a new *Borgward BIV, Ausführung C*. Portions of the drive train have been removed. Repair work is also being conducted within the vehicle itself. (Harbort)

Repair work on the steering brakes. (Harbort)

Repairs are also conducted on the *Sturmgeschütze*. (Harbort)

Cigarette break! (Harbort)

Above and above right: In the billets of the battalion maintenance platoon in the Warsaw suburb of Wlochy. (Harbort)

Right: Track links of different types have been added to the hull and superstructure to provide additional armor protection. (Fiedler)

Maintenance work on a *BIV*, *Ausführung C*. An access hatch over the gearbox has been opened. (Fiedler)

Above left: Additional demolition-charge carriers have been parked in the open. Note the mounting bracket for additional track links on the sidewall of the engine compartment. (Fiedler)

Above: A large tripod constructed out of wood serves as a provisional crane. Using the pulley, the engine-compartment cover is removed from the vehicle. (Harbort)

Left: The tank mechanic can now continue his work unimpeded. Access to the engine and drive train is made easier. (Harbort)

Engine repair! The amount of repair work on the vehicles increased during the operations in Warsaw. (Fiedler)

In lieu of the 3-ton prime mover, *Panzer-Abteilung (Fkl) 302* has already received a few *Sd.Kfz. 251, Ausführung D* halftracks to be used to move spare demolition charges. (Fiedler)

Unfortunately, no other information is available concerning this demolition-charge carrier. It is not known what purpose the tubes fastened on the sidewalls had. Note the modified exhaust system placement on the vehicle's rear. (Tamine)

The same vehicle from another perspective. (Fiedler)

A new operation starts. The engine of the demolition-charge carrier is started with a crank. (Jenckel)

Top right and below: The *BIV* driver and some combat engineers and signals mechanics check out the demolition-charge carrier one last time. (Paust)

Right: A radio-control section moves through the destroyed city to its intended attack position. (Jenckel)

Destroyed houses and craters are all that can be seen! (Jenckel)

The demolition-charge carrier follows its command and control vehicle. (Jenckel)

The attack position has been reached. The *BIV* driver activates the radio-control receiver and then climbs out of the *BIV*. (Jenckel)

After that, the gun commander starts the radio-controlled operation of the demolition-charge carrier. (Jenckel)

Detonation!

After the demolition-charge carrier has returned it receives a new demolition-charge container and is reloaded with the prepared charge. (Jenckel)

During breaks in operations less-experienced crews are trained on operating the command transmitter. (Jenckel)

A *BIV*, *Ausführung C* of the *2./Panzer-Abteilung (Fkl) 302*. (Jenckel)

Left: One of the three *Panzerbefehlswagen IV's* of the staff of *Panzer-Abteilung (Fkl) 302*. (Schatz)

Opposite page top: The combat elements of the battalion were billeted in the second half of August in the center of the city in the vicinity of the Saxony Palace. The *Leitpanzer* and the demolition-charge carriers moved to their operational areas from there. (Fiedler)

Opposite page bottom: A *Panzerbefehlswagen IV* and two assault guns of *Panzer-Abteilung (Fkl) 302*. (Fiedler)

Below: An unusual *Zimmerit* application can be seen on this *Panzerbefehlswagen IV*. (Fiedler)

Right: A frontal view of the *BIV*, *Ausführung C*. The radio-control frequency code of *NU* has been painted on the demolition-charge container. (Harbort)

Opposite page top: Another command and control assault gun moves forward. Only a few of the assault guns of the battalion had the *Saukopf* gun mantlet. (Harbort)

Opposite page bottom: A mascot has been affixed to the gun barrel of this assault gun of the *3./Panzer-Abteilung (Fkl) 302*. Harbort)

Right: Every block of houses is destroyed where rebels are suspected. (Harbort)

Left: In addition to the demolition-charge carriers, *Sturmpanzer IV "Brummbären"*, *Sturmtiger*, the 54 cm mortar *"Karl"* and heavy artillery and mortars are employed in the operation. (Fiedler)

The crew of assault gun *210*. (Harbort)

Opposite page, bottom left: During the German occupation "Pilsudki Square" was renamed as "Adolf Hitler Square". Pilsudski Square was the front grounds of the Saxony Palace. The rider monument is in the foreground; in the background can be seen the columned entryway and passage to the Saxony Gardens. (Harbort)

Opposite page, right: Taking a break underneath the rider monument. Vehicles have been parked in the niches of the entryway. In front of them is a *VW 82*. (Harbort)

Below and top of next page: A disabled *Sturmgeschütz* is loaded onto a lowboy (*Sonderanhänger 116*). (Werner)

Werner, the driver, in front of his *Sturmgeschütz* on the lowboy. This assault gun has been outfitted with steel return rollers. (Werner)

The move-out begins. The lowboy is being towed by an 18-ton prime mover (the final version of the *Sd.Kfz. 9*). (Werner)

An assault gun has slipped into the moat of an old fort during the fighting. (Werner)

A few *BIV* drivers and combat engineers of the *3./Panzer-Abteilung (Fkl) 302* wait for the next operation in a *Sd.Kfz. 251*. (Dubberstein)

Elements of the *3./Panzer-Abteilung (Fkl) 302* cross a bridge over the Weichsel while moving to another area of operations. (Dubberstein)

After reaching the assembly area the crews await their orders. (Dubberstein)

Numerals *301* and *322* can be seen on the rear of the vehicles. (Dubberstein)

The *Sturmgeschütze* of the battalion were frequently employed in support of infantry. (Winkler)

This *Sturmgeschütz* has been christened *Wartburg*, an association with its garrison city of Eisenach. Note the unusual machine-gun shield. (Werner)

An assault gun has advanced as far as a barricade. (Brandner)

The crew of this *Sturmgeschütz* takes a lunch break. Replacement track links have been added to the superstructure. (Jenckel)

The mounting for the radio-control antenna has been affixed to this assault gun, however, the antenna itself is missing. A piece of metal has been added to protect the gunner's sights.(Jenckel)

Get ready to move out! (Jenckel)

An assault gun works its way through the Warsaw city center. Extreme alertness was demanded since there were sharpshooters everywhere. (Harbort)

By the end of August 1944 the resistance in the old city of Warsaw has been crushed. (Fiedler)

Polish *Hiwis* assist in loading up this assault gun with ammunition. (Hitzfeld)

The assault guns of the *4./Panzer-Abteilung (Fkl) 302* had an unusual, meandering camouflage scheme. One can see the numeral on the next-to-last side skirt. The last vehicle is a *Sturmpanzer IV "Brummbär* of *Sturmpanzer-Kompanie z.b.V. 218*. (Jenckel)

The elimination of the Polish resistance in the suburbs of Mokotow and Zoliborz started in the first half of September 1944. (Wilhelm)

Above and 2 photos opposite page: This *BIV* has apparently become disabled through a mechanical problem. The demolition-charge container is still on the vehicle. Combat engineers and flamethrower teams advance through artificial smoke. (Wilhelm)

Above: A *BIV* is guided forward via radio control. (Wilhelm)

Opposite page top: *Leitpanzer 112* stops at a road obstacle and covers the advance of the infantry. (Wilhelm)

Opposite page bottom: Smaller road obstacles cannot stop the assault guns. (Wilhelm)

Right: The *Leitpanzer* follows at a safe distance. (Wilhelm)

An assault gun has bottomed out in a crater. *Leitpanzer 413* assists in its recovery. (Münch)

This *BIV*, *Ausführung C* has dropped its demolition charge. When it automatically pulled back, it backed into a street lamppost. (Leon)

Radio-control operations were conducted until the rebels surrendered on 2 October 1944. (Fiedler)

This page and opposite page top: After the combat phase of the operations in Warsaw ended, the battalion was transported by rail to the East Prussian border area and offloaded at Tilsit. In the meantime, the battalion had been redesignated as *Panzer-Abteilung 302 (Fkl)*. The armored vehicles were checked out and prepared for future operations. (Fiedler)

During the fighting from 18-25 October 1944 the assault guns were the primary combat vehicles employed. The demolition-charge carriers were staged in areas in the woods. (Müntz)

Above: A 1-ton prime mover of the *2./Panzer-Abteilung 302 (Fkl)*. (Tamine)

Above: Camouflaged in an ambush position! (Esser)

Right: A short break in the action at a farmstead! (Fiedler)

The assault guns were employed continuously. The *2./Panzer-Abteilung 302 (Fkl)* had painted over its old tactical insignia — a parallelogram and circle. This company also had a few *Sturmgeschütze* with the *Saukopf* gun mantlet. (Fiedler)

Behind the command and control vehicle is a *BIV*. The *BIV* lacks its track on the right-hand side. To the left in the picture is a 15 cm *sIG auf GW 38t "Bison"*. (Fiedler)

Above and opposite page top: The personnel of the maintenance platoon get the opportunity to repair damage on the armored vehicles. (Fiedler)

Occasional counterattacks were launched after the initiation of the Russian offensive on 13 January 1945. (Fiedler)

After the *Sturmgeschütze* were lost, the crews were employed as infantry. (Esser)

2.2.15: The 4./Panzer-Abteilung (Fkl) 301

As part of the formation of *Panzer-Abteilung (Fkl) 302* it was intended that the *4./Panzer-Abteilung (Fkl) 301* would be removed from that unit and incorporated into the new battalion. The commander of *Panzer-Abteilung (Fkl) 301*, *Hauptmann* von Abendroth, suggested using the *1./Panzer-Abteilung (Fkl) 301* instead of the *4./Panzer-Abteilung (Fkl) 301*. However, this idea was rejected by higher headquarters.

Thus, when *Panzer-Abteilung (Fkl) 301* was transported to Lemberg to join *Heeresgruppe Nordukraine* on 5 June 1944, the *4./Panzer-Abteilung (Fkl) 301* stayed behind at its garrison in Bertangles near Amiens.

Oberleutnant Witter remained company commander, while the 1st Platoon was led by *Leutnant* Hamann and the 2nd Platoon by *Leutnant* Glässel.

After the Allied landing in Normandy, the *4./ Panzer-Abteilung (Fkl) 301* was moved on 9 June 1944 via Amiens — Paris — Melun — Alencon to Belun (Etampes), the unit arriving on 14 June. As the Allies enjoyed total air superiority, the transfer had to be conducted at night and it was not possible to move the entire unit at one time. For these reasons it was not until 16 June that all elements of the company arrived in the new billeting area.

During the night of 20 June 1944 positions were occupied north of St. Martin de Fontenay. After heavy enemy artillery bombardments the positions four kilometers south of Caumont were abandoned.

On 21 June 1944 the company received orders to move into the Reims area for the purpose of consolidating with *Panzer-Abteilung (Fkl) 302*, then being established. The unit moved out on 28 June but, on 30 June, the order was rescinded and the company, by then in the Belun area, was ordered back into the combat zone.

As a result of battalion organizational changes, the company's incorporation into *Panzer-Abteilung (Fkl) 302* was no longer necessary; it was turned into a separate remote-control unit. The designation *4./Panzer-Abteilung (Fkl) 301* was retained.

In the first half of July 1944 the company took part in the construction of a blocking position in the Jurques — Villers Bocage — St. Martin area. On 26 July the unit moved into its new quartering area and was attached to the *9. SS-Panzer-Division "Hohenstaufen"*.

At the end of July the US 1st Army breached the German front line in the St. Lô area. Soon afterwards the British attacked near Caumont in the direction of Falaise.

The company carried out its first *BIV* mission between St. André-sur-Orne and May-sur-Orne at approximately 0700 hours on the morning of 31 July 1944. One platoon was committed against enemy positions near Hill 72, the other against the town of Verriéres. While a few demolition-charge carriers were guided into the most forward enemy positions southeast of Verriéres and the town and were detonated, several of the *BIV's* committed against Hill 72 broke down and were left on the battlefield. Unfortunately, the operation did not achieve the hoped-for success and the company was released from its attachment to the *9. SS-Panzer-Division "Hohenstaufen"* on 1 August 1944.

In order to prevent the undamaged vehicles from falling into enemy hands in the area of Hill 72, a detachment was formed which blew them up during the night of 2 August 1944. The participants in this operation were decorated by a unit of the *1. SS-Panzer-Division "Leibstandarte Adolf Hitler"* located nearby.

On 7 August 1944 the *4./Panzer-Abteilung (Fkl) 301* was committed in support of the *89. Infanterie-Division*. The first casualties were recorded during an operation near Bretteville — Potigny on 8 August. The leader of the 1st Platoon, *Leutnant* Hamann, was captured. In addition, the gun crews suffered fatal casualties. On 9 and 10 August the company participated in attacks by the *12. SS-Panzer-Division "Hitlerjugend"*.

Gunter Hoffmann (†), then an *Unteroffizier*, described this action:

> Our company's assault guns and demolition-charge carriers were positioned in the main line of resistance. I was the driver of one of the *BIV's* and was located approximately 300 meters behind the command and control assault gun with my vehicle, which was well camouflaged. It was intended to employ my vehicle. Our engineers had already armed the vehicle, which meant that the detonators had been inserted into the charge. One was for the command "Total Ignite" and the other for the radio command "Jettison". In addition, two detonators had been installed on the retaining cables for the purpose of severing them, and the safety bolts had been removed from the charge retaining arms.
>
> Filled with anticipation, my gaze remained fixed on my control tank. My commander flashed a green light, the signal to begin the operation.
>
> My vehicle and I shot through the terrain at maximum speed, and I stopped in front of my command and control vehicle, 150 meters from the main line of resistance. In front of me, approximately 800 meters away, I saw the armada of enemy tanks. I set the safety switch to 100 meters and pushed the high-frequency switch forward. My vehicle reacted immediately to the radio signal "Forward". Now I had to get out as quickly as possible. In a matter of seconds, the vehicle reached 60 kilometers an hour. I tumbled into the mud and saw my *BIV* racing toward the enemy.
>
> I sought cover on my command and control vehicle. The commander was standing in the turret hatch, the command set hanging in front of his chest. I was able to see my *BIV* clearly as it raced along on a zigzag course. Such a vehicle is difficult to hit. The target was a Sherman tank. Twenty meters to the right of the Sherman was a command vehicle, a half-track. My *BIV* stopped between the two vehicles, the "Jettison" command was issued, the retaining cables were blown and the Ekrasit charge dropped.
>
> At the same instant I switched my vehicle into reverse gear and rushed back at maximum speed. About five seconds later we saw a huge explosion. The Sherman was left lying on its side and all that could be seen of the command vehicle was burning wreckage. Meanwhile my faithful *BIV* had arrived back with us. The vehicle had been hit numerous times by machine-gun fire. Under heavy enemy fire I jumped into my vehicle. Once I reached the rear position it was fitted with a new demolition charge. During the move back I realized that I was sitting on something hard. It turned out to be an English hand grenade which was not armed. My vehicle saw action two more times that day. On the last operation, I issued the command "Total Ignite" which destroyed my faithful vehicle.

On 10 August the company moved through Falaise to

Livarot, where the trains were ambushed by an English patrol. The combat elements were encircled on 19 August. On 26 August the Seine was crossed at the village of La Maillerraye-sur-Seine north of Rouen. The retreating unit was attacked by partisans while moving through Yverville three days later.

The retreat continued through Amiens, Albert, Bapaume, Cambrai, Charleroi, Namur, Hasselt, Sittart and Nimwegen. The *4./Panzer-Kompanie (Fkl) 301* arrived in Arnhem on 7 September 1944. After a brief stay there, the unit entrained in Arnhem on 15 September and was transported via Wesel to Eisenach for rest and refit.

Following arrival at the garrison of *Panzer-Versuchs- und Ersatz-Abteilung 300 (Fkl)*, the members of the company were sent on leave. When they returned to their unit at the beginning of October 1944, training resumed at the Berka Troop Training Area and the garrison at Eisenach.

In the weeks that followed, company personnel were assigned additional special duties such as building barracks and providing an honor guard at the swearing-in of *Volkssturm* units.

On 2 January 1945 the company received orders to proceed to the Grafenwöhr training area, where it was to become the *1./Panzer-Abteilung (Fkl) 303*, a new unit then being established. Once renamed, the *4./Panzer-Abteilung (Fkl) 301* was disbanded.

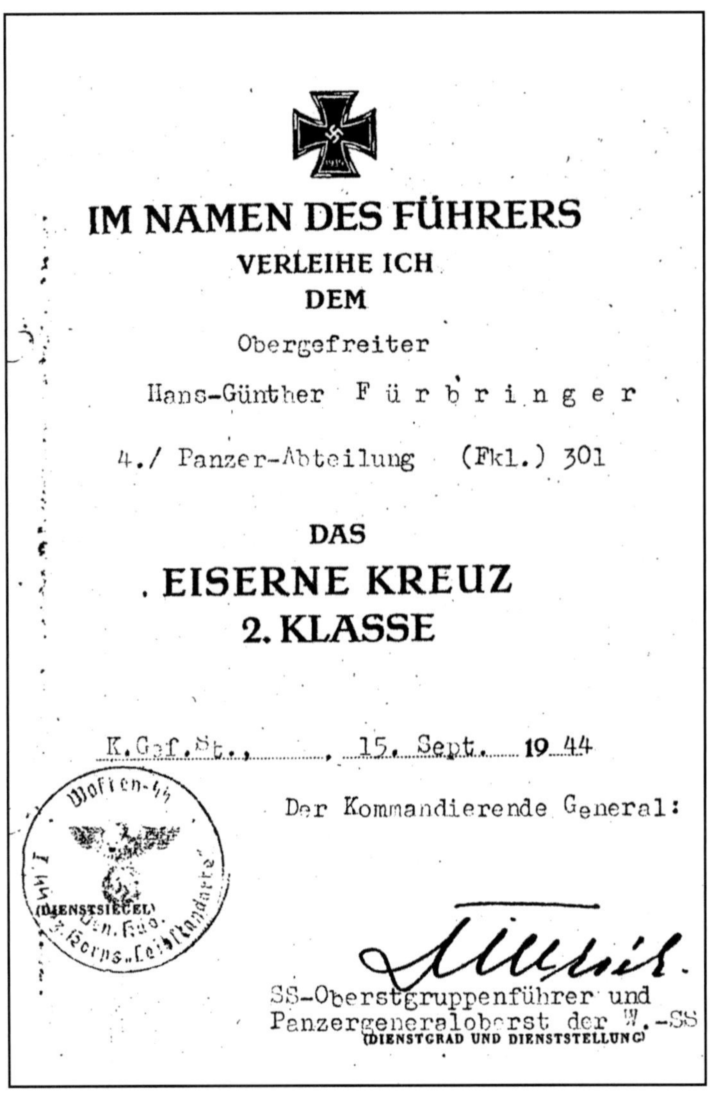

IM NAMEN DES FÜHRERS
VERLEIHE ICH
DEM
Obergefreiter
Hans-Günther Fürbringer
4./ Panzer-Abteilung (Fkl.) 301
DAS
EISERNE KREUZ
2. KLASSE

K.Gef.St., 15. Sept. 1944

Der Kommandierende General:

(DIENSTSIEGEL)

SS-Oberstgruppenführer und
Panzergeneraloberst der W.-SS
(DIENSTGRAD UND DIENSTSTELLUNG)

Besitzzeugnis

Dem Obergefreiten
(Dienstgrad)
Hans-Günther Fürbringer
(Vor- und Familienname)
4./Pz.Abt.(Fkl) 301
(Truppenteil)
wurde das
Panzerkampfwagenabzeichen
in Silber
verliehen.

Rgt.St.Qu., den 30.Okt.1944
(Ort und Datum)
(Unterschrift)
SS-Sturmbannführer u.
Regiments-Kommandeur.
(Dienstgrad und Dienststellung)

2223 a. Kroll & Straus, Berlin SO 36

Two award documents for *Obergefreiter* Hans-Günther Fürbringer of the *4,/Panzer-Abteilung (Funklenk) 301*. **Left**: The Iron Cross, Second Class (Note that it is signed by *SS-Oberstgruppenführer* Sepp Dietrich.) **Right**: The *Panzerkampfwagenabzeichen* in Silver.

Gliederung der Panzer - Abteilung (Fkl) 301 (Stand 07.02.1944)

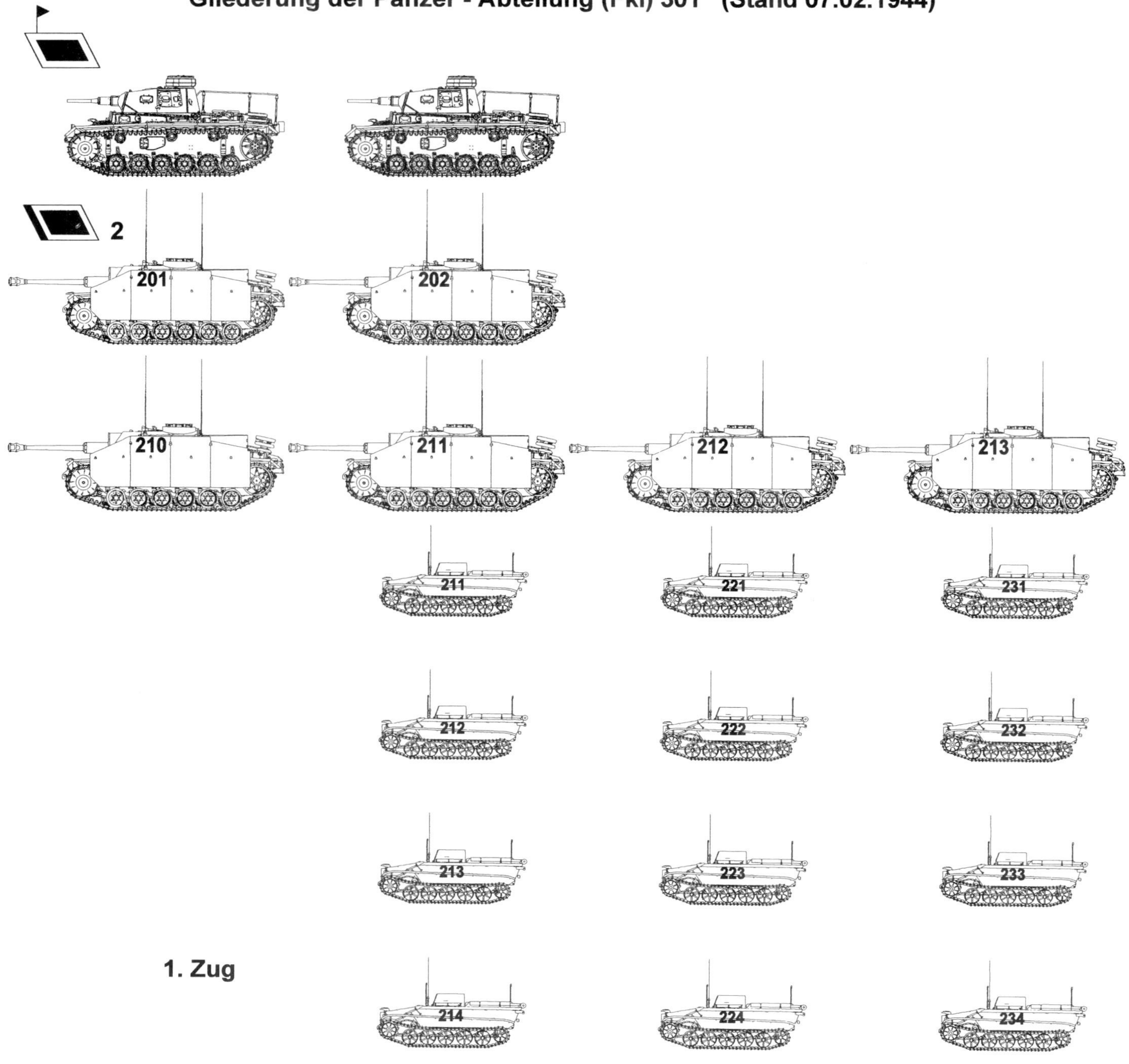

Gliederung der Panzer - Abteilung (Fkl) 301 (Stand 07.02.1944)

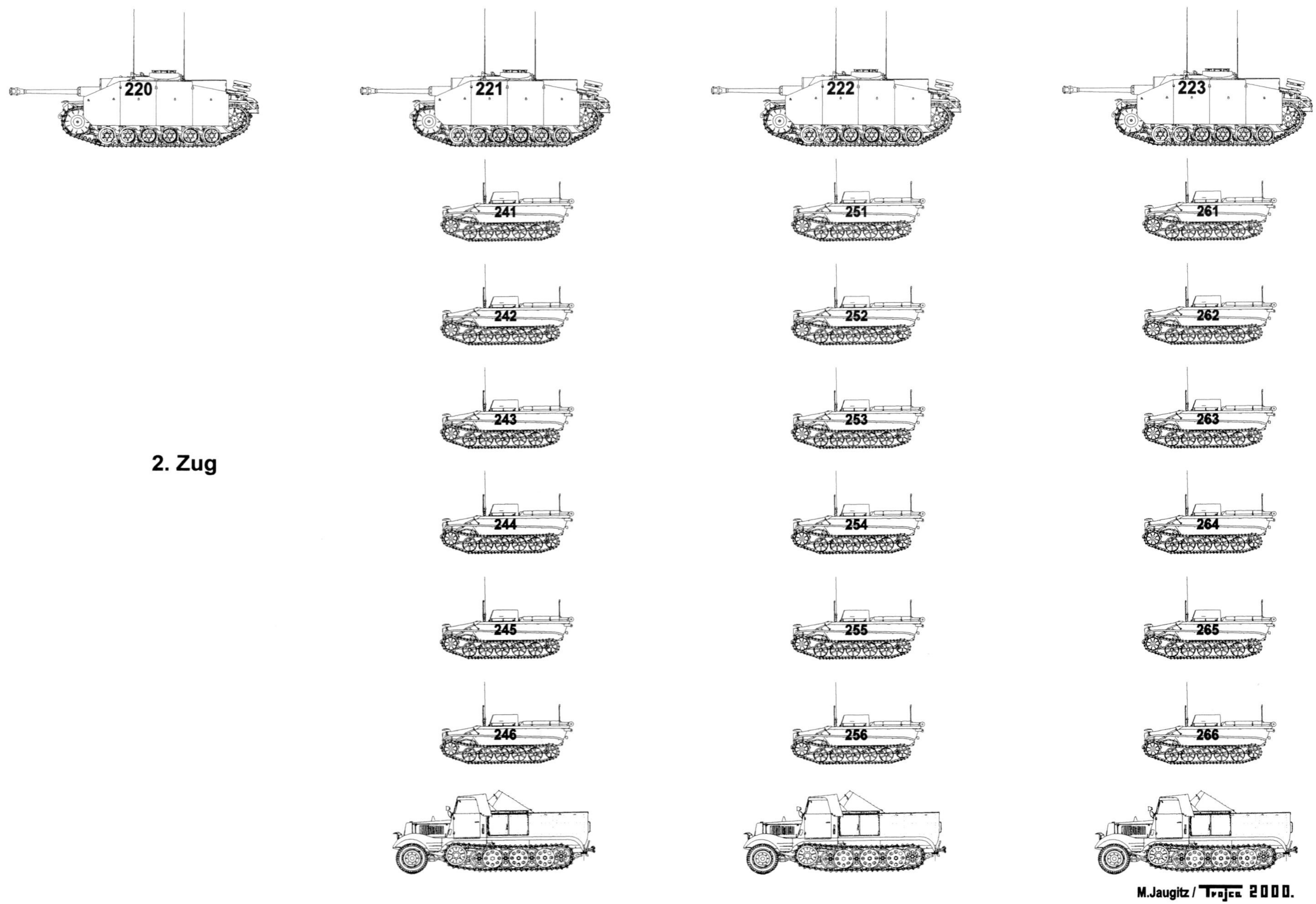

Gliederung der Panzer - Abteilung (Fkl) 301 (Stand 07.02.1944)

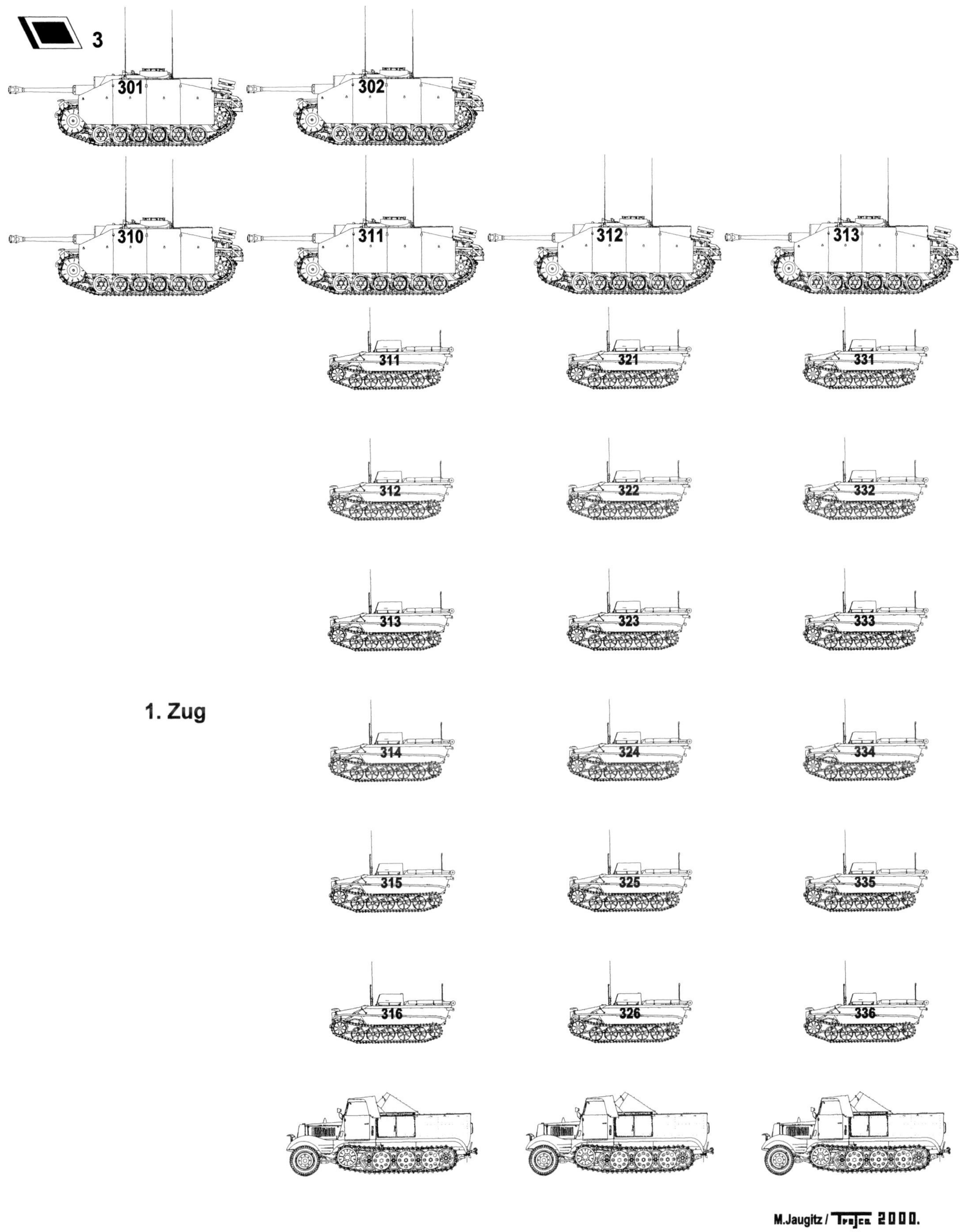

Gliederung der Panzer - Abteilung (Fkl) 301 (Stand 07.02.1944)

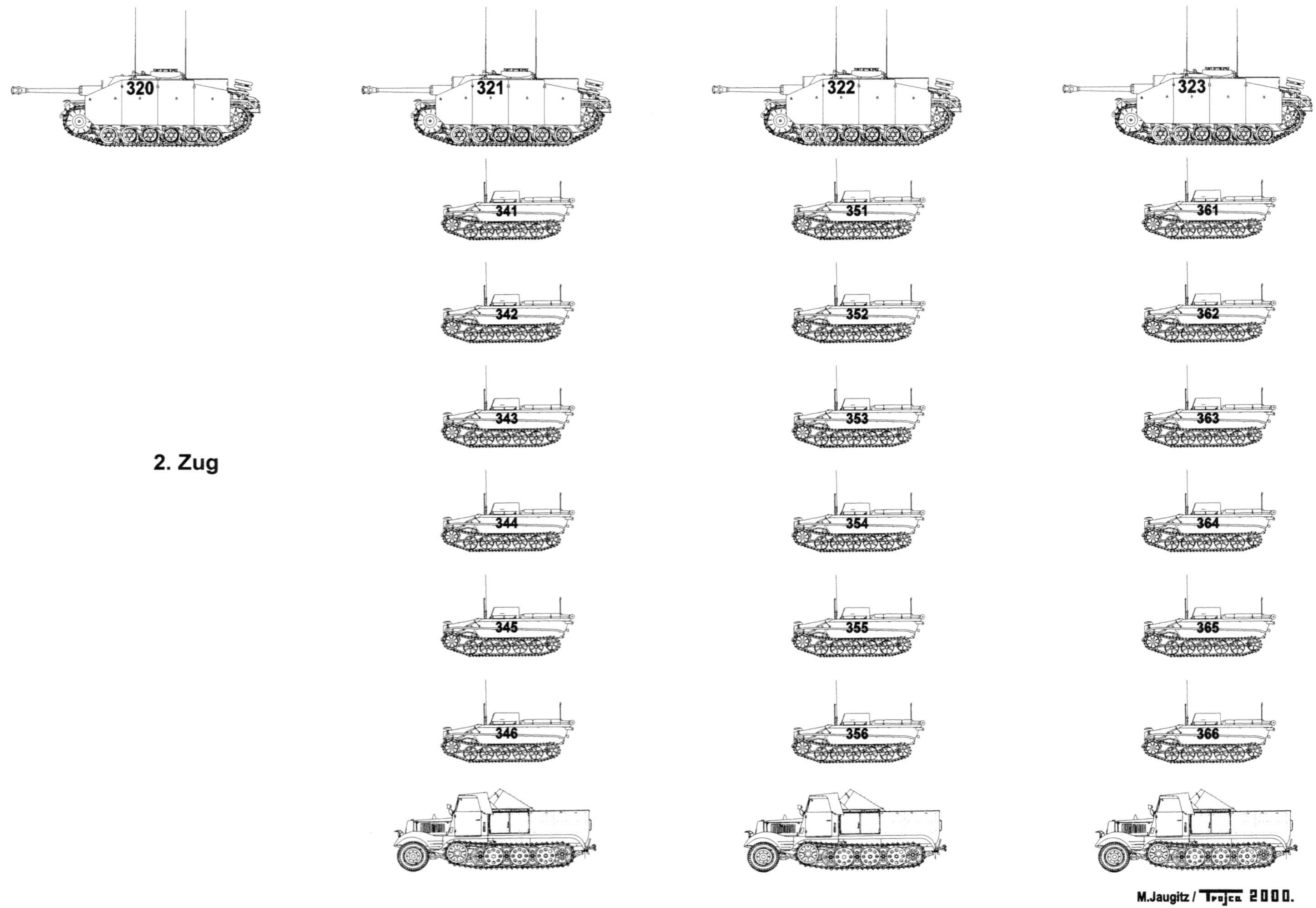

Gliederung der Panzer - Abteilung (Fkl) 301 (Stand 07.02.1944)

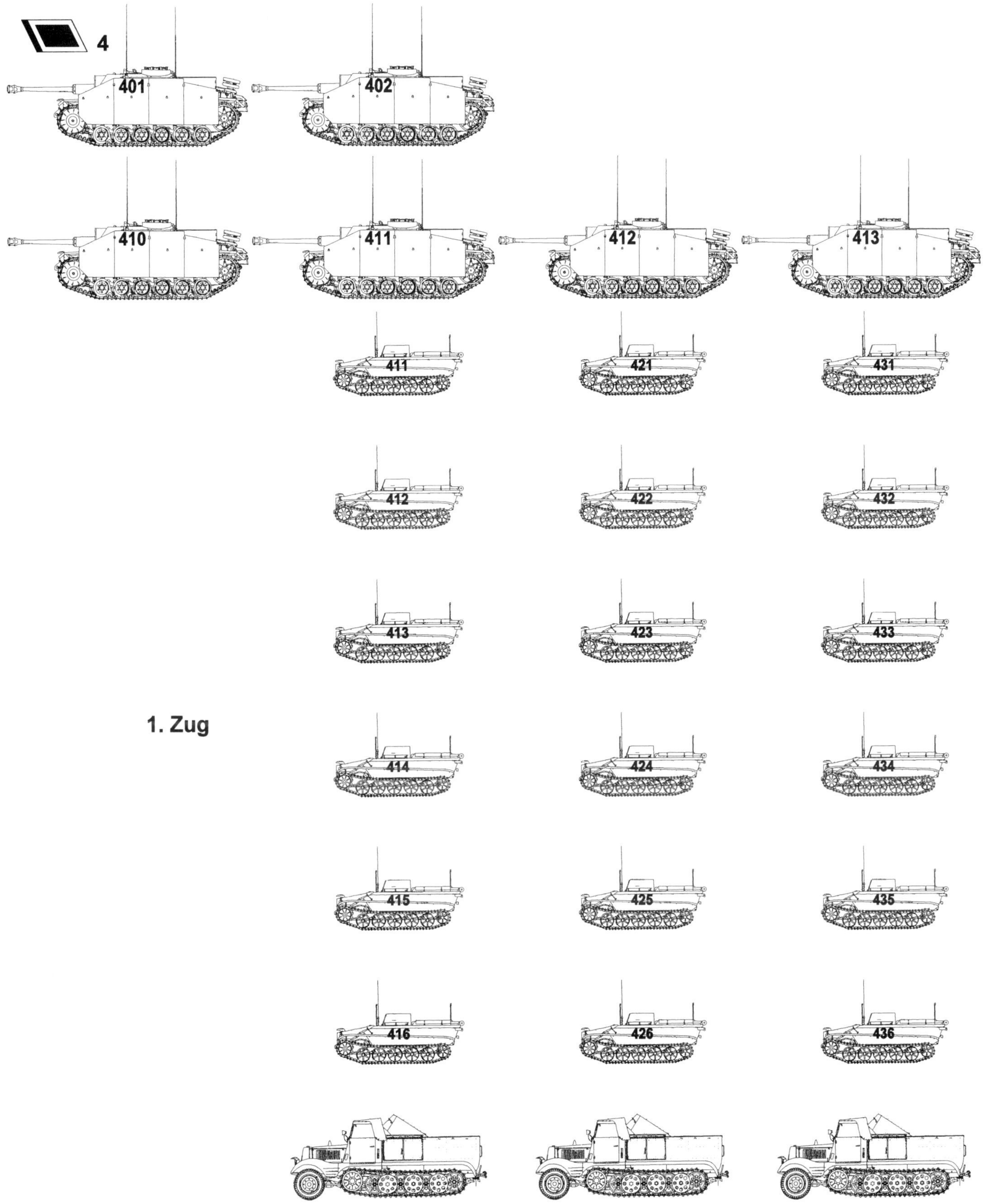

M.Jaugitz / Trojca 2000.

Gliederung der Panzer - Abteilung (Fkl) 301 (Stand 07.02.1944)

2. Zug

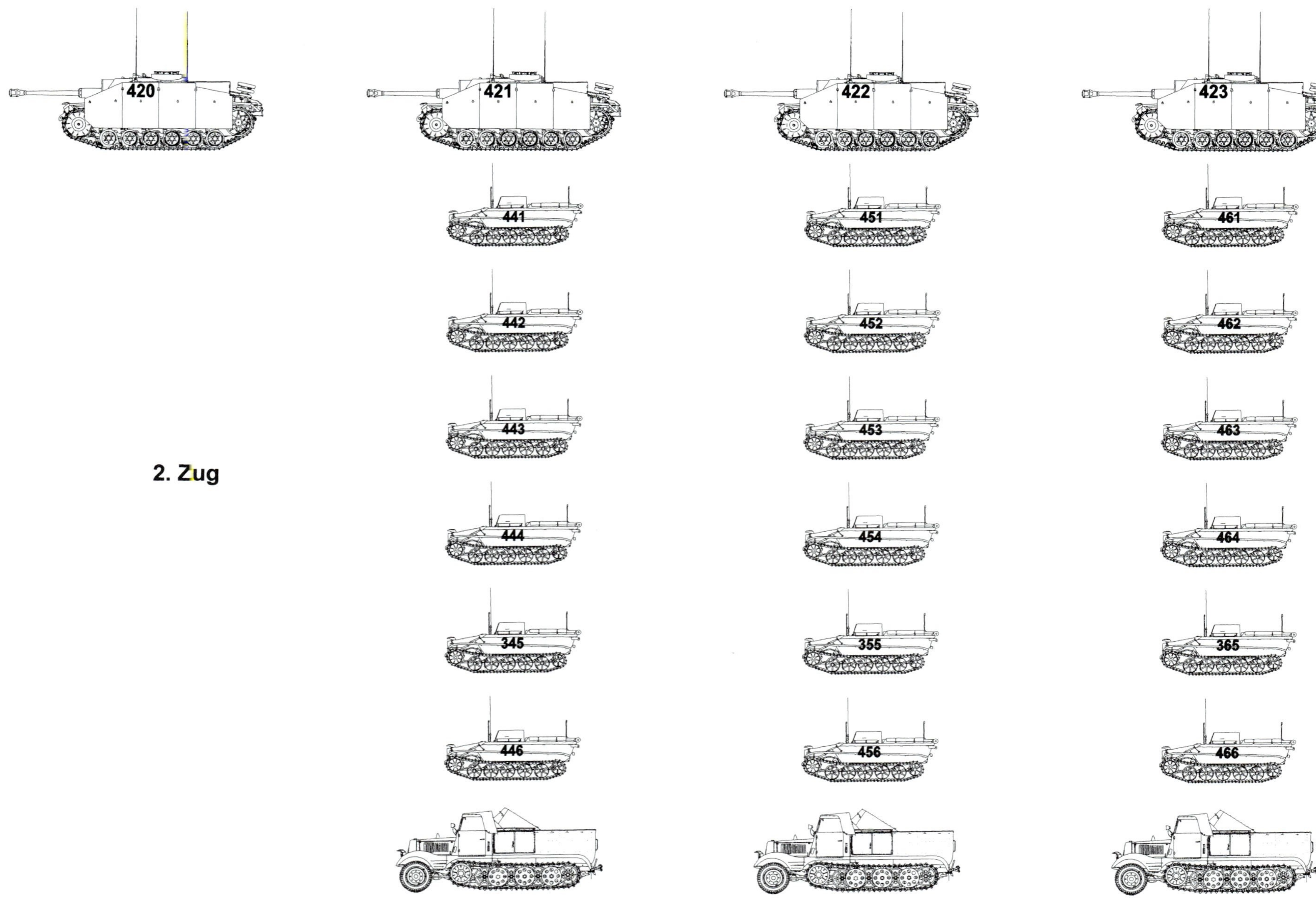

M.Jaugitz / Trojca 2000.

A *BIV* and a 3-ton prime mover of the *4./Panzer-Abteilung (Fkl) 301* in the area of Caumont. (Fürbringer)

The company was placed under the operational control of the *9. SS-Panzer-Division "Hohenstaufen"* on 26 July 1944. A member of the division poses next to a demolition-charge carrier. A few commanders of the *Waffen-SS* showed great interest in radio-control operations and requested similar formations be constituted within the *Waffen-SS*. (König)

Operations with the *89. Infanterie-Division* took place on 7 August 1944. Here: *Sturmgeschütz III 421*. This vehicle has also been equipped with a deep-water exhaust stack. (ECPA DAT 2695 L 09 and 10)

Demolition-charge carriers were employed again on 9 and 10 August 1944. On those dates they were with the *12. SS-Panzer-Division "Hitlerjugend"*. Note the tactical insignia — a parallelogram and a *4* — on the forward fender of the assault gun. (ECPA DAT 2695 L 11 and 16)

2.2.16: Panzer-Kompanie (Fkl) 319

Following the defeat of the German forces in the battles in northern France in summer 1944, many *Wehrmacht* units retreated to the German border, most of them in disarray. The armored and motorized units had lost the bulk of their equipment in countless battles and in the flight from the Allies.

In Germany new units were formed for the purpose of halting or delaying the Anglo-American advance.

One such unit was *Panzer-Kompanie (Fkl) 319*, which was formed by the *Panzer-Versuchs- und Ersatz-Abteilung 300 (Fkl)* on order of the *AHA* (*I Nr. 39923/44 geheim*, dated 28 August 1944). The company, which was created by combining personnel from the *2.*, *3.* and *4./Panzer-Versuchs- und Ersatz-Abteilung 300 (Fkl)*, was commanded by *Oberleutnant* Schlüter.

The number assigned to the company did not follow the established sequence as there already was a *Panzer-Kompanie 318*, although it was not a remote-control unit. The company was a separate unit equipped with older tanks and employed on the Eastern Front in the anti-partisan role. It was intended to equip the unit with captured T 34 tanks.

After *Panzer-Kompanie (Fkl) 319* was hastily equipped with vehicles, it entrained in Eisenach on 28 August 1944 and was transported through Gießen and Koblenz west along the Mosel River. Meanwhile, the situation at the front in northern France had changed again. The Allies had breached the Meuse — Mosel line established by the Germans by 31 August and were driving northeast. Brussels, the Belgian capital, was lost on 3 September.

Panzer-Kompanie (Fkl) 319 was redirected and sent back to the *LXXXI. Armee-Korps* in the Liège area where, together with other German armored units, it was to oppose the advancing Allied troops and blow vital bridges.

On 7 September US forces breached the German front line near Hasselt and advanced through the Meuse Valley in the direction of Liège. *Panzer-Kompanie (Fkl) 319* was committed immediately, suffering its first casualties. An assault gun was lost on 8 September. *BIV* demolition-charge carriers were also used to destroy bridges over the Meuse between Liège and Verviers.

Together with the battered *116. Panzer-Division*, the remote-control unit withdrew into the Aachen area. After six weeks of operations, the remnants of the company were moved back to Sindorf near Elsdorf for rest and expedited maintenance. For a time, consideration was given to incorporating the company into the *116. Panzer-Division*.

At the beginning of October the unit had four serviceable assault guns and 14 demolition-charge carriers. The *Oberquartiermeister West* in Wiesbaden made available 43 *BIV's*, however, only 23 could be made operational. After delivery to the unit, the vehicles' radio and mechanical systems were supposed to be checked in the reconstitution area. On 24 October the company was allocated four new *Sturmgeschütz III's* which reached the unit on 9 November.

On 26 October 1944 *Leutnant* Keitel assumed command of *Panzer-Kompanie (Fkl) 319*. Platoon leaders remained unchanged, with *Leutnant* Schmidt (1st Platoon) and *Leutnant* von Hoven (2nd Platoon). The remote-control officer was *Leutnant* Lindmüller.

By the beginning of November the company's complement of armored vehicles had risen to nine serviceable *Sturmgeschütz III's* and 26 *BIV's*. Furthermore, 10 more *BIV's* were undergoing short-term repairs.

On 4 November 1944 *Leutnant* Keitel informed the *LXXXI. Armee-Korps* that the remote-control unit would be fully operational on 11 November 1944.

Following lengthy preparations, the Americans launched a major offensive on 16 November 1944, achieving a three- to four-kilometer-wide penetration at Immendorf. Geilenkirchen was lost on 18 November, after which the US forces advanced toward Würselen and Stolberg.

On 19 November *Panzer-Kompanie (Fkl) 319* was moved back into the combat zone and attached to *Panzer-Abteilung (Tiger/Fkl) 301*. Together the two elements were supposed to establish an antitank position between Barmen and Bourheim on 20 November. On the 21st, following a shortening of the front, there was fighting near Merzenhausen in which several assault guns were heavily damaged. The company also recorded seven killed on this day. On 23 November radio-controlled tanks were successfully used operationally for the first time.

The days that followed witnessed further combat operations in the Eschweiler — Linnich — Jülich area. However, *Panzer-Kompanie (Fkl) 319* only saw action in the conventional armor role. After twelve days of fighting, the American forces had driven 15 kilometers into the German defenses by 28 November. In the days that followed, the Allies gained further ground between Inde and Ruhr in spite of heavy losses. On 29 November, the company commander, *Dr.* Keitel, recently promoted to *Oberleutnant*, led an attack against Lochberg and entered the town. However, lacking infantry support, the assault guns were soon forced to retire.

After the start of the Ardennes Offensive there was a halt in the fighting on the Ruhr front. There were further withdrawals to the east. The damaged assault guns were repaired and on 8 January 1945 the company had five assault guns and 34 demolition-charge carriers on hand.

On 14 January 1945 the Army General Staff ordered *Panzer-Kompanie (Fkl) 319* be integrated into *Panzer-Abteilung 303 (Fkl)*. For this purpose the company was to proceed to the Grafenwöhr training area and not Eisenach.

The company entrained in Horrem near Cologne at the beginning of February 1945 for transport to Grafenwöhr. With its incorporation into *Panzer-Abteilung (Fkl) 303*, *Panzer-Kompanie (Fkl) 319* was disbanded.

Panzerkompanie (Fkl) 319 (Stand 01.09.1944)

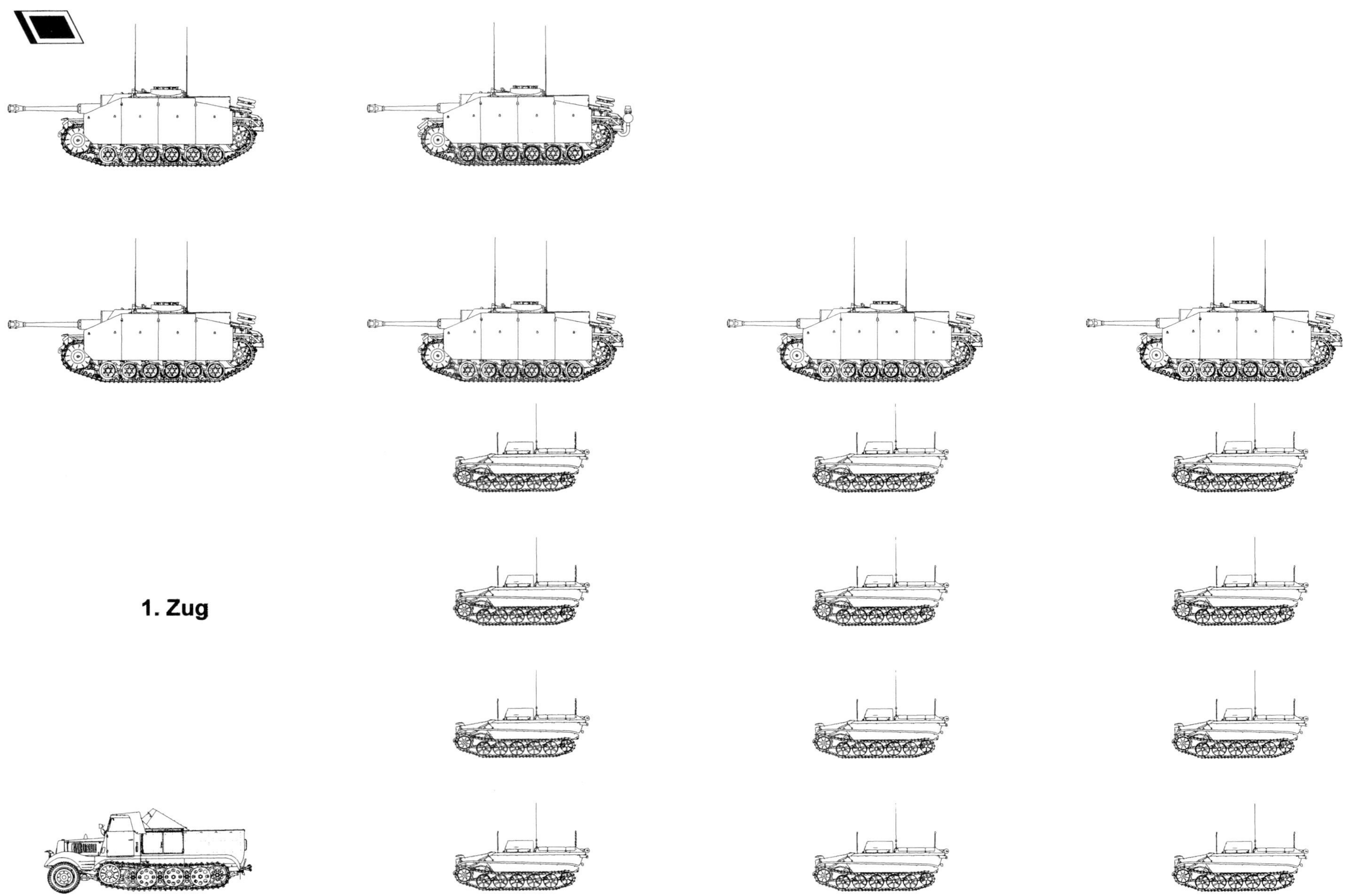

M.Jaugitz / Trojca 2000.

Panzerkompanie (Fkl) 319 (Stand 01.09.1944)

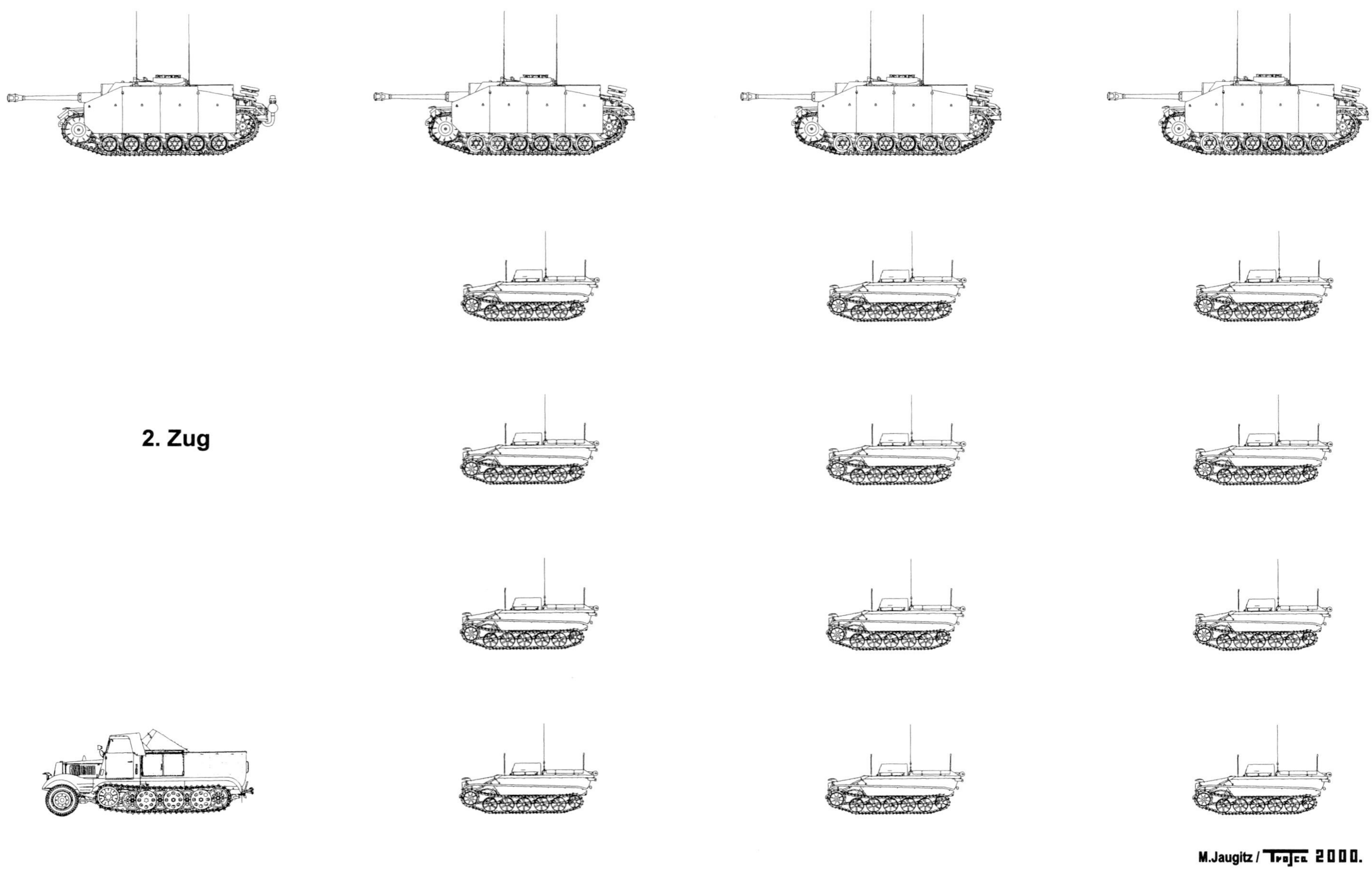

Panzerkompanie (Fkl) 319 (Stand 01.09.1944)

Sondergeräte - Reserve

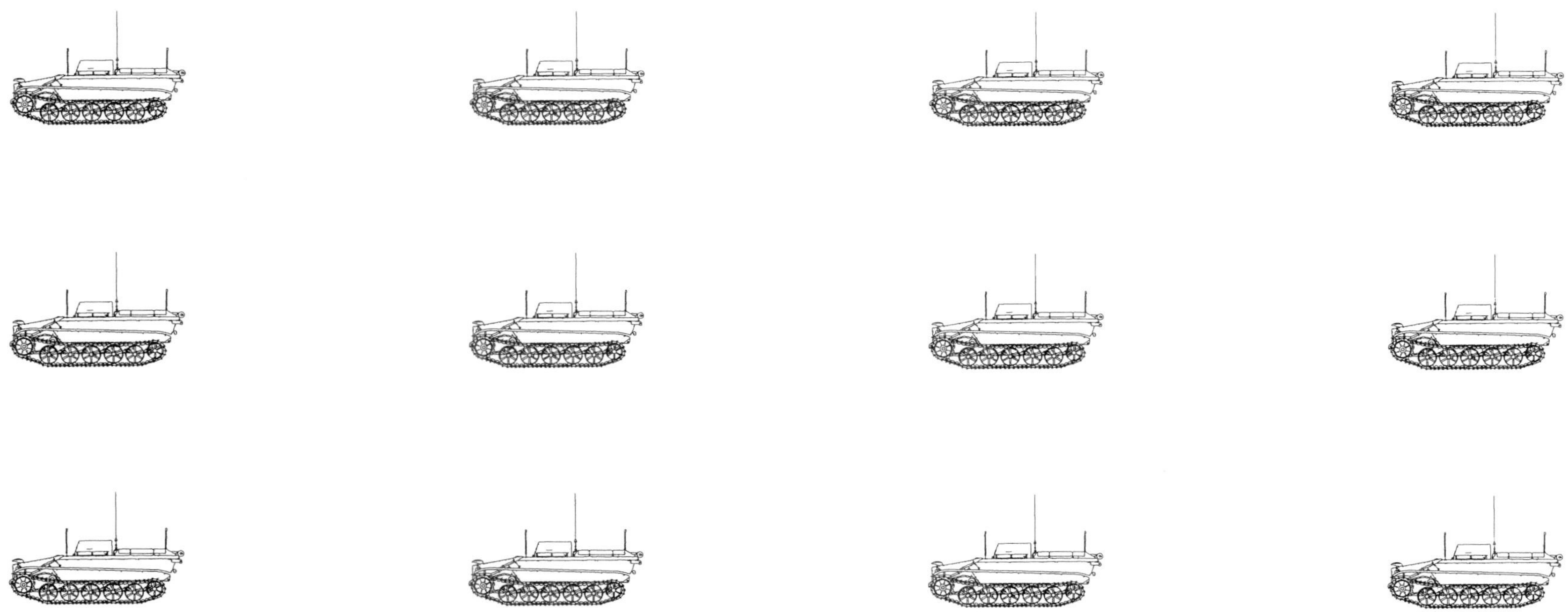

M.Jaugitz / Trojca 2000.

Top Row: *Panzer-Kompanie (Fkl) 319* was moved from Eisenach to the German/Belgian border on the Western Front at the end of August 1944. (Cornelius)

Left: This *Sturmgeschütz III, Ausführung G* was based on the chassis of the *Panzer III, Ausführung M.* (Cornelius)

Bottom row: After the German formations had pulled back over the Meuse, the bridges between Lüttich and Verviers were blown. A few of them were destroyed by *Panzer-Kompanie (Fkl) 319.* (Cornelius)

A few *NSU-Kettenkräder* were also issued to the company.

Oberleutnant Keitel took over the command of the company on 26 October 1944. (Keitel)

Several *BIV's* were lost in the fighting for Aachen

2.2.17: schwere Panzer-Abteilung (Tiger / Fkl) 301

On order of the *AHA Stab I* (*Nr. 40843/44 geheim*), *Panzer-Abteilung (Fkl) 301* was renamed and reorganized as *schwere Panzer-Abteilung (Tiger/Fkl) 301* on 2 September 1944. The unit had been withdrawn from the Eastern Front at the end of August and transferred to the Grafenwöhr training area for refit.

The battalion consisted of a headquarters, a headquarters company, three tank companies, a supply company and a maintenance company. On 27 August the Inspector-General of Armored Troops ordered further changes in the unit's equipment.

The battalion was issued new vehicles at the Grafenwöhr training area. It was issued 66 *Borgward BIV* demolition-charge carriers (*Ausführung C*) and 31 *Tiger I* tanks. The battalion never reached its authorized strength of 32 *Tiger I's*. The majority of the heavy tanks were repaired vehicles or came from an armor unit in the process of reequipping.

Deliveries were very slow. The battalion received its first 11 tanks on 4 September 1944, followed by 10 more in September and another 10 at the end of October, the latter coming from *schwere SS-Panzer-Abteilung 503*. On 23 October 1944 the battalion received seven *Sd.Kfz. 251's*, whose intended role was the transporting of demolition charges for the *BIV's*.

In October 1944 the battalion's officer complement was as follows:

Battalion Commander: *Hauptmann* Krämer
Headquarters Company: *Oberleutnant* Rohmeyer
Remote-Control Officer: *Oberleutnant* Stuke
Battalion Surgeon: *Stabsarzt Dr.* Wick
1./schwere Panzer-Abteilung (Tiger/Fkl) 301: *Oberleutnant* Endemann
2./schwere Panzer-Abteilung (Tiger/Fkl) 301: *Oberleutnant* Meyer
3./schwere Panzer-Abteilung (Tiger/Fkl) 301: *Oberleutnant* Zywietz
Supply Company: *Oberleutnant* Sickendick
Maintenance Company: *Oberleutnant* Bayert

During its stay at Grafenwöhr the combat elements were trained on the *Tiger I* and the *Borgward BIV, Ausführung C*. The former members of the supply and maintenance sections of the tank companies were consolidated in the supply company and prepared for their new duties. The maintenance company had to familiarize itself with new recovery systems, especially the unit's two *Bergepanthers*. In spite of the general fuel shortage, the battalion conducted several successful exercises in the training areas.

On 31 October 1944 *schwere Panzer-Abteilung (Tiger/Fkl) 301* entrained at Grafenwöhr and was transported northwest.

After a two-day trip the battalion detrained in Erkelenz, after which it was attached to the *LXXXI. Armee-Korps*. The battalion split up and occupied quarters in different towns; the *1. Kompanie* in Lich, the *2. Kompanie* in Stetternich and the *3. Kompanie* in Mersch-Pattern. The headquarters company moved into the forest near Stteternich and the supply company was quartered at Niederaussem. The maintenance company remained far behind in Frechen near Cologne.

It was also planned that the battalion would be attached to the *116. Panzer-Division*. There followed regular trips to the front to scout the area of operations and visit neighboring units. On 14 November 1944, during a visit to *schwere Panzer-Abteilung 506*, the battalion command was surprised by the start of the American offensive on the Ruhr front. All of the villages were bombarded by enemy artillery, while hundreds of bombers literally "plowed up" the area.

In the evening the battalion headquarters was summoned to the corps. The commanding general, *General der Infanterie* Köchling, briefed the headquarters staff on the situation. *Hauptmann* Krämer requested that the battalion be employed as a complete entity. On 15 November the battalion moved through Steinstrass and Obersiers to Pattern, which had so far escaped serious damage. The combat elements remained there just one day. On 16 November the battalion was withdrawn into the Lich area. On 18 November *schwere Panzer-Abteilung (Tiger/Fkl) 301* was ordered to construct a blocking position between Jülich and Ruhrdorf in the area of the *246. Volksgrenadier-Division*.

In two march serials, the battalion rolled over the rail bridge near Jülich and the Ruhrdorfer bridge. Both bridges were already under artillery fire. Moreover, oncoming vehicles blocked the *Tigers'* path. After a long road march, the battalion arrived in the area of the *246. Volksgrenadier-Division*. In the evening a staff conference was held in the division command post to decide the plan of attack.

On the morning of 19 November 1944 the just-arrived separate *Panzer-Kompanie (Fkl) 319* was attached to *schwere Panzer-Abteilung (Tiger/Fkl) 301*. While the battalion's first two companies attacked with the attached *Panzer-Kompanie (Fkl) 319*, the *3. Kompanie* remained in reserve in Barmen. The *1. Kompanie* soon contacted the enemy, knocking out four Sherman tanks. There was no further fighting on this day. In the evening the companies were pulled back, the *1. Kompanie* to Merzenhausen, the *2. Kompanie* to Pattern and the 3. *Kompanie* to Koslar.

On 20 November 1944 American forces attacked Merzenhausen. The assault guns of *Panzer-Kompanie (Fkl) 319* suffered heavy losses. The *1. Kompanie* destroyed 12 Sherman tanks and the *3. Kompanie* also had contact with the enemy. Not until later did headquarters receive the report that the *2. Kompanie* was heavily engaged. *Oberleutnant* Meyer, the company commander, had been killed.

The defensive struggle continued on 21 November. *Leutnant* Nebe, who had been placed in temporary command of the *2. Kompanie*, was killed in a tank engagement. Following this, the two surviving *Tigers* were moved back to Inden under the command of *Leutnant* Jakob.

On 22 November it became clear that the situation had become untenable. During a conference at corps, the battalion commander received orders to establish a bridgehead near Jülich. All operational *Tigers* were combined in one tank company. *Oberleutnant* Zywietz assumed command.

On 23 November 1944 the newly-formed tank company and *Panzer-Kompanie (Fkl) 319* committed radio-controlled demolition-charge carriers for the first time. Since it appeared that further remote-control missions offered little hope of success and the massed employment of *BIV's* was not possible, the demolition-charge carriers were pulled back to Kirchherten on 24 November. In Jülich *Hauptmann* Krämer was briefed on the situation by *Oberst* Tolsdorff, commander of the *340. Volksgrenadier-Division*. Seven *Tigers* subsequently moved across the Ruhr bridge in the direction of Bourheim. However, persistent enemy artillery fire prevented the planned attack on Koslar. The *Tigers* pulled back toward Jülich. While on the move the tanks were attacked by enemy fighter-bombers, but there were no losses.

The battalion engaged enemy tanks again on 25 November. Approximately 30 American tanks were knocked out. However, the battalion also sustained losses, including *Leutnant* Jakob of the *2. Kompanie*.

On the following day the *Tigers* were positioned in readiness in the forest near Hambach. The combat elements were attached to the *3. Panzer-Grenadier-Division*. During the night the order was rescinded and the *Tigers*, under the command of *Oberleutnant* Zywietz, were moved to Pier, arriving on 27 November. After his recent heavy losses, the enemy undertook no attacks on that day or on 28 November.

On 29 November the enemy resumed his attacks in the Inden area. He succeeded in establishing a small bridgehead at the rail line between Inden and Pier. Allied troops took Lochberg, in the area of the *3. Fallschirm-Jäger-Division*. There *Panzer-Kompanie (Fkl) 319* undertook a counterattack. Together with *schwere Panzerjäger-Abteilung 519*, the *Tigers* of *schwere Panzer-Abteilung (Tiger/Fkl) 301* were to attempt to crush the bridgehead. Meanwhile the *246. Volksgrenadier-Division* had been moved into the area.

The attack, which began at about 0300 hours on 30 November, was halted soon after it began when the battalion commander was put out of action by enemy artillery fire. At about 0800 hours on the morning of 2 December a *BIV* was sent against the dairy, an isolated building between Inden and Pier. In command of the command and control tank was *Oberfeldwebel* Kastner. Unfortunately the demolition charge carrier got stuck in a ditch in front of the dairy. *Oberfeldwebel* Kastner was killed a short time later when his Tiger was engaged by enemy antitank guns. By the end of November *schwere Panzer-Abteilung (Tiger/Fkl) 301* had sustained 22 personnel losses. In addition, *Leutnants* Grams and Möser, both platoon leaders, were killed in the early days of December.

The fighting died down somewhat between 3 and 5 December. The tank company, which was positioned near Mariaweiler, was divided up in anticipation of an enemy attack. The US forces launched another attack on 6 December. The *Tigers* destroyed 11 Shermans. However, two *Tigers* which became bogged down had to be blown up.

The American forces resumed their attack on 8 December 1944. The Ruhr bridgehead at Jülich had to be abandoned. For the most part the enemy attack was repulsed. The front was quiet on 9 December. However, on 10 December the enemy attacked between Altdorf and Großheu and achieved several penetrations west of Düren.

Schwere Panzer-Abteilung (Tiger/Fkl) 301 and the attached *Panzer-Kompanie (Fkl) 319* had already been ferried to the east bank of the Ruhr during the night of 11-12 December. The formation was supposed to be committed in the area around Birkesdorf to prevent the enemy from taking possession of the *Autobahn* bridge.

The first American tanks appeared in front of the *Autobahn* bridge at about 1055 hours on 12 December. The *LXXXI. Armee-Korps* again issued an urgent order for the Ruhr bridges to be blown, so that neither infantry nor tanks could cross the river. Subsequent demolition efforts by the combat engineers were not very successful, however.

The necessary follow-up demolitions were ordered in the afternoon. *Schwere Panzer-Abteilung (Tiger/Fkl) 301* was ordered to employ demolition-charge carriers to destroy the *Autobahn* bridge at Birkesdorf. The first attempt at about 1700 hours, involving three *BIV's*, was a failure. One *BIV* was knocked out by artillery fire, another overturned. Only one reached its target. At about 1825 hours another attempt was made, this time with two *BIV's*. It, too, was a failure.

Finally, during the night, personnel of *schwere Panzer-Abteilung (Tiger/Fkl) 301* succeeded in placing several demolition-charge containers on the bridge. These were wired together and, on the night of 13 December, they were detonated, destroying the bridge.

This action was described by *Oberleutnant* Stuke, the remote-control officer:

One morning — on or about 12 December 1944 — a senior engineer officer— I believe it was an *Oberst* from the army headquarters — sought out our battalion commander. Upset, he said that it had been discovered that the preparations for blowing the Autobahn bridge had not been properly carried out. It would be a catastrophe if the bridge fell into enemy hands intact. He was on a desperate search for someone who could help prevent the bridge and Autobahn from offering the Americans an "open invitation" into the Cologne area. There were no combat engineers available, and no other unit was prepared to take on such a mission.

I was there when the *Oberst* made his plea for help. The discussion went back and forth. There was agreement when I proposed a solution and volunteered to carry it out. Preparations for the mission were to be made the same night. Meanwhile the *Oberst* wanted to determine the exact timing of the demolition so that he could brief the unit in whose sector the bridge was as to the demolition plan and schedule.

And this is how it went:

There were seven volunteers for the mission: five drivers for the tracked vehicles, *Gefreiter* A. of the signals section and *Unteroffizier* B. While the vehicles were being fitted with demolition charges and prepared for action, three of us (*Unteroffizier* B, *Gefreiter* A and I) reconnoitered the terrain near the mission area as best we could in daylight. We found a place about 100 meters from the bridge from where we would detonate the demolition charges using heavy field cable and a car battery. We also looked for a place where we could position our small column under some cover near the "scene of the crime."

The *Oberst* had meanwhile informed his headquarters about his conversation with us and gave the rigid deadline of "late this evening." Our column moved out while it was still daylight (the *Oberst* planned to follow later), so as to reach the departure point for our tracked vehicles before it got dark. From there they would proceed up the gently-rising *Autobahn* to the bridge over the Ruhr.

But first we had to wait for a while, as it didn't seem dark enough. So far the operation had proceeded without incident. The critical part began about twenty minutes later. The first tracked vehicle (also carrying automobile batteries and field cable) moved off. *Unteroffizier* B. was supposed to start the remaining vehicles after six minutes intervals. *Gefreiter* A., the signals man, and I walked ahead for the first twenty paces. At the place where the ignition was to take place, we removed the batteries and stowed them in vegetation next to the water drainage beside the Autobahn.

Gefreiter A. secured the end of the field cable to the vegetation, shouldered the backpack and we set off toward the Ruhr. We weren't too confident, especially since the tracked vehicle made a considerable amount of noise. But the Americans seemed unaware of our presence. To anticipate your question: Not a shot was fired during all our preparations. Of course, we could not have known that then. At the point where the bridge crossed the water, we laid down the places for the demolition charges in a zigzag pattern over the width of the roadway. All the driver had to do was activate a lever and the charge fell harmlessly to the ground. When all of the charges were in place and the cable connections had been made, all of the charges would be armed simultaneously.

The first tracked vehicle headed back. The others moved forward one after the other. Altogether the work across the Ruhr took a long 45 minutes. On the way back *Gefreiter* A. again carried the drum on his back, in order to connect the second wire to the ignition battery. We simply couldn't believe that everything had gone so smoothly. While *Gefreiter* B. worked with the cable ends at the battery, I walked to where *Unteroffizier* A. was waiting with the drivers and I told him that everything had gone perfectly. I believe that they wanted to observe the fireworks from there. I made my way back to *Gefreiter* A.

Not long after that the *Oberst* arrived and gave the green light for the detonation. There was a tremendous bang, the earth shook and pieces of stone flew about. But had the bridge been rendered impassable? We waited a while to see if there would be any reaction from the American side. When there was none, we walked toward the detonation site. We heard the sound of rushing water, softly at first but louder as we drew nearer. And then we could see that all of the drivable portion of the bridge had fallen into the water. It was dark and not a word was spoken, but it was still obvious that a great weight had been lifted from the shoulders of our engineer *Oberst*.

In spite of the bridge demolitions, American forces succeeded in occupying the west bank of the Ruhr from Linnich to Düren by 13 December 1944. Following the successful demolition of the *Autobahn* bridge, *schwere Panzer-Abteilung (Tiger/Fkl) 301* and the attached *Panzer-Kompanie (Fkl) 319* were moved to Arnoldsweier as corps reserve.

On 15 December 1944 *schwere Panzer-Abteilung (Tiger/Fkl) 301* received word that it was to be attached to the *9. Panzer-Division*. The unit assembled in Niederaussem and prepared to entrain, which was scheduled for 20 December.

There, on 16 December, the battalion received news of the start of the German Ardennes Offensive. The *9. Panzer-Division* was part of the *5. Panzerarmee* and, in the second half of December, it was committed to relieve the *2. Panzer-Division* fighting near Celles. On 22 December members of the headquarters of *schwere Panzer-Abteilung (Tiger/Fkl) 301* went into the Eifel to establish contact with the division headquarters. By that time the plan to move the *Tigers* there by rail was no longer possible as the rail lines had been destroyed. On 25 December the battalion was released from its attachment to the division. The *Tigers* were unloaded and *schwere Panzer-Abteilung (Tiger/Fkl) 301* was moved to Nörvenich where it became army-group reserve of the *LXXXI. Armee-Korps* and occupied quarters.

During the first half of January 1945 the battalion carried out repairs to its tracked and wheeled vehicles and conducted a reorganization of its personnel. At that time the battalion had 27 *Tigers*, 2 *Bergepanther*, 11 *SPW's* and 52 *BIV's (Ausführung C)*. These figures include vehicles undergoing repairs.

On 16 January 1945 the Allies launched "Operation Blackcock" in the northern sector of the Ruhr front. Field Marshall Montgomery wanted to take the "Heinsberger Triangle" and the sector of front between Linnich and Roermond. After a heavy bombardment of the towns there, units of the British XII Corps attacked the German main line of resistance with strong tank support. The villages had to be wrestled from the defenders one by one.

On 16 January 1945 *schwere Panzer-Abteilung (Tiger/Fkl) 301* was placed on alert and rushed from Nörvenich northwest to the *178. Infanterie-Division*. Even though it moved by night, the unit was still attacked by enemy fighter-bombers. The battalion prepared for operations at Annenbosch (Holland). In the days that followed, the *Tigers* saw action in the towns of Königsbosch, Diergardt, Braunsrath, Obspringe and Waldfeucht.

In his book *The Battle of the Rhineland*, author R.W. Thompson described the *Tigers'* part in the operations at Waldfeucht on 20 January 1945:

> The pressure was maintained by day and night. On 20 January the Light Highland Infantry took Bocket with little effort and opened the way for the attack on Waldfeucht and the Echterbosch Forest. The enemy seemed more confused by the frequent snowstorms than by the British infantry, which was experienced in mountain operations, and it became clear that he had lost his composure. A battalion-strength counterattack against Bocket supported by *Tiger* tanks failed. Three companies of the 5th KOSB (The King's Own Scottish Borderers) took Waldfeucht in a night assault, and the fourth company moved into Echterbosch.
>
> Yet it was in these two villages that the 5th KOSB would fight the fiercest battle of "Operation Blackcock" on 21 January. The night of 20-21 January was bitterly cold. The entire region lay under deep snow, and the village of Waldfeucht with its medieval gate looked almost peaceful. In the early morning hours of 21 January the 5th KOSB quietly occupied all-round defensive positions. There was an eerie stillness. Dawn hadn't broken when enemy mortar fire heralded an attack, and at half past six two *Tiger* tanks appeared in the gray light. Two small groups of infantry followed. Four of our own tanks were quickly put out of action by enemy mortar fire, but the antitank gunners held their fire until the *Tigers* appeared in the village street one hundred meters away. Only then did the crews of the six-pounders open fire, destroying both enemy tanks with direct hits. This was only the beginning, however, a minor prelude to a big attack.
>
> The enemy mortar fire intensified. Six *Tiger* tanks and at least four *Panthers*, all carrying mounted infantry, rushed in from the north. Fifteen self-propelled guns, which were in close contact with the infantry, began blasting the buildings into rubble, filling the narrow streets with flying splinters and a maddening din.

As the German front line could no longer be held, the bridgehead near Heinsberg was abandoned. Using the last of its fuel, *schwere Panzer-Abteilung (Tiger/Fkl) 301* moved back to Kirchherten where it was assigned as army reserve. Since there was a general fuel shortage at that time, the *Tigers* occupied specially-prepared earth-walled bunkers. Those tanks which were still operational were split into two tank companies. One was led by *Oberleutnant* Endemann, the other by *Leutnant* Schreiner.

After a few quiet days, the next Allied attack began on 24 February. Under cover of an artillery barrage, the enemy had succeeded in establishing a bridgehead at Erkelenz. On 25 February *schwere Panzer-Abteilung (Tiger/Fkl) 301* was alerted and once again attached to the *LXXXI. Armee-Korps*. The tank companies rolled to the front. *Leutnant* Schreiner led eight *Tigers* in the direction of Lich, while *Oberleutnant* Endemann took the remainder to Mersch-Pattern. The companies were able to knock out several American tanks, however losses were high. The company commander of the *1./schwere Panzer-Abteilung (Tiger/Fkl) 301*, *Oberleutnant* Endemann, was among those killed.

At the end of February the unit had just five operational Tigers; ten more were undergoing repairs.

In March the remaining operational *Tigers* were committed individually, supporting small *Kampfgruppen* in their struggle against the advancing enemy. The tanks covered withdrawal movements at Niederaussem, Stommeln and Sinnersdorf. While the *Tigers* of the tank companies were constantly employed, the rest of the battalion succeeded in reaching the right bank of the Rhine.

On 6 March 1945 the battalion command post moved to Bergisch-Gladbach. The tanks assembled in the Gummersbach area, where the maintenance company was also located. On 15 March many members of *schwere Panzer-Abteilung (Tiger/Fkl) 301* received decorations for their actions in the previous weeks. All of the tanks had been consolidated in one tank company under the command of *Leutnant* Schreiner.

On 18 March 1945 *Leutnant* Schreiner arrived in Wermelskirchen, where the *Tigers* were assembled after the completion of repair work. In the evening the tank company rolled through Leichlingen to the Düsseldorf *Autobahn*. On 18 March they reached Mettmann, where the *Tigers* were dispersed on the grounds of a large farm. The tank company was made army reserve. During the night of 24/25 March the company moved to Selbeck; on 27 March it moved to Mühlheim-Dümpten. On 28 March the tank company had seven serviceable *Tigers*. On order of the battalion, three *Tigers* were to remain in Dümpten, while the other four under the command of *Leutnant* Schreiner were to proceed in the direction of Gelsenkirchen-North. There was no contact with the enemy.

After a lengthy withdrawal movement, the four *Tigers* arrived in Lütgendortmund on 31 March 1945. *Leutnant* Schreiner scouted in the direction of Unna and Soest. After brief skirmishes near Dortmund the four *Tigers* returned to Lütgendortmund. The other three *Tigers* under *Leutnant* Schwickert engaged elements of the US 2nd Armored Division "Hell on Wheels" in Dümpten.

On 10 April 1945 demolition-charge carriers were supposed to again be used to destroy important bridges over the Ruhr. The order was rescinded, however, and the *BIV's* were sent back to the supply company.

On 12 April 1945 the *Tigers* moved to Gevelsberg. The *Tigers* again contacted the enemy near Hasslingshausen. Tank commander Palitz destroyed an enemy tank. The American forces had discovered the German tanks and attempted to destroy them with antitank guns. On 16 April the *Tigers* arrived in Sprockhövel. The battalion's command post was also there. With the situation hopeless and almost no fuel left, the battalion commander, *Hauptmann* Krämer, ordered the Tigers blown up on 17 April 1945.

The majority of the battalion's demolition-charge carriers had been blown up during the retreat.

Many members of the battalion tried to make their way east in small groups. The majority of them were captured by American forces and imprisoned in the notorious camp at Sinzig.

Gliederung der schweren Panzer - Abteilung (Tiger / Fkl) 301 (Stand: Oktober 1944)

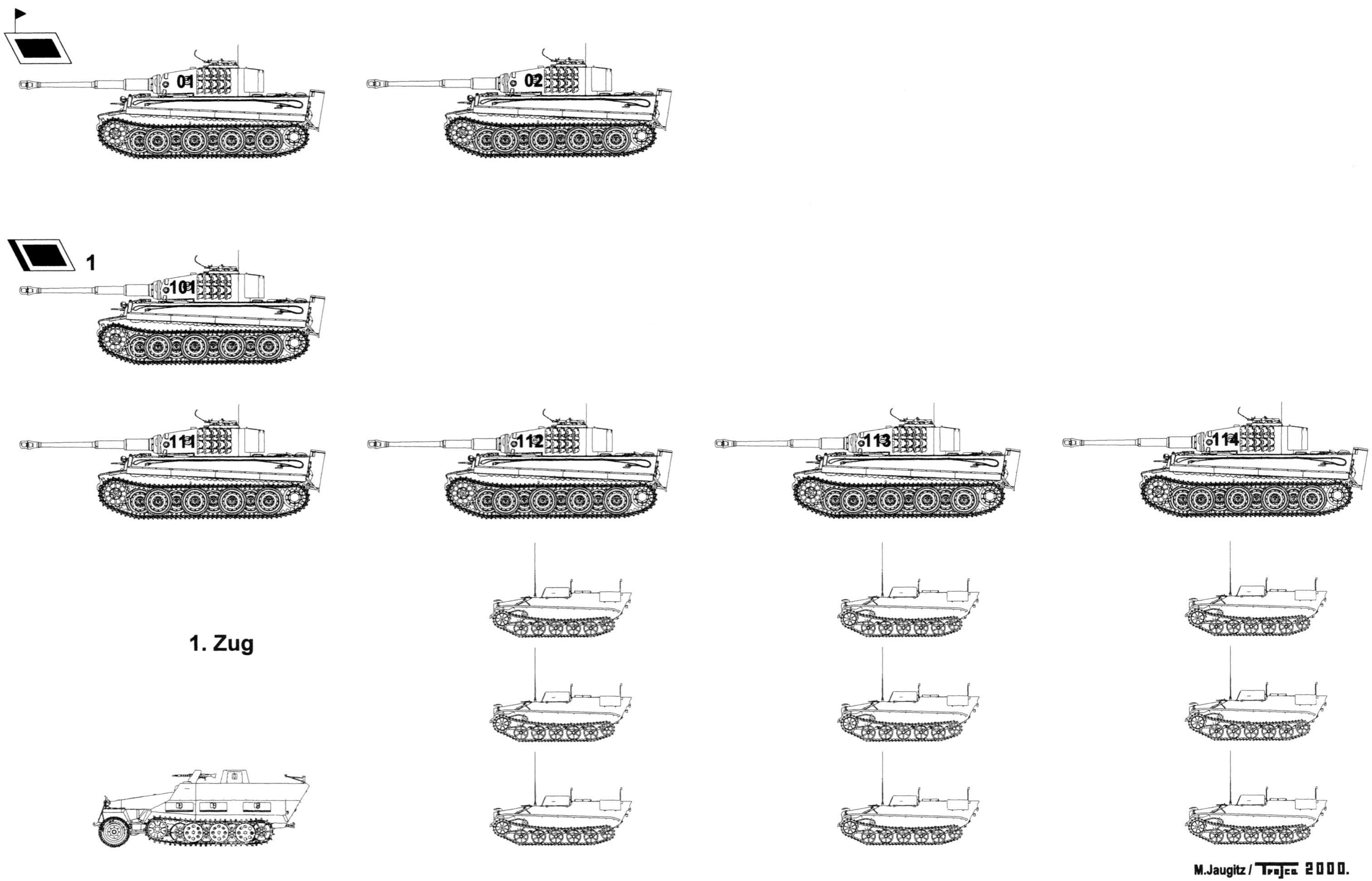

Gliederung der schweren Panzer - Abteilung (Tiger / Fkl) 301 (Stand: Oktober 1944)

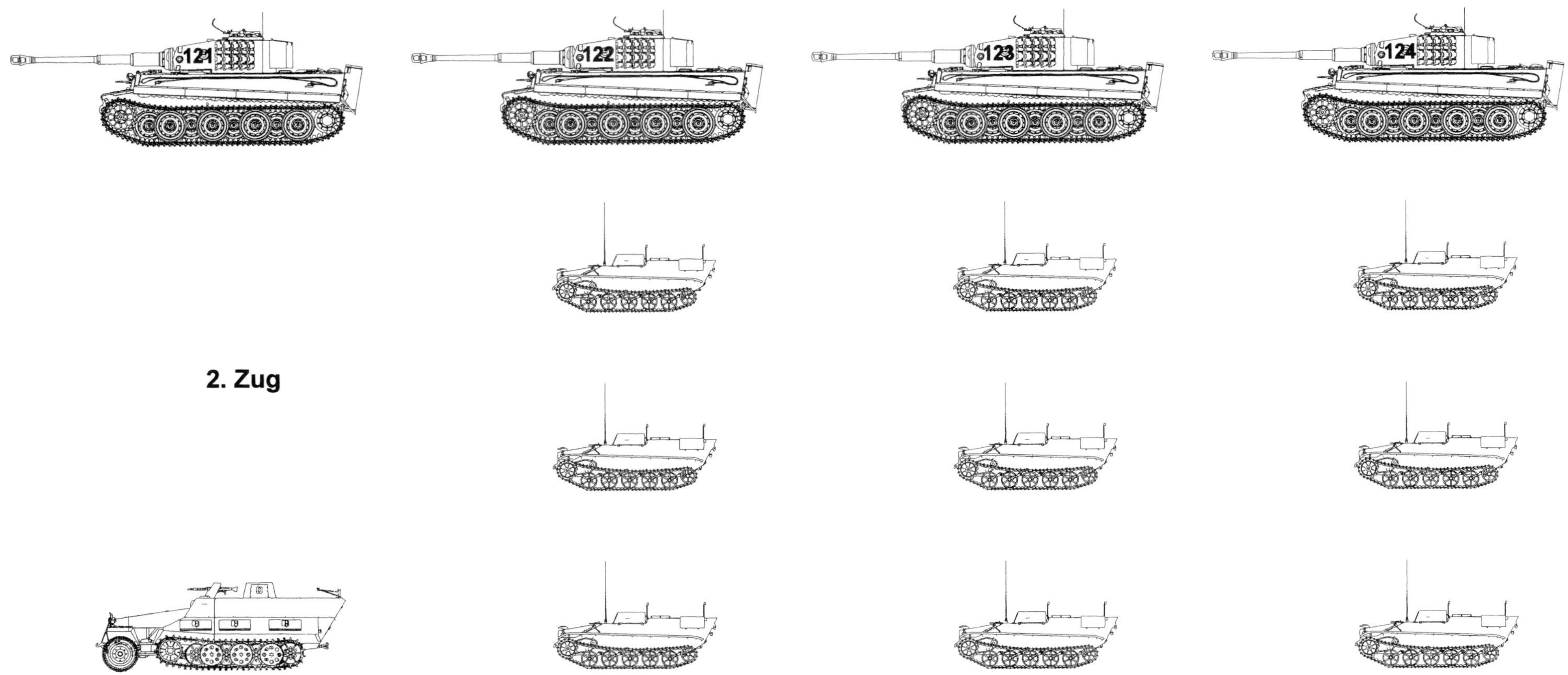

M.Jaugitz / Trojca 2000.

Gliederung der schweren Panzer - Abteilung (Tiger / Fkl) 301 (Stand: Oktober 1944)

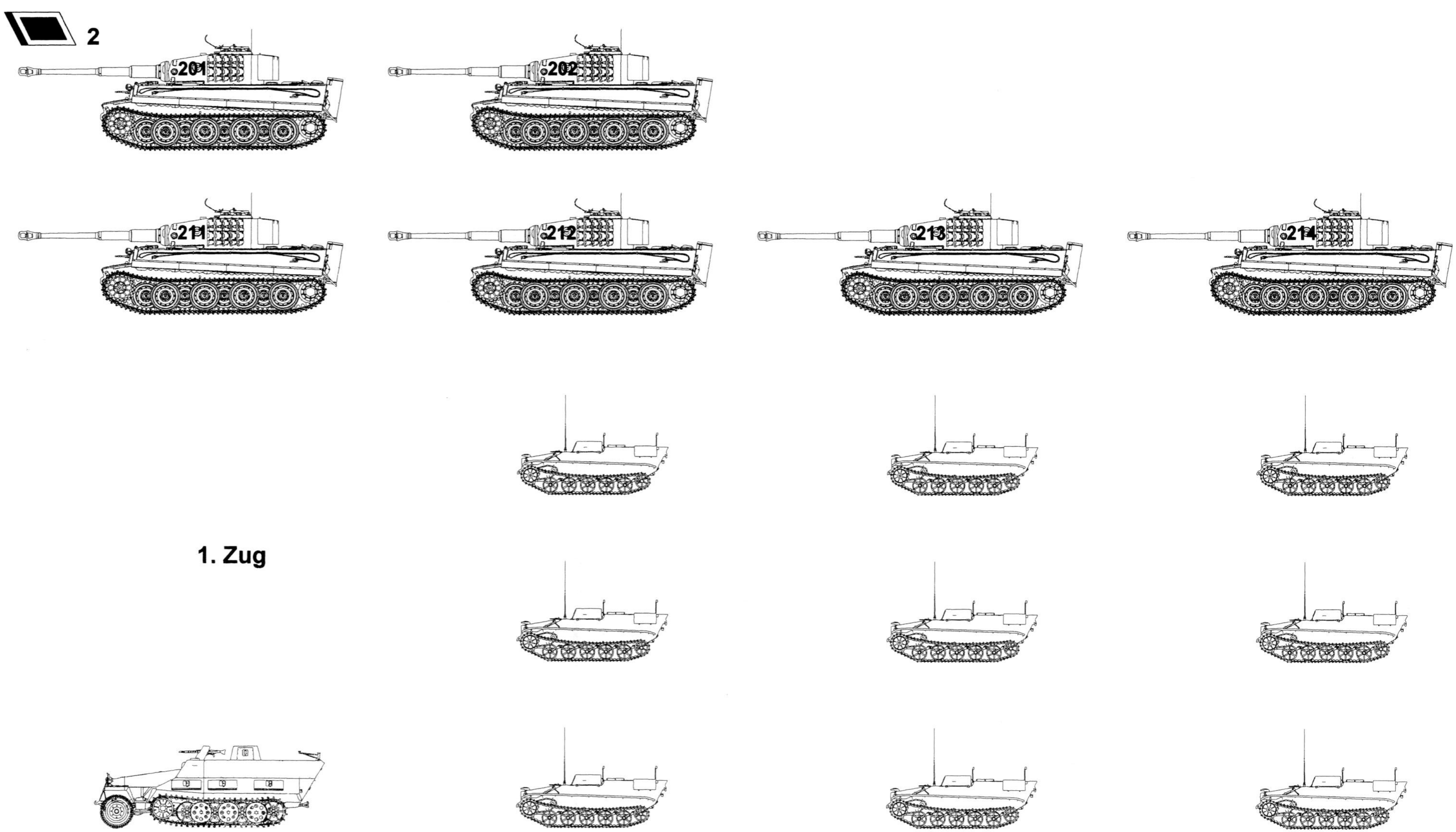

M.Jaugitz / Trojca 2000.

Gliederung der schweren Panzer - Abteilung (Tiger / Fkl) 301 (Stand: Oktober 1944)

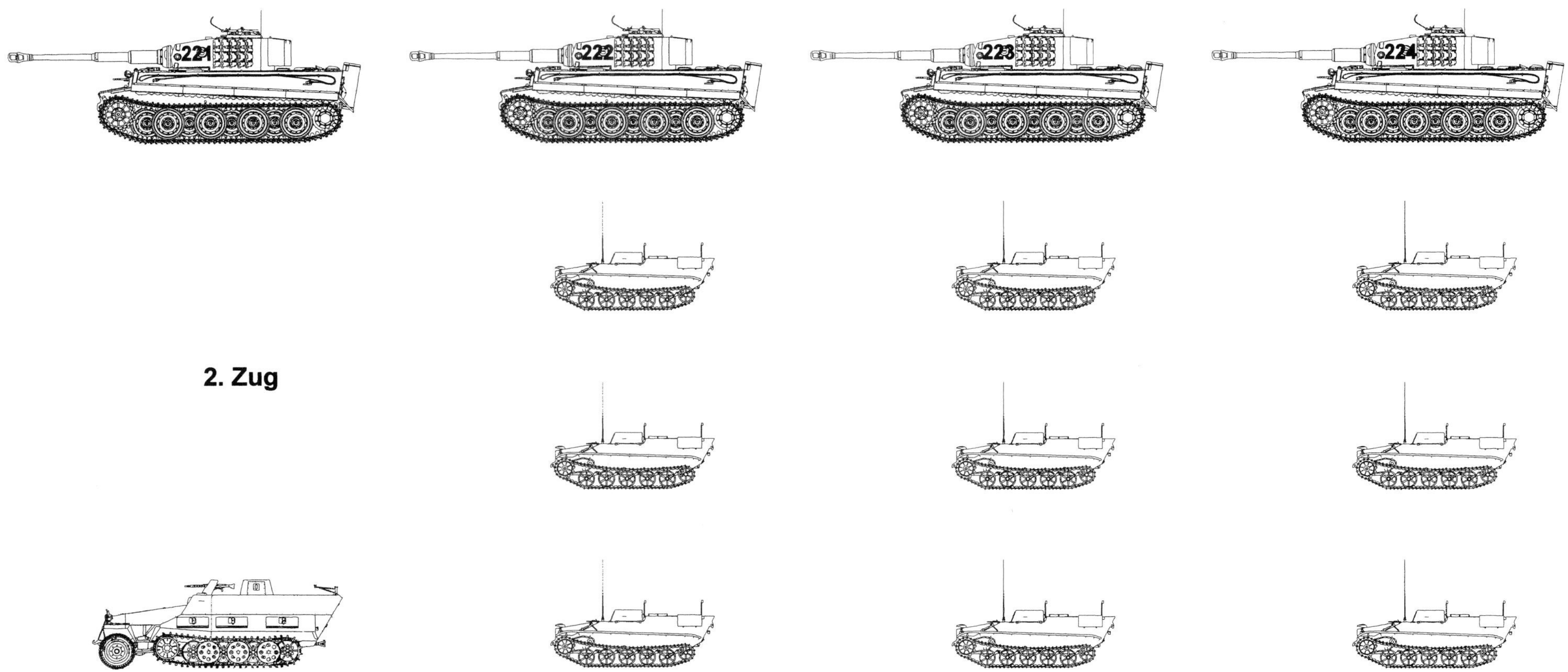

M.Jaugitz / Trojca 2000.

Gliederung der schweren Panzer - Abteilung (Tiger / Fkl) 301 (Stand: Oktober 1944)

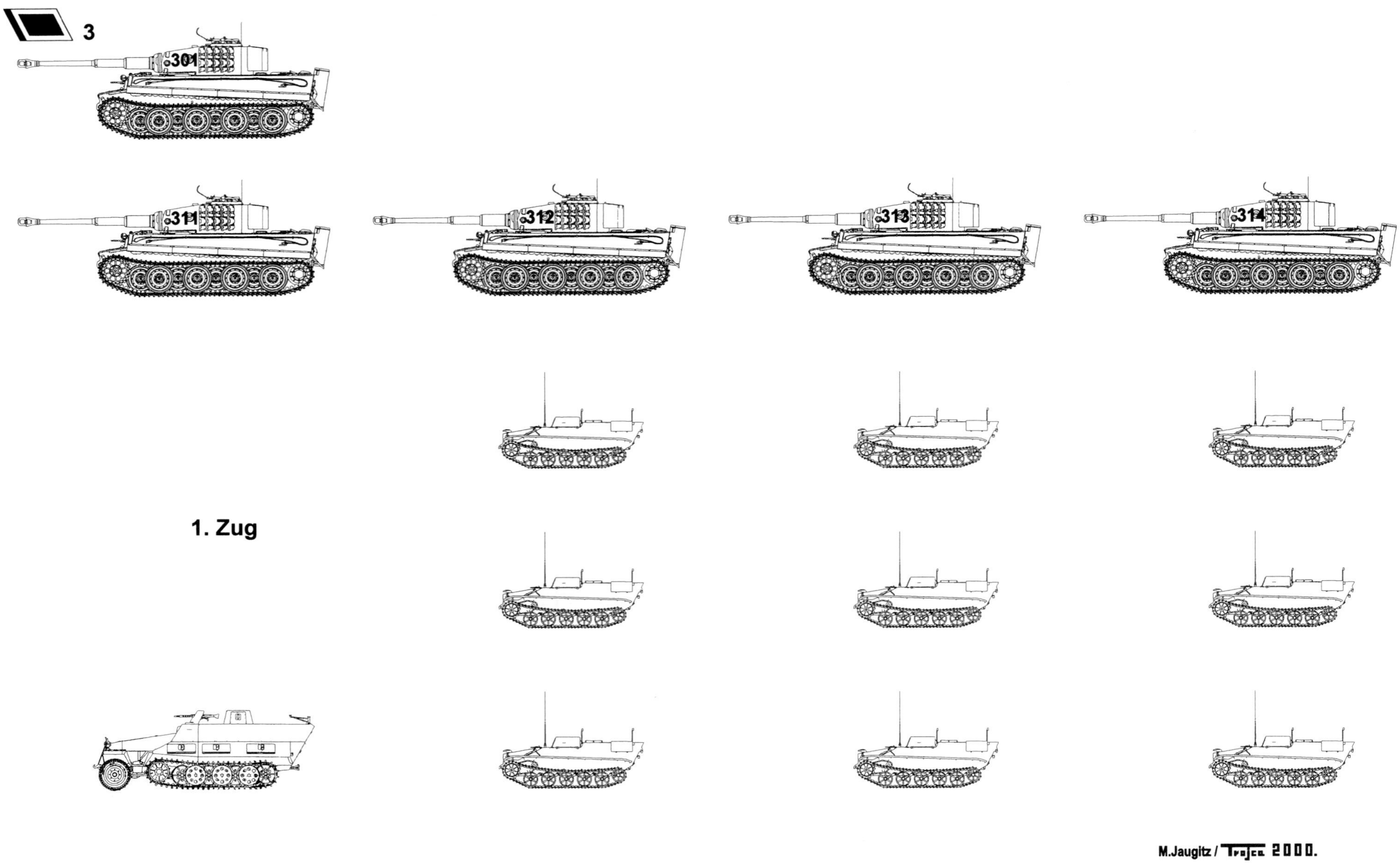

Gliederung der schweren Panzer - Abteilung (Tiger / Fkl) 301 (Stand: Oktober 1944)

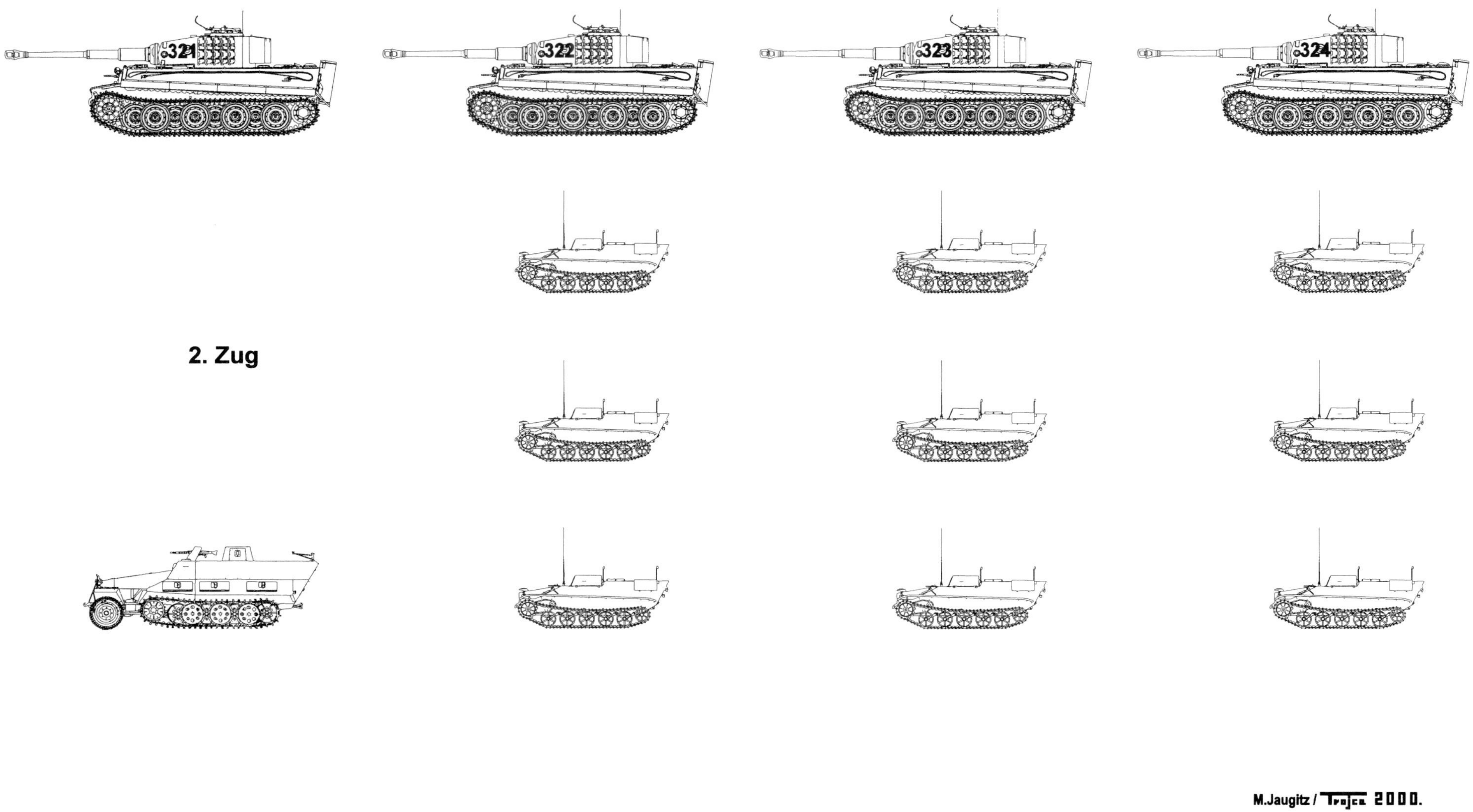

Gliederung der schweren Panzer - Abteilung (Tiger / Fkl) 301 (Stand: Oktober 1944)

Sondergeräte - Reserve
bei der Versorgungs - Kompanie

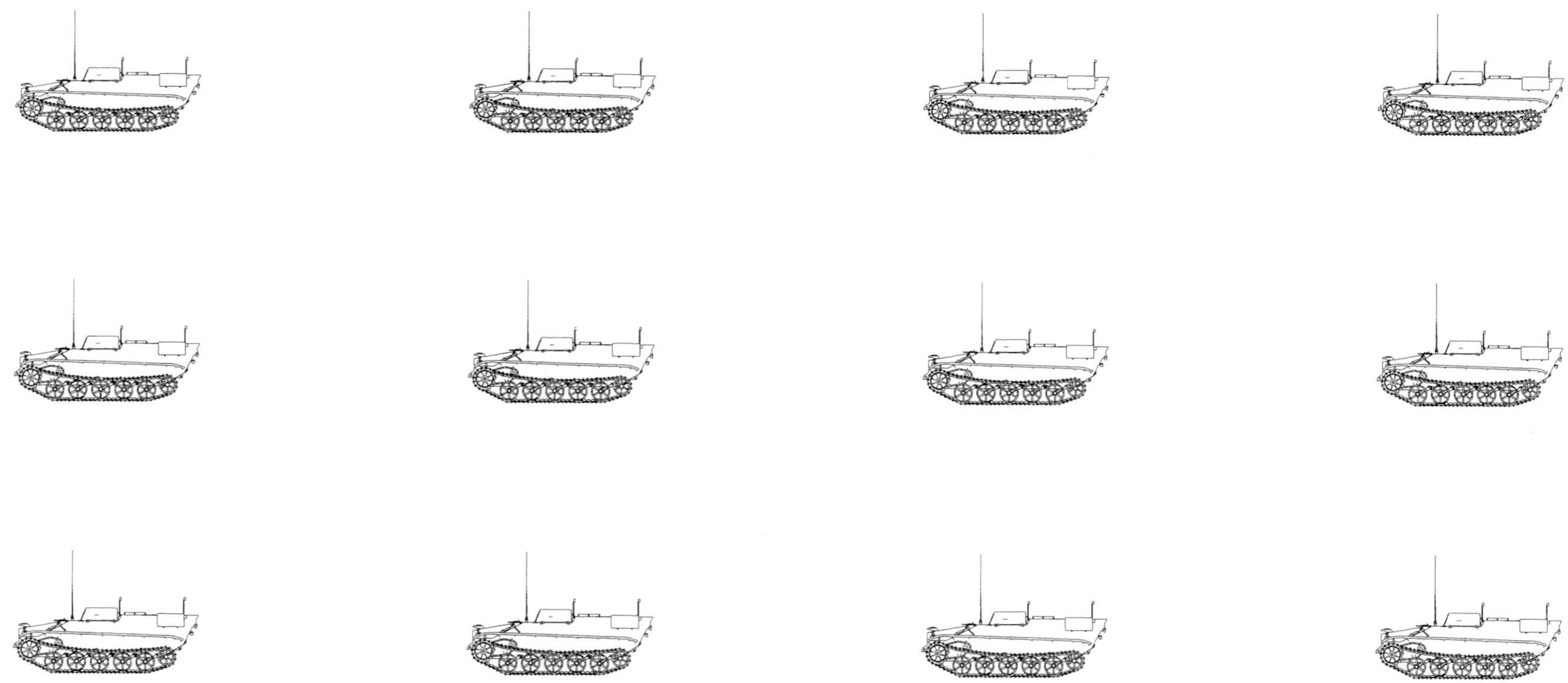

M.Jaugitz / Trojca 2000.

The newly formed *Panzer-Abteilung (Tiger/Fkl) 301* received 31 *Tiger I's* of different varieties at the Grafenwöhr Training Area. (Höland)

The crew of *Unteroffizier* Braun get the *Tiger* operationally ready. The outside roadwheels have been removed on this *Tiger*. (Höland)

Left: No turret numbers have been applied as of yet. Note the smoke-grenade launcher next to the Bosch blackout-drive light. (Höland)

Opposite page top and bottom: The *Tiger* of tank commander Haneke of the *3./Panzer-Abteilung (Tiger/Fkl) 301* in the brush at the Grafenwöhr Training Area. The turret numeral *323* was not applied until later. (Haneke)

Below: The crews were training on the heavy tank in September and October 1944. This *Tiger* has steel roadwheels. (Schreiner)

Tiger 112 in the billeting area at Grafenwöhr. (Münch)

The battalion was loaded on trains on 31 October 1944 and sent to the northwest. (Höland)

The battalion was unloaded at Erkelenz on 2 November 1944. The tanks were immediately camouflaged. (Höland)

Tiger 113 was moved to Pattern in the middle of November 1944 along with the rest of the combat elements. Note the jerry cans fastened to the rear of the turret. (Höland)

This page and top of next next page: Several *BIV*, *Ausführung C* demolition-charge carriers were lost in the fighting on the Ruhr front. One of these carriers was recovered almost undamaged by the US forces. It was later extensively evaluated. (Jentz)

Tiger 113 is recovered by two *Bergepanthers* after it had received hits on the track and main gun. (Höland)

Tiger 111 in December 1944 at Niederaussern. The battalion made preparations for the planned move to the Ardennes. (Specht)

The tank commander of *Tiger 111*, *Unteroffizier* Specht, is in the commander's hatch. (Specht)

Above: After being alerted in January 1945, *Panzer-Abteilung (Tiger/Fkl) 301* was moved to the German/Dutch border area. The *Tiger* of tank commander Sartory still has the old-style running gear. (Sartory)

Top Left: These two *Tigers* were lost during the fighting in the area around Waldfeucht on 21 January 1945. (Tank Museum)

Left: The battalion commander, Krämer (in the field-gray cap), with a few officers. The adjutant and later commander of the *Einsatzkompanie*, *Leutnant* Schreiner, wears camouflage coveralls in *Waffen-SS* pattern. (Schreiner)

2.2.18: Panzer-Abteilung (Fkl) 303

On 2 January 1945 the *Organizations-Abteilung* of the Army General Staff ordered the formation of *Panzer-Abteilung (Fkl) 303*. The battalion, to be based at Eisenach, would have 45 assault guns. The *4./Panzer-Abteilung (Fkl) 301* and the *4./Panzer-Abteilung (Fkl) 302* were to be used in the formation process, which was supposed to be completed by 31 January 1945.

On or about 10 January the two companies were transferred to the training area at Grafenwöhr, after which they were renamed. The *4./Panzer-Abteilung (Fkl) 301* became the *1./Panzer-Abteilung (Fkl) 303* and the *4./Panzer-Abteilung (Fkl) 302* became the *2./Panzer-Abteilung (Fkl) 303*.

After arriving at the training grounds in mid-January, the men were trained in the use of the *Panzerfaust* and *Panzerschreck* antitank weapons. It was also planned to equip the unit with the *BIV, Ausführung C*, demolition-charge carriers. Test and training drives were begun, however, these were soon called off.

On 20 January the *AHA* ordered *Panzer-Kompanie (Fkl) 319* withdrawn from the Ruhr front and transferred to Grafenwöhr. The company was renamed and was incorporated into *Panzer-Abteilung (Fkl) 303* as its *3. Kompanie*.

On 14 February 1945 *Panzer-Abteilung (Fkl) 303* received orders to move to the Hindenburg Garrison in Potsdam. The unit departed for its destination on 15 February, leaving its demolition-charge carriers behind at Grafenwöhr.

While the unit was still en route to Potsdam, the *AHA* ordered a change in equipment on 17 February. The battalion was to be equipped with 31 assault guns only. All remote-control equipment was dispensed with and the suffix *(Fkl)* was dropped from the unit designation.

In February 1945 the battalion's complement of officers was as follows:

Battalion Commander: *Hauptmann* Faßbeck
Headquarters Company: ?
Battalion Surgeon: *Oberarzt Dr.* Jokszies
1./Panzer-Abteilung 303: *Hauptmann* Witter
2./Panzer-Abteilung 303: *Leutnant* Hollensteiner
3./Panzer-Abteilung 303: *Oberleutnant* Keitel
Supply Company: *Hauptmann* Bachmann
Maintenance Platoon: *Leutnant* Wilkniß

This latest order was modified just four days later. Instead of *Sturmgeschütz III's* the battalion would now be equipped with *Panzer IV's*. Furthermore, it was ordered that the battalion would first be renamed *Panzer-Abteilung "Döberitz"* and then, after its incorporation into the *Panzer-Division "Schlesien"*, would be designated as *Panzer-Abteilung "Schlesien"*.

The unit received its armored vehicles in Potsdam on 26 February. The *1. Kompanie* was equipped with *Jagdpanzer IV* tank destroyers armed with the 75-mm *KwK 42 L/70*, while the *2.* and *3. Kompanien* were issued *Panzer IV, Ausführung J* battle tanks. After the tanks had calibrated their guns at Berlin-Krampnitz, the battalion entrained in Potsdam on 28 February and was transported into the Frankfurt/Oder area. After detraining in Pillgram, the unit occupied quarters near Jakobsdorf.

The battalion changed its billets several times. On 7 March it entrained again and was transported to Stettin, where it detrained on 9 March.

On 10 March the tanks and tank destroyers began moving into the assembly area in the Retzowfelde — Ferdinandstein area east of the Oder. On 11 March at about 1000 hours an attack was launched against the Soviet positions, however, it was halted by fire from antitank guns. In the following days (until 22 March) the armored vehicles saw action daily in both the defensive and offensive roles in the Sinzlow — Podejuch — Finkenwalde — Altdamm area.

After a brief rest period, the battalion entrained in Stettin on 25 March 1945 and was transported to Eberswalde. There, the remnants of *Panzer-Divisionen "Schlesien"* and *"Holstein"* were consolidated to form *Panzer-Regiment 118* within the reformed *18. Panzer-Grenadier-Division*.

On 17 April the *18. Panzer-Grenadier-Division* departed for the Müncheberg area east of Berlin. On 18 April the tank crews were engaged in fierce fighting. In spite of the massed German defenses, superior forces of the Red Army broke through the German positions between Seelow and Wriezen and, by 20 April, advanced as far as Müncheberg. The battered *Panzer-Abteilung "Schlesien"* made a fighting withdrawal toward Straußberg. In the days that followed, the unit withdrew via Fredersdorf and Hirschgarten to Berlin.

Small numbers of tanks from *Panzer-Abteilung "Schlesien"* saw action in the Battle of Berlin in the Mahlsdorf, Neukölln, Charlottenburg and Wilmersdorf sectors. In spite of the heavy fighting in the streets and houses, the maintenance platoon, which was located not far from the technical institute, succeeded in repairing a number of damaged tanks.

"Fortress Berlin" surrendered on 2 May 1945. Several members of *Panzer-Abteilung "Schlesien"* managed to make their way through enemy lines to the west.

An account of the final battle for Berlin by former *Unteroffizier* Carl M. Wenzel (tank commander of tank *"224"*, of the *2./Panzer-Abteilung "Schlesien"*) provides a vivid description of the final three weeks of the war:

Account of the Final Operations and Capture of the Crew of Tank *"224"* (Field Post Number 17589)

Crew:

Commander: *Unteroffizier* Carl M. Wenzel
Gunner: *Obergefreiter* Helmuth Klindt
Loader: *Gefreiter* Horst Ziegner
Driver: *Obergefreiter* Rudolf König
Radio operator: *Obergefreiter* Hans Brodersen

The last operation of the unit — Field Post Number 17589 — began with its alert late in the evening of 16 April 1945 in its billeting area at Eberswalde. The unit conducted a road march into the area east of Berlin and went into action on 18 April.

The attack, which was carried out in the early morning hours of 18 April, cost the division heavy losses because the Russians had lured us into a trap. We all got out in one piece except the tank commander, who received a minor wound, a small piece of shrapnel in his neck. In the afternoon we were encircled, however, we succeeded in breaking through the Russian ring and making it to our own lines.

That evening a minor engine defect forced us to return to our support base for repairs.

The next day we were ordered to secure in Müncheberg and the surrounding area. In the evening we pulled back to a rest position in some woods near Hohenstein. The next morning we were committed to a counterattack, however, it failed to achieve the desired results as the enemy was too strong. During the night we withdrew closer to Berlin.

On 21 April 1945 we were with the supply section having a mechanical problem fixed. We pulled back further via Straußberg, Fredershof and Hirschgarten. On account of the strong Russian pressure, we moved our support base into Berlin during the night. On 22 April we were at the Spaeth Nursery and during the course of the afternoon we became operational. We were supposed to report to the battalion command post in Mahlsdorf, however, the Russians were already there. We looked for the battalion without success. Having failed to find the tank companies, we reported to the supply company in Grunewald the next morning.

On 23 April, while moving to the new battalion command post in Neukölln, our engine failed again on Gneisenaustraße. It was evening by the time the maintenance section had been summoned and the repair completed. During the night we reported to the division command post in the town hall of Neukölln. On 24 April we were sent to a new area of operations on Kaiser-Friedrich-Straße. From there we proceeded to the Kurfürstendamm. We were committed on the Kurfürstendamm, Hubertusallee and Warmbrunnerstraße from 24 to 26 April. Early on 27 April we withdrew and moved to our new area of operations near Fehrbelliner Square.

Our vehicle developed a steering gear problem forcing us to proceed to the maintenance section, which was at the technical institute. During the night we took up our position on the corner of Berliner Straße and Mannheimer Straße. During the period from 27 April to 1 May we were in continuous action in the area of the Fehrbelliner Square. The square was encircled on the evening of 1 May.

We succeeded in breaking out and reached the remnants of our unit by the radio tower on Suarezstraße. During the night of 1-2 May we guarded the Heerstraße. On the morning of 2 May we began our breakout from Fortress Berlin toward the west in hopes of reaching the American lines at the Elbe. In the early morning hours of 3 May the commander of our Kampfgruppe gave the order to prepare to blow up our vehicles. We were ordered to try to get through to the German main line of resistance in small groups. As a result, we headed west on 3 May. We got as far as a few kilometers south of Rathenow. There, on the morning of 8 May, we were captured by the Russians in a barn.

Gliederung der Panzer - Abteilung (Fkl) 303 (Stand 01.03.1945)

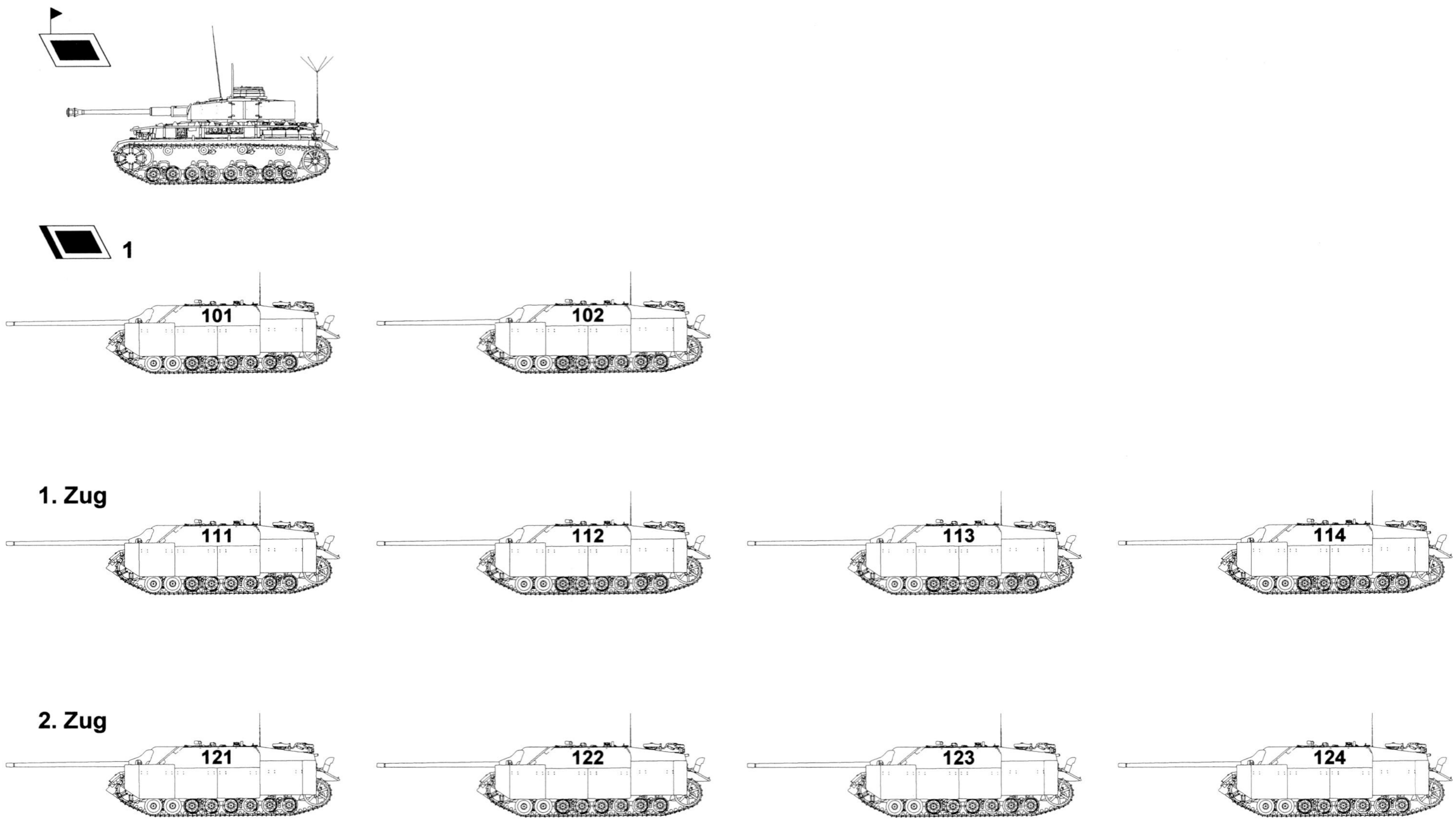

M.Jaugitz / Trojca 2000.

Gliederur g der Panzer - Abteilung (Fkl) 303 (Stand 01.03.1945)

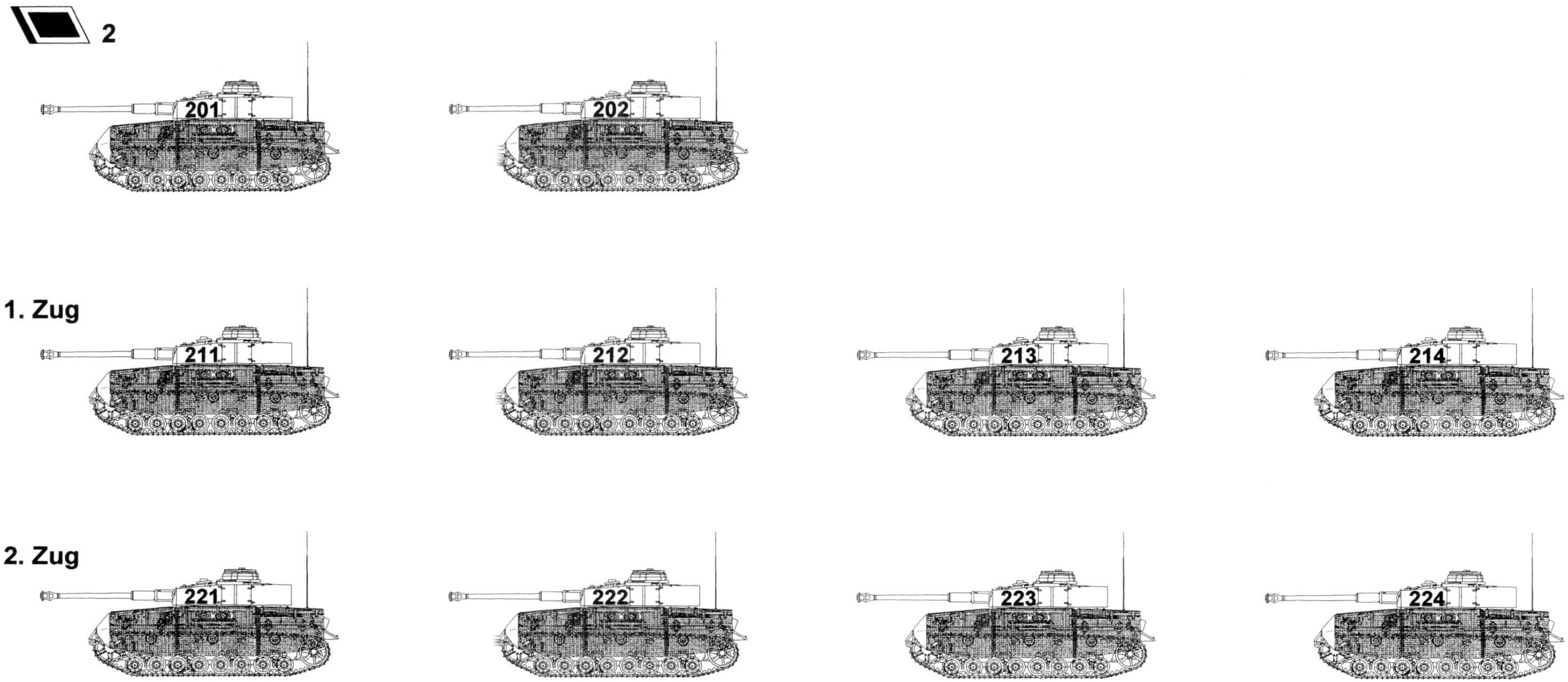

M.Jaugitz / Trojca 2000.

Gliederung der Panzer - Abteilung (Fkl) 303 (Stand 01.03.1945)

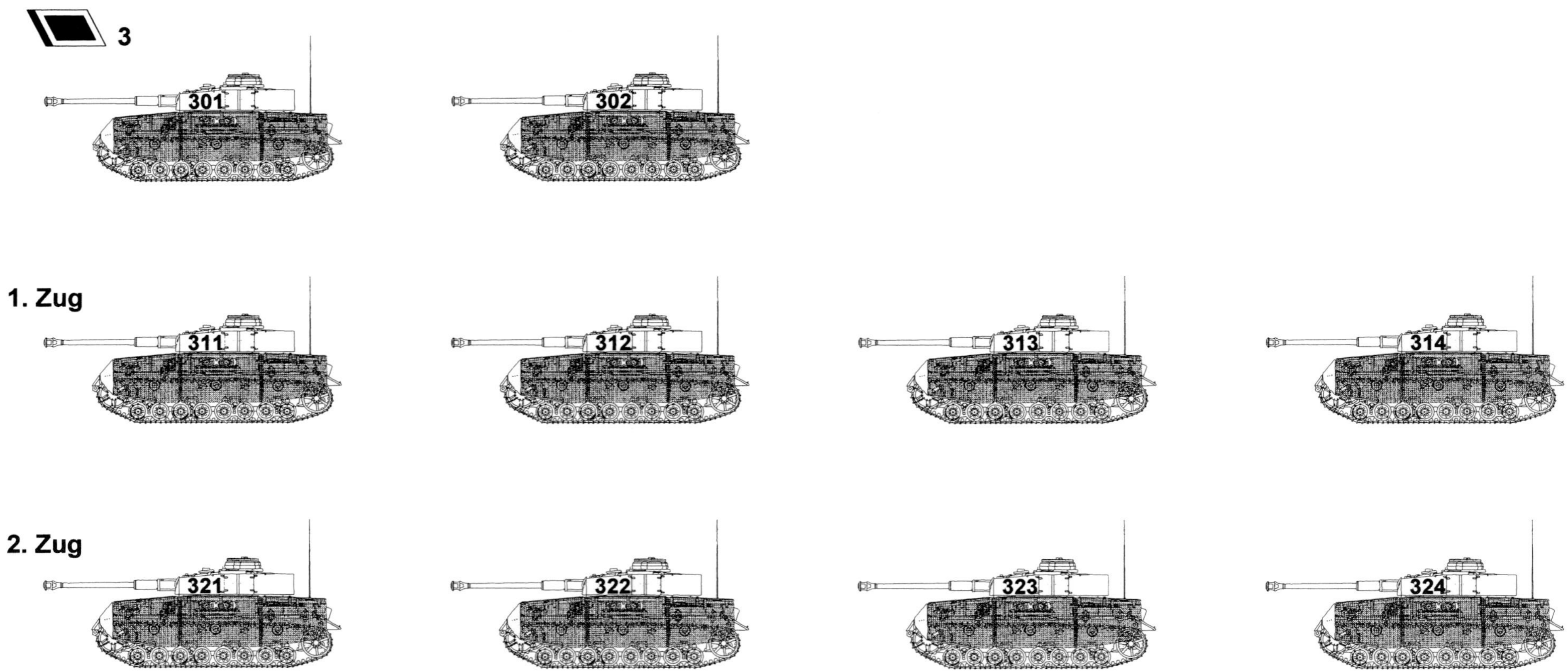

M.Jaugitz / Trojca 2000.

This page and top of next page: In January 1945 portions of the newly formed *Panzer-Abteilung (Fkl) 303* reached the Grafenwöhr Training Area. The personnel were trained on the *Panzerfaust* and the *Panzerschreck*. (Wenzel)

The battalion received its armored vehicles after being moved to Potsdam at the end of February 1945. The *2./* and *3./Panzer-Abteilung (Fkl) 303* received the final version of the *Panzer IV, Ausführung J*. (Wenzel)

The battalion was moved to the Frankfurt/Oder area on 28 February 1945. A few officers of the battalion supervise the rail-loading operation. (Wenzel)

2.2.19: Fkl-Panzer-Zug / Panzer-Abteilung (Fkl) 303

On 18 February 1945 the Inspector-General of Armored Forces directed radio-controlled demolition-charge carriers be committed with the *25. Panzer-Grenadier-Division*. *IN 6* was to put together the remote-control platoon from *Panzer-Abteilung (Fkl) 303* under the command of a veteran officer. The unit in charge of the formation process was the *Panzerjäger-Ersatz-Abteilung 3* in Potsdam. A written order was not issued until 23 February (Source: Bundesarchiv / Militärarchiv Freiburg):

—

REFERENCE: *OKH/Gen.Insp.d.Pz.Tr./Org. II No. 3659/45* secret of 18 February 1945

SUBJECT: Formation of *Panzer-Zug (Fkl) 303*

1.) Panzer-Zug (Fkl) 303 is to be formed by *W.Kdo. III* with support from *AHA/IN 6* in Potsdam as soon as possible.

2.) It is intended that *Panzer-Zug (Fkl) 303* be employed with the *25. Panzergrenadier-Division (Heeresgruppe Mitte)*. Recall by *GenStdH/Op.Abteilung*.

3.) Organization and strength: Tank platoon of four assault guns with 12 demolition-charge carriers in accordance with *KStN* and *KAN 1171 f* of 1 June 1944. Further: Tank and signals maintenance section.

4.) Available for the formation are:

Four *Sturmgeschütze (Fkl)* of the former *Panzer-Abteilung (Fkl) 303* en route to Potsdam.

Crews, necessary radio-control equipment plus vehicle maintenance section, which *AHA/IN 6* is having sent to Potsdam from the *Panzer-Versuchs- und Ersatz-Abteilung 300* in Eisenach.

5.) There will be no further allocations of equipment.

6.) *Panzer-Versuchs- und Ersatz-Abteilung 300* (*W.Kdo. IX*) is the designated replacement unit.

7.) Report completion of the formation process to

OKH/GenStdH/Op.Abteilung
OKH/GenStdH/Org.Abteilung
Gen.Insp.d.Pz.Tr.
AHA/Stab I a (1)
OKH/ObdE/AHA/STAB Ia (1)
No. 11238/45 geh. Of 23 March 1945

For the Commander:
(signature illegible)

—

Leadership of the platoon was given to *Leutnant* Rudolf Andreas, who hastily assembled his unit from the noncommissioned officers and enlisted men sent by the replacement battalion in Eisenach. As a result, it was possible to dispatch the remote-control platoon (consisting of three *Sturmgeschütz III's* and twelve *Borgward BIV, Ausführung C* demolition-charge carriers) to *Heeresgruppe Weichsal* at Wriezel on 22 February 1945. The *25. Panzer-Grenadier-Division* sent a telex message to confirm the unit's arrival.

It is not known if the remote-control platoon saw any committment in the last days of February. The *25. Panzer-Grenadier-Division* was still reporting the platoon at full operational strength in its vehicle summary dated 2 March 1945. Fourteen *BIV, Ausführung C* demolition-charge carriers were delivered from the beginning until the end of March 1945.

Since the beginning of February 1945 the *25. Panzer-Grenadier-Division* had been engaged in the defensive fighting around Küstrin. Following an unsuccessful attempt to relieve the garrison of Küstrin, the division was pulled out of the line and transferred to Wriezen for a quick refit near the front at the end of March. The division reported 16 demolition-charge carriers written off as total losses during the period 1 to 31 March 1945. Most of these vehicles had been lost in the fighting around Küstrin.

Since the remote-control platoon had proven successful in combat, the Inspector-General of Armored Forces issued an order to increase the strength of the small unit on 23 March 1945. (Source: Bundesarchiv / Militärarchiv Freiburg)

—

Telex Telex No. 5232 24 March 1945

To: (Various staffs)

Inspector-General of Armored Forces intends to strengthen the *Einsatzzug (Fkl)* of *Panzer-Abteilung 303* by sending personnel, weapons and armored vehicles to the *25. Panzer-Grenadier-Division*.

Panzer-Versuchs- und Ersatz-Abteilung 300 (Eisenach) is responsible for providing same.

Execution:

1.) Personnel to be provided by the *ObdE*.

Needed are:

6 Crews for assault guns (including 1 officer)
1 Supply Sergeant
14 *BIV* drivers
1 Vehicle mechanic (noncommissioned officer)
4 Vehicle mechanics (enlisted men)
3 Signals mechanics

Later issuance of the *"Springer"* demolition-charge carrier to be taken into consideration.

2.) Equipment

2 assault guns and 5 *BIV C's* are being allocated by the Inspector-General of Armored Forces from the stocks of *Panzer-Versuchs- und Ersatz-Abteilung 300*.
4 assault guns and 5 *BIV C's* are being allocated by the Inspector-General of Armored Forces from new production.

Associated ammunition, weapons and equipment.

Additionally, maintenance repair equipment as per *D 623/11* and remote-control special equipment to be provided by the *ObdE*

Necessary wheeled vehicles to be provided by the 25. Panzergrenadier-Division.

3.) *ObdE* is being requested for accelerated deliveries to the *Einsatzzug* of *Panzer-Abteilung (Fkl) 303*.

4.) Request orders.

The Inspector-General of Armored Forces

AbteilungOrg.II Nr. F 1289/45 geheim dated 23 March 1945

The Chief of Staff
(Thomale)
Generalleutnant

—

Following the arrival of reinforcements, the unit was reorganized and later incorporated into the *25. Panzer-Grenadier-Division* as the *4./Panzerjäger-Abteilung 25*. As part of the division it participated in the fighting withdrawal through Brunow-Kruge to Eberswalde. Together with other decimated units, it formed *Gruppe Steiner*. There followed counterattacks on Liebenwalde and the Ruppin Canal as well as defensive fighting around Oranienburg. The remnants of the *25. Panzer-Grenadier-Division* subsequently withdrew through Mecklenburg via Lake Malchin to Lake Schwerin where, in the first days of May, it surrendered to American forces.

The author has in his possession two letters from former members of the *Einsatzzug* of *Panzer-Abteilung (Fkl) 303* describing this period:

Former *Oberfähnrich* Johannes Pfleiderer (†) wrote:

In March 1945 I joined the replacement battalion in Eisenach from the officer candidate school in Wischau. After serving as Platoon leader in *3./Panzer-Abteilung 300* for a few days, I was detached to the *"Funklenk-Zug 303 — Andreas"*. I took a detail to the equipment depot at Döberitz, where I was issued six assault guns and five *Borgward BIV's*. We arrived in Werbig on or about Easter, probably on 4 April 1945. *Leutnant* Andreas was wounded. From the platoon a company was formed which was designated as the *4./(Fkl) Pz.Jäger.Abteilung 25*. *Leutnant* Frowein assumed command.

We were located in the Glanz Estate in Platkow, approximately 10-15 kilometers behind the front, where we functioned as the local reserve. The *BIV's* were guided from the assault guns and were used to blow up bridges and other similar objects during the withdrawal in the direction of Berlin.

After several quiet days, a fierce artillery duel began on the morning of 14 April 1945. The Soviet troops were advancing from the Oder west toward Berlin. For us there followed two weeks of mad retreats on roads clogged by refugees. Usually we drove the entire night. The Russian units first advanced to Seelow, but then they advanced north of the city toward Bernau and Oranienburg instead of heading for the center of Berlin. We used the *BIV's* to blow bridges, but I cannot remember when and where they were used. They did not play any decisive role.

Once, near Neustrelitz, there was a two-day break, during which the *4. Kompanie* was pulled out of the line to tend to its vehicles. We were running out of fuel and ammunition. I remember that we had to blow up some vehicles for that reason. Then at an abandoned airfield we captured some high-percentage alcohol. This we used to fuel our assault guns, but the engines only ran for a short time. When they became hot our new fuel evaporated. We therefore had one assault gun tow another. Every few kilometers we had to switch them around. It was a very laborious way to proceed. *Leutnant* Frowein was wounded sometime during this period. I assumed command of what was left of the company.

On the morning of 3 May 1945, after having driven all night, we were about to turn into a village when loud noises alerted us to the presence of Soviet troops. We were able to withdraw and several hours later came upon an American unit on the road to Schwerin.

We were captured. We sank two of the remaining assault guns in the lake. The other gave up the ghost soon after we had set out on the embankment. During the course of the day countless German soldiers arrived and gathered in a large square in the center of the town.

Further information was provided by Walter Jazbec, former *BIV* driver and finally loader in the assault gun commanded by *Oberfähnrich* Pfleiderer:

We arrived at the Oder in March, more or less a company. There were *BIV's* with us. I arrived as a *BIV* driver but then I was made a loader-radio operator in one of the six assault guns we picked up in March. It was brand-new, came from Döberitz and had a machine-gun with telescopic sight. The assault-gun commander was *Herr* Pfleiderer. I believe we were also the platoon leader's vehicle. Driver was Walter Czepluch; gunner Heinz Cornelius.

When the Russian offensive began, we were outside of Seelow. From then on our assault gun was in action almost constantly. We were committed in Müncheberg, Sieversdorf and Buckow on Lake Schermützel. At Lake Schermützel we still had *BIV's* with us. During the night we moved into an assembly area, heavily wooded. We were suddenly attacked and fired on. We climbed into our assault guns and got out of the woods. Then we realized that we had left two *BIV's* behind. At first light we drove back into the position and recovered the BIV's.

The next places we were employed were Straußberg and the sector between Straußberg and Wriezen. This is where we saw our last operation. I have no direct knowledge of *BIV* operations. We lay under fire at a railway line in the sector mentioned above. Transmission damage! Our vehicle refused to move from the spot. We summoned another assault gun and had it tow us into cover behind a hill. By radio we summoned a recovery vehicle, which soon arrived. We were towed to the maintenance company. The front was so fluid that we were constantly being moved west. Near Templin-Lychen the prime mover left us on the open road to go get another broken-down vehicle. *Oberfähnrich* Pfleiderer tried to find another vehicle to tow us. But the Russians were quicker and we blew up the assault gun. We headed west on foot and outside of Parchim near Goldenstädt we were taken prisoner by the Americans. An operational *BIV* from our platoon also arrived there. I believe the driver's name was Gehrmann.

Fkl - Panzer - Zug / Panzer - Abteilung (Fkl) 303 (Stand 20.02.1945)

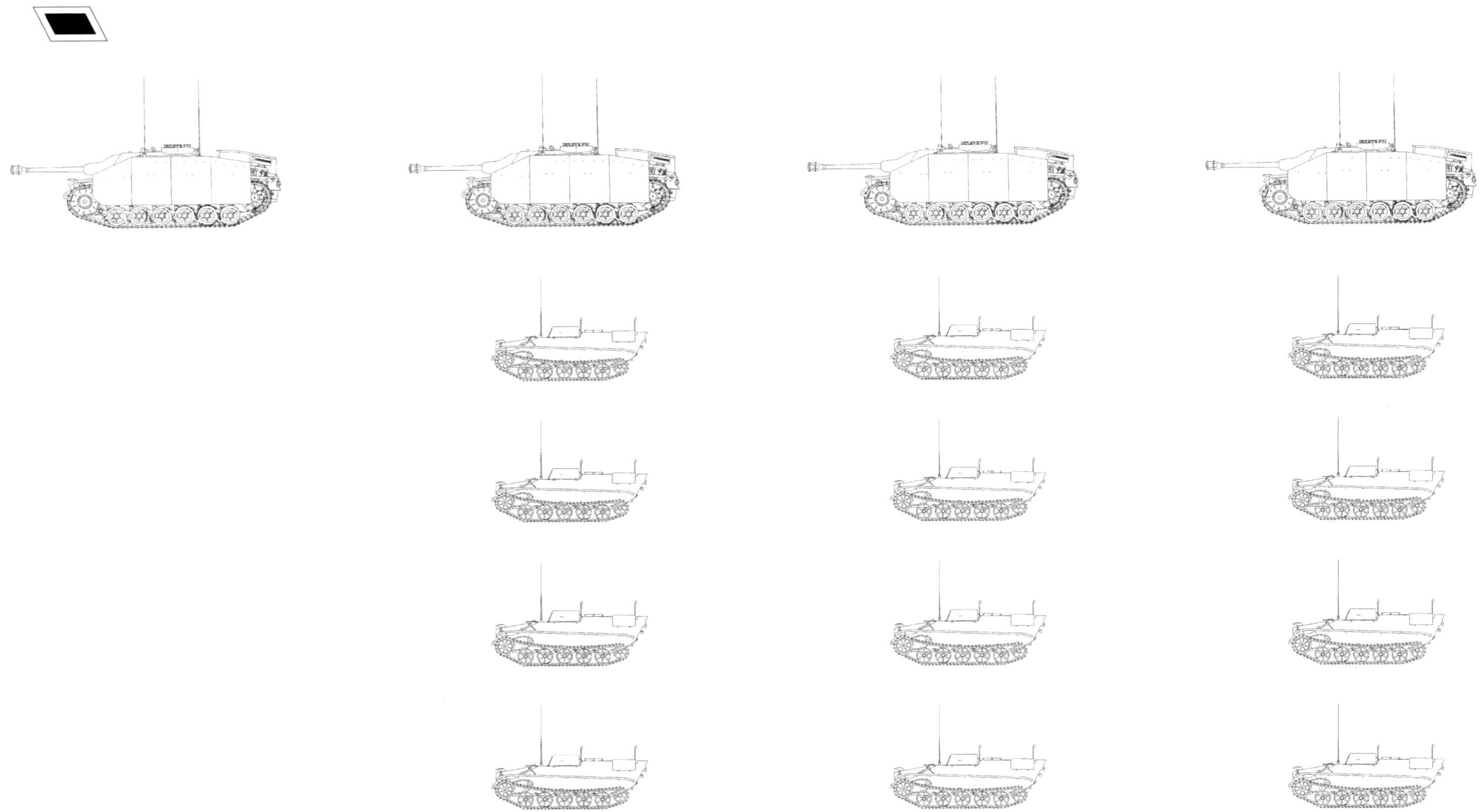

M.Jaugitz / Trojca 2000.

2.3: The Replacement and Training Formations

2.3.1: Panzer-Feldersatz-Kompanie (FL)

When *Panzer-Abteilung 300* was transferred to the Eastern Front, the need arose for a unit which could provide personnel to replace those killed, wounded or captured during operations.

On 18 May 1942 the *AHA* ordered *Panzer-Ersatz-Abteilung 5* in Neuruppin to form *Panzer-Feldersatz-Kompanie (F.L.)* for the battalion.

In accordance with *Kriegsstärkennachweisung 1161* of 8 May 1942, the replacement company was to consist of the following personnel:

Company Personnel

Officers: 1

Noncommissioned officers: 17

Enlisted: 21

Replacement Personnel

Officers: 2 (Platoon Leaders)
Noncommissioned officers: 10 (vehicle commanders (also remote-controlled section leaders)
Enlisted: 148 (gunners, loaders, radio operators, drivers, combat engineers, tank mechanics, signals mechanics and *BIV* drivers)

Total

Officers: 3

Noncommissioned officers: 27

Enlisted: 169

Since the unit had only a few vehicles, it had to rely on the motor pool of *Panzer-Ersatz-Abteilung 5.*

On 1 November 1942 the remaining personnel of *Panzer-Feldersatz-Kompanie (F.L.)* were transferred to the newly formed *Panzer-Ersatz-Abteilung 300* where they served as the cadre for the *2./Panzer-Ersatz-Abteilung 300* (see Chapter subsection 2.3.3).

2.3.2: Panzer-Versuchs- und Ausbildungs-Kompanie (FL) 300

After *Panzer-Abteilung 300* had been transferred from Cottbus to the Eastern Front for employment in the Crimea, the question arose, which unit in Germany would take over the testing of demolition-charge carriers and train specialists in the use of wire- and radio-controlled special vehicles.

An *AHA* directive *(AHA I a II 15843/42)* dated 29 May 1942 called for the formation of *Panzer-Versuchs- and Ausbildungs-Kompanie (FL) 300* by *Panzer-Ersatz-Abteilung 5* in Neuruppin. *Panzer-Ersatz-Abteilung 5* was also responsible for ensuring that sufficient qualified personnel were available for *Panzer-Versuchs- und Ausbildungs-Kompanie (FL) 300.*

On 3 October 1942 the *AHA* ordered that the company's personnel and equipment were to be used in the formation of *Panzer-Ersatz-Abteilung 300* (see Chapter subsection 2.3.3).

2.3.3: Panzer-Versuchs- und Ersatz-Abteilung 300 (Fkl)

On 3 October 1942 the *AHA* ordered the formation of *Panzer-Ersatz-Abteilung 300 (Fkl) (I a VII 31475/42 g.).*

Formation of the battalion, which was to consist of a battalion headquarters, a replacement company and a testing and instructional company, had been completed at Neuruppin by 1 November 1942. The personnel and equipment of *Panzer-Versuchs- und Ersatz-Kompanie 300 (Fkl)* and those members of *Panzer-Ersatz-Abteilung 5* who were in *Panzer-Feldersatz-Kompanie (FL)* were to be incorporated into the replacement battalion.

At the end of the year *Panzer-Ersatz-Abteilung 300* was renamed and reorganized. Its new designation was *Panzer-Ersatz-Abteilung (Fkl) 300.* The unit consisted of a headquarters and four companies. The *1. Kompanie* was organized as a testing and instructional company, the *2.* and *3. Kompanien* were training companies, and the *4. Kompanie* was a replacement company. At the same time the unit was ordered to move to Eisenach. The *6./Panzer-Ausbildungs-Abteilung 18* was used for the formation of the training companies.

On 9 March 1943 the *AHA* ordered another redesignation and the battalion was renamed *Panzer-Versuchs- und Ersatz-Abteilung 300 (Fkl).*

On 18 March 1943, a few days after he became Inspector-General of Armored Forces, *Generaloberst* Guderian visited the unit at its garrison at Eisenach.

At this time both the separate Funklenk-Kompanie 312 and Kompanie Abendroth were attached to Panzer-Versuchs- und Ersatz-Abteilung 300 (Fkl).

In the summer of 1943 the battalion's officer complement was as follows:

Battalion commander: *Major* Weicke

1./Panzer-Versuchs- und Ersatz-Abteilung 300 (Fkl): *Oberleutnant* Fischer

2./Panzer-Versuchs- und Ersatz-Abteilung 300 (Fkl): *Oberleutnant* Bauche

3./Panzer-Versuchs- und Ersatz-Abteilung 300 (Fkl): *Oberleutnant* Schlüter

4./Panzer-Versuchs- und Ersatz-Abteilung 300 (Fkl): *Leutnant* Glaessel

The battalion was equipped with a very mixed complement of vehicles. Both older versions of the *Panzerkampfwagen III* and various variants of the *Sturmgeschütz III* were used as command and control vehicles. Demolition-charge carriers no longer fit for operational employment use were employed for driver and operator training.

Later on, *Panzer-Versuchs- und Ersatz-Abteilung 300 (Fkl)* also included the *Panzerkampfwagen IV*, the *Panther* and prototype radio-controlled vehicles in its complement of vehicles.

The *2.* and *3. Kompanien* were responsible for the training of personnel. This focused on the training of special drivers for the remotely controlled demolition-charge carriers and specialists for maintaining and repairing the special radio equipment. The *4. Kompanie* was the replacement company and was responsible for immediately making good losses suffered by units at the front. The *1. Kompanie* had a special mission. It consisted of three platoons under the command of *Leutnant* Dannenberg (1st), *Leutnant* Wizgall (2nd) and *Leutnant* Haas (3rd). Each platoon of this testing unit had a different function.

For this reason, the *1./ Panzer-Versuchs- und Ersatz-Abteilung 300 (Fkl)* was frequently visited by representatives of *Inspektion 6 (IN 6)* of the *Allgemeines Heeresamt* (*AHA* = General Army Office) and the *Heereswaffenamt* (*WaA* = Army Ordnance Office).

In the autumn of 1944 *Hauptmann* Sahmel of *Abteilung WAPrüf 6* (Section for the Development and Evaluation of Tanks and Motor Vehicles) became commander of *Panzer-Abteilung 302 (Fkl)*. He had been involved in the development of the gasoline-powered *Goliath (Sd.Kfz. 303)*. It was also to the advantage of the *1./Panzer-Versuchs- und Ersatz-Abteilung 300 (Fkl)* that the Army Ordnance Office's motor transport test station (abbreviated *Verskraft*) had a workshop and small test stretches on the training area at the Berka garrison.

Among the tasks assigned to the *1./Panzer-Versuchs- und Ersatz-Abteilung 300 (Fkl)* were the development of an unmanned battlefield reconnaissance vehicle equipped with a television camera and testing of amphibious demolition-charge carriers with floating charges. In addition, experiments were conducted with guiding a demolition-charge carrier by radio from an aircraft.

Leutnant Wizgall played a significant part in the development of the *Sd.Kfz. 304 Springer* demolition-charge carrier.

Early on the company was equipped with a Tatra vehicle equipped with an experimental model of a night-vision device. Based on this, development work proceeded on the infrared device and an image sensor. *Leutnant* Haas was responsible for testing and evaluation of the system, both as a driving aid and a targeting device for night fighting.

At the beginning of 1945 *Versuchszug Leutnant Haas* (Experimental Platoon *Leutnant* Haas) was formed with four *Panthers* equipped with image-sensor devices. The platoon was sent to Hungary and was employed in the fighting at Lake Balaton. Commanding the *Panthers* were *Leutnant* Haas, *Feldwebel* Boll, *Feldwebel* Deichfuß and *Unteroffizier* Cibis. The experimental platoon achieved a number of kills with the new equipment and was much feared by the enemy side — mainly because of its night operations.

At the end of March 1945 the majority of the personnel of the *2.*, *3.* and *4./Panzer-Versuchs- und Ersatz-Abteilung 300 (Fkl)*, together with most of their wheeled and tracked vehicles, were attached to the newly formed *Panzer-Ausbildungs-Verband Thüringen* (Armored Training Unit Thuringia).

After arriving at the front, *Panzer-Ausbildungs-Verband Thüringen* was used to reconstitute the strength of the *2. Panzer-Division*. Many former members of *Panzer-Versuchs- und Ersatz-Abteilung 300* saw action again northwest of Frankfurt between Fulda and Schlüchtern using barely operational armored vehicles.

(Raum für weitere amtliche Eintragungen, insbesondere für Bedingungen der Erlaubnis oder Ausdehnung der Erlaubnis nach Ergänzungsvorschriften).

Nur gültig für die Dauer des Dienstverhältnisses bei der Wehrmacht. (Bei der Entlassung abzugeben.)

Wehrmacht-Führerschein

für

den Bruno Müntz
(Vor- und Zuname)

Truppenteil usw.: Panzer-Schütze 3./Pz.Vers.u.Ers.Abt.300(Fkl)

Standort Eisenach

geboren am 4. August 1924

Der Bruno Müntz
(Vor- und Zuname)

erhält die Erlaubnis, nach bestandener Prüfung ein Kraftfahrzeug mit Antrieb durch Verbrennungsmaschine der Klasse - 2 - zu führen.
nur für Vollketten-Kfz. gepanzert bis 15 to.

Eisenach, den 18. Mai 1943

Panzer-Versuchs-u. Ers.-Abteilung 300

(Eigenhändige Unterschrift des Inhabers)

Driver's License for *Panzerschütze* Bruno Müntz of the *3./Panzer-Vesuchs- und Ersatz-Abteilung 300 (Funklenk)*. He has been authorized to drive fully tracked vehicles up to 15 tons.

Above: After the battalion was formed, a large portion of the armored vehicles of *Panzer-Abteilung 301* were transferred into it. (Geissel)

Above and below left and right: Training is also conducted on the proper handling of smoke and thermal charges for conducting individual antitank fighting. Here: A Molotov Cocktail is used against a T 34. (Susenbeth)

Experienced tank commanders trained the new recruits. Here: *Oberfeldwebel* Harbort conducts combat training. (Susenbeth)

Training on the main gun, fieldcraft and the radio equipment were part of the training schedule in both training companies. Note the difference on the *Panzer* collar tabs of the recruits. (Susenbeth)

Above and top right: During the annual Armed Forces Day the armored vehicles were put on display for the Eisenach populace. This was an especially interesting event for the children and young people. Here: A *Sd.Kfz. 250* and a modified *Panzer III, Ausführung G* (Chassis number 65075). (Susenbeth)

Opposite page: The recruits were trained on the *BIV's* at the nearby Berka Training Area. They first had to be tested to receive the appropriate tracked driver's license. "Driving School" signs have been mounted on the demolition-charge carriers. (Lock)

Above: *Leitpanzer 213* has almost sunk up to its return rollers in the mud. (Lock)

Left: Training to be a tank driver was considerably more extensive. In addition to actual training in the tank hull, there was also the requirement to react to critical situations. (Lock)

Right-hand images: In the first months of its existence, *Panzer-Versuchs- und Ersatz-Abteilung 300 (Fkl)* had two *Sturmgeschütz III, Ausführung D* vehicles in its inventory. Both vehicles were painted a sand color. A mount for the radio-control antenna can be seen on the right-front of the fighting-compartment roof. (Jung)

Below: A *Panzer III* of the battalion moves cross-country during a training exercise.

Additional views of the two *Sturmgeschütz III*, *Ausführung D* vehicles discussed on the previous page. (Jung)

Above: A section has moved out for an exercise. The demolition-charge carriers follow the command and control vehicle. *Leutnant* Glässel is in the hatch of the assault gun. (Jung)

Top right: Group photo after the successful completion of the driving school. (Jakob)

Right: The battalion commander, *Major* Weicke (middle), with a few officers. Left — *Hauptmann* von Abendroth; right — *Major* Weicke and *Oberleutnant* Nolte. (Schatz)

Panzer-Versuchs- und Ersatz-Abteilung 300 (Fkl) was visited on numerous occasions by high-ranking officers of the *Allgenmeines Heeresamt* and the *Heereswaffenamt*. At this demonstration the complete palette of demolition-charge carriers was put on display. (Susenbeth)

Above: A radio-controlled *BIV* demonstrates its functioning through the use of smoke dischargers. The demonstration is critically observed. (Susenbeth)

Below: The general has the officers and specialists brief him with additional information on the radio-control weapons. (Susenbeth)

The company commander of the *1. (Versuchs-)Kompanie*, *Oberleutnant* Fischer, with his platoon leaders, *Leutnants* Wizgall and Dannenberg. Note the tactical insignia of the *1./Panzer-Versuchs und Ersatz-Abteilung 300 (Fkl)* on the right fender of the *Kfz. 16*. (Susenbeth)

Oberst Mildebrath of *Inspektion 6* of the *Allgemeines Heeresamt* is given a demonstration of the operational possibilities of radio-controlled vehicles. (Susenbeth)

A few officers of the battalion used the opportunity to fire machine guns. (Susenbeth)

Swearing in new recruits at the garrison of *Panzer-Versuchs- und Ersatz-Abteilung 300 (Fkl)*. Some of the assault guns have the mounts for the side skirts. (Susenbeth)

The battalion received the *Panzer V "Panther"*, *Ausführung D* in June 1943. (Jakob)

The same *Panther* in front of the vehicle garages. Next to it is the experimental version of the *Borgward BIV*, *Ausführung C*. Note the cover on the engine compartment. (Jakob)

The chassis numeral 211038 has been painted on the gun mantlet and the side of the turret on another *Panther*. (Jakob)

Above and next page: One of the three *Panthers* was later used as a driving school vehicle. It was converted to use propane gas in that capacity. The closure over the machine-gun port and the antenna mount on the turret roof indicate this was a command tank. (Jakob)

Panzer-Versuchs- und Ersatz-Abteilung 300 (Fkl) also had a *Panzer III, Ausführung H* in its inventory. (Wimmer)

Part of the training schedule included convoy practice in the area of the Eisenach garrison. (Wimmer)

Recruits on a command and control vehicle. The tactical insignia of the *3./Panzer-Versuchs- und Ersatz-Abteilung 300 (Fkl)* has been painted on the front slope. (Jakob)

Left and below: This vehicle has also been converted to use propane gas. The bottles have been mounted on the rear deck. Jakob)

Left and next page: Training *BIV* drivers in the field. (Jung)

PV-394

Taking a break at the training area. (Jung)

This *BIV* was used for transporting food and other containers. (Jakob)

Occasionally, a live charge was ignited in order to accustom soldiers to it. (Hanke)

Whenever the units bivouacked at the Berka Training Area, guard posts were set up for the *BIV's* and the other armored vehicles. No other information is available on the *Panzer IV*, *Ausführung G*. (Jakob)

Above and next four pages: In addition to the *Sturmgeschütz III*, *Ausführung G*, the battalion also had two *Sturmgeschütz III*, *Ausführung F* vehicles for training drivers, loaders and gunners. One of the vehicles had the chassis number 91253. Some of the assault guns were also converted to use propane. (Schwarz)

Exchanging war stories during a break in the action (Cornelius)

Above: The successor to the battalion commander, *Major Weicke*, was *Hauptmann* Wollschläger. Until he was wounded in February 1943, *Hauptmann* Wollschläger was the commander of *schwere Panzer-Abteilung 502*. (Wizgall)

Left: *Leutnant* Wizgall was the platoon leader of the 2nd Platoon of the *1. (Versuchs-)Kompanie*; *Leutnant* Haas was the platoon leader of the 3rd Platoon. (Wizgall)

Oberleutnant Bauche was the company commander of the *2./Panzer-versuchs- und Ersatz-Abteilung 300 (Fkl)*. (Jung)

Above and below: A *VW-Schwimmwagen* (*Sondertyp 129*) was employed with *Panzer-Versuchs- und Ersatz-Abteilung 300 (Fkl)* in 1944. A portion of the special equipment has been removed. (Susenbeth)

Hauptmann Sahmel (at the time of the picture still with *IN 6*) and *Leutnant* Wizgall during a test on the Baltic Coast. The experimental company had the task of outfitting a *BIV* with a "swim" set and conducting tests as an amphibious demolition-charge carrier. (Wizgall)

The tests were conducted in close cooperation with the *NSU* company. To the right in the picture is *Dr.* Froede, the designer of the *NSU-Kettenkrad*. (Wizgall)

Trials of the radio-controlled *NSU-Kettenkräder*. *Leutnant* Wizgall can be seen in the image above. (Wizgall)

Left and below left: *Leutnant* Haas in a *NSU-Kettenkrad* during a demonstration at the Eisenach garrison. The picture of the interior shows details of the steering mechanism. (Susenbeth)

Below: *Leutnant* Wizgall during a demonstration for higher political leaders. To the left of *Leutnant* Wizgall is Propaganda Minister Goebbels. (Susenbeth)

Above: An experimental model of the *Goliath* is demonstrated for Hitler. (Kugler)

Top right: A *BIV* with a television camera is next to a 3-ton *Opel Blitz*. The *Seedorf P* television receiver is located in the built-up area. (Susenbeth)

Right: The *BIV* with the television camera was also demonstrated to Hitler and senior generals. (Trenkle)

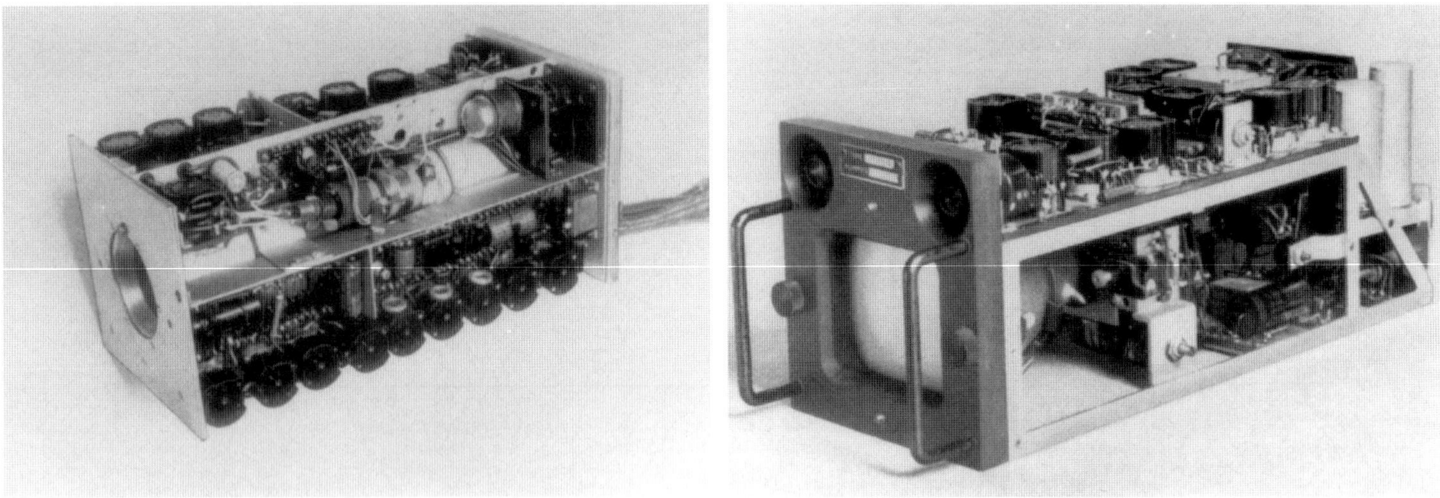

Details of the *Tonne P* television camera. (Trenkle)

This *BMW 326* cabriolet was also outfitted with a night-vision device. (Münch)

The first experiments with night-vision devices were conducted by *Panzer-Versuchs- und Ersatz-Abteilung 300 (Fkl)* in the spring of 1943. A Tatra 97 staff car was used as the basis for the trials. (Susenbeth)

The same vehicle in 1944. Note the ventilation slits for the 4-cylinder Boxer engine in the trunk. (Susenbeth)

Personnel were detached from *Panzer-Versuchs- und Ersatz-Abteilung 300 (Fkl)* in 1943 for the trials on the *Sprengboote* (demolition boats). Steering of these boots was accomplished with components employed on the *BIV*. (Hauptmann)

Elements of *Panzer-Versuchs- und Ersatz-Abteilung 300 (Fkl)* were integrated into *Panzer-Ausbildungs-Verband Thüringen*. The battalion was employed to the north-west of Frankfurt/Main. (Regenberg)

2.4: Other Special Units

2.4.1: Munitionsschlepper-Kompanien 801-806

The infantry began demanding an armored ammunition carrier as early as 1937. The Carl F.W. Borgward Company was tasked with its development and designed a heavy tracked carrier weighing 3.5 tons and armored against small-arms fire. A 55 horsepower, Type 6 M 2.3 RTBV engine produced by Borgward was intended to deliver the required performance of 30 kilometers per hour. The armored ammunition carrier, which was capable of carrying an internal load of 500 kilograms, could also tow an ammunition trailer carrying an additional 500-kilogram payload.

A pre-production series was built, of which 20 examples were used for service trials and three by *WaPrüf 6*. An order was placed for another five, of which three were to be delivered with riveted hulls. The armored ammunition carrier was given the designation *VK 301/302*.

In addition to the pre-production series, Borgward was issued a contract to build 100 vehicles. Deliveries were supposed to begin at the end of 1940, however, production was very slow and was halted at the end of 1942 after the delivery of 45 vehicles.

These armored ammunition carriers were sent to the infantry school at Döberitz where they were intended to form the basis of a new armored ammunition carrier company.

Contracts were issued for development of a 50-mm self-propelled antitank gun based on the *VK 302* as well as a self-propelled carriage for a 105-mm *Leichtgeschütz 350* recoilless rifle. Two prototypes of the self-propelled antitank gun were built, however, these did not prove a success in service trials. Only one prototype of the recoilless rifle self-propelled carriage was built.

gepanzerte Munitionsschlepper-Kompanie 801

On 26 March 1943 the *AHA* issued an order (*AHA I a (1) No. 10215/43*) for the formation of *gepanzerte Munitionsschlepper-Kompanie 801*. It was intended to incorporate the company into the *78. Sturm-Division* as an experimental company. In fact, however, *gepanzerte Munitionsschlepper-Kompanie 801* was assigned to the *1. Infanterie-Division* in June 1943.

In order to assess the operational use of the special vehicles as quickly as possible, a detailed report from the 1. Infanterie-Division was requested. (Source: National Archives, USA)

—

REFERENCE: *Gen.St.d.H./Op.Abt. III/6105/43 geheim*, dated 11 June 1943.

SUBJECT: Attachment of *gepanzerte Munitions-Schlepper-Kompanie 801*

The above-cited order attached the *gepanzerte Munitions-Schlepper-Kompanie 801* to the *1. Infanterie-Division* for service trials.

The Borgward Company was tasked with the construction of the armored ammunition carrier and produced 45 examples. The specifications requested were not met, however, and production was halted.

In order to acquire operational experience in the potential, methods of employment and command and control relationships relating to the armored ammunition carrier, the company was equipped with the 45 existing *Borgwards*. It should be mentioned that no delivery of spare parts for disabled carriers was possible.

The armored ammunition carriers are to be employed in such a manner that the questions asked in the attached questionnaire can be answered unequivocally.

Interim Report Dated 15 September 1943 is enclosed as an attachment.

1 enclosure

DISTRIBUTION:

Organizations-Abteilung

Gen.Qu.

AHA/In 2

—

Questionnaire

Introduction

The armored-ammunition-carrier company is solely combat-support unit and not a combat-arms formation.

The Borgward armored ammunition carrier has a limited performance and development was not completed. It is no longer being produced. Therefore, mechanical trials are not to be carried out.

The purpose of the service trials with the armored-ammunition-carrier company is to gain experience in ammunition transport within the infantry regiment on the battlefield between division supply columns or forward ammunition issue points and the infantry battalions and companies.

The armored ammunition carrier which is to enter service in the future is to be a fully tracked vehicle with the following performance specifications: gross weight 6,000 kilograms, payload up to 2,000 kilograms plus 4 personnel, speed up to 30 kilometers an hour, armored against small-arms fire and shrapnel.

The trials are to answer the following questions:

A) Organization:

How many armored ammunition carriers are needed for each infantry battalion? (Anticipated about 6)
In addition, how many armored ammunition carriers are needed for use by the regiment? (Anticipated about 7)

B) Personnel Complement:

How many personnel will each platoon need?
Which company should the platoons be attached to?
Can the regiment's maintenance section maintain the ammunition carriers?

C) Concentrated Use of the Armored Ammunition Carriers:

Will concentrated use by a battalion cause problems?
What about supply?

D) Armament:

What armament do the platoons require?

E) Delivery of Ammunition:

Is the delivery of ammunition to the infantry's weapons in the front line possible?

F) Operations:

First day of operations on ___
Number of days of operations: ___
Number of missions ___
Company committed as an organic whole?
Committed by platoon?
Single armored ammunition carriers committed?
Day missions?
Night missions?
Successful on the battlefield?
Type of delivery of ammunition and equipment to the troops on the battlefield?

G) Road March:

1) Lessons learned during the march?
2) Speed?
3) Night marches?
4) Did the company provide its own security?
5) Were security elements provided by the division?
6) Was there fighting with enemy or partisans?

H) Missions:

1) How many kilograms of the various types of ammunition were requested?
2) Transport of equipment?
3) Movement of personnel?
4) Movement of wounded?
5) Could the company provide a complete basic load of ammunition to a regiment of the division employed at the front?

J) Special Technical Lessons Learned:

K) Interim (Result) Assessment:

L) Reductions to the Tables of Organization:

How many ammunition vehicles, horses and soldiers could be eliminated given the organization proposed under A)

	In positional warfare	In mobile warfare
Infantry Battalion		
Infantry Company		
Machine-Gun Company		
Regiment		
Infantry Gun Company		

M) Other Suggestions:

Unfortunately, no information has surfaced to date to indicate the results of the report. According to information from the *1. Infanterie-Division* the vehicles were a success in frontline operations. In the summer of 1944 the ammunition-carrier company was sent back to the infantry school at Döberitz and temporarily disbanded. The unit's experienced personnel were intended to be used to form new armored ammunition carrier companies.

An order issued on 6 August 1944 (*AHA/Stab I Nr. 33790/44 geh.*) instructed the infantry school at Döberitz to form *gepanzerte Munitionsschlepper-Kompanien 801* to *806* using the former members of the "old" *gepanzerte Munitionsschlepper-Kompanie 801*.

Instead of the *VK 301/302*, the new companies would be equipped with converted *Borgward BIV, Ausführung A* demolition-charge carriers which were no longer usable in their intended role.

Like its predecessor, the reconstituted *gepanzerte Munitionsschlepper-Kompanie 801* was assigned to the *1. Infanterie-Division*. In autumn 1944 the unit was engaged in defensive fighting in the East Prussian/Baltic border region and, after operations at Schloßberg, it withdrew to Königsberg. While part of the division was evacuated by sea from Pillau, the rest was taken prisoner by the Soviets in the Samland.

gepanzerte Munitionsschlepper-Kompanie 802

After being issued with *Borgward BIV* armored ammunition carriers, *gepanzerte Munitionsschlepper-Kompanie 802* was assigned to the *170. Infanterie-Division*. After a fighting retreat, the division withdrew from the Suwalki area into the Königsberg area on the Baltic coast in late summer. By then it had been reduced to a *Kampfgruppe*. On 13 April 1945 *gepanzerte Munitionsschlepper-Kompanie 802* was disbanded and its remaining personnel were attached to the *AOK Ostpreußen*.

gepanzerte Munitionsschlepper-Kompanie 803

Gepanzerte Munitionsschlepper-Kompanie 803 was the third armored ammunition carrier company to be formed and was attached to the *28. Jäger-Division*. There is an interesting report on the company for the period January to March 1945. (Source: Bundesarchiv/Militärarchiv Freiburg)

—

28. Jäger-Division location classified, 12 March 1945

To

OKH/Gen.d.H./Gen.d.Inf.

b. Chef Gen.St.d.H.

As requested, the division encloses the requested situation report on *gepanzerte Munitionsschlepper-Kompanie 803*.

During the course of the offensive and defensive fighting in East Prussia in January of this year the armored ammunition carriers had to be blown up one after another as fuel was issued exclusively for assault guns and command purposes. Loading aboard trains was also not possible.

On 29 January a horse-drawn transport column was formed from personnel in the unit and those attached to it.

As a result of the order by the *4. Armee* to comb the supply units for excess personnel and sharply curtail their strength to the benefit of the infantry, 6 noncommissioned officers and 19 enlisted men were transferred to the infantry at the beginning of January. On 11 March 1945 the company was disbanded. The rest of the company's personnel:

4 noncommissioned officers (including 1 medic) and

14 enlisted men

were incorporated into units of the commander of the *Division Nachschubtruppe 28* since they were not fit for combat duties.

(signature illegible)

—

gepanzerte Munitionsschlepper-Kompanie 803 location unknown, 8 March 1945

To

OKH/Gen.d.H./Gen.d.Inf.

b. Chef Gen.St.d.H.

Report
for the period from 4 January to 8 March 1945

On 4 January 1945, by order of the *28. Jäger-Division*, command of the *gepanzerte Munitionsschlepper-Kompanie 803* was transferred to *Oberleutnant* Hartwich.

At that time the company's personnel and materiel strength was as follows:

1. Personnel:

   1 officer; 9 noncommissioned officers; and, 36 enlisted men

2. Vehicle situation: 15 Sd.Kfz. 301's of which 13 were serviceable, 1 in short-term repairs and one conditionally operational due to a cracked cylinder block.

   1 medium motorcycle; 2 assigned trucks; and, 1 small field kitchen stove.

3. Weapons and equipment:

   Seitengewehr 84/98: 19
   Karabiner 98k: 37
   Karabiner 43: 6
   M.Pi. 38/40: 3
   Rifle grenade launchers: 3
   Flare pistols: 6
   Binoculars 10 x 50: 4
   Binoculars 6 x 30: 6
   March compass A: 11
   Flashlights: 20
   Cleaning device 34: 23
   Small receiver: 1

Operations: On 12 January 1945 2 platoons = 10 *Sd.Kfz.* to *Infanterie-Regimenter 83* and *49*

Performance:

By night: 14 trips with 8.5 tons of small-arms ammunition

By day: 8 trips with 5.6 tons of small-arms ammunition

When the withdrawal movements began in mid-January 1945, the 10 carriers in service were loaded with ammunition and equipment. Because of the fast withdrawal which was ordered from the Goldap area, these had to be blown up or otherwise destroyed, as follows:

On 22 January 1945: Vehicle chassis number 360214 by order of the *28 Jäger-Division Ib* because of lack of spare parts. Hand tools and spare tracks had to be destroyed because of a shortage of transport.

On 23 January 1945: Vehicle chassis number 360214 by order of the company commander because it was unserviceable and repair behind the main line or resistance was impossible.

On 24 January 1945: Due to lack of fuel vehicles 360258, 360400, 360438, 360460, 360465, 360468, 360507 and 360575 were destroyed by order of the commander of the division supply units after consultation with the division.

On 26 January 1945: By order of the commander of division supply units all secret files and documents burned.

On 28 January 1945: Vehicles chassis number 360509, 360564, 360343 and 360395 at the *Jäger-Regiment 49* location.

On 4 February 1945: Due to lack of fuel vehicle chassis number 360710 was destroyed by order of the commander of the division supply units after consultation with the division.

Personnel changes: To bring the infantry up to strength, by order of the division,

on 7 February 1945 4 noncommissioned officers and 14 enlisted men
on 3 March 1945 2 noncommissioned officers and 5 enlisted men

were transferred to combat-arms units.

Organization and Status of the Company on 8 March 1945

Actual strengths:	Officers	NCOs	Enlisted
	1	3	17
Departed:			
Hospital:	—	1	3
Leave:	=	=	2
Present:	1	2	12

Weapons and equipment:

Karabiner 98k: 16
Karabiner 43: 3
Seitengewehr 84/98: 2
M.Pi. 38/40: 2
Rifle grenade launchers: 3
Flare pistols: 5
Cleaning device 34: 12
March compass A: 6
Binoculars 10 x 50: 2
Binoculars 6 x 30: 3
Flashlights: 13
Field kitchen stove: 1
Small receiver: 1

The missing weapons and equipment were given to the noncommissioned officers and enlisted men transferred to the combat-arms units troops.

Vehicles:

1 medium motorcycle
2 trucks attached from the division.

The vehicle list and documents for the carriers are in possession of the company.

By order of the division, the remnants of the company, plus personnel from other units, volunteers and horse-drawn vehicles, were used to form a transport column with a capacity of 20 tons on 29 January 1945.

The operations described at the beginning were conducted under the most difficult road and terrain conditions. The carriers proved their worth. No large-scale operations were undertaken because of the fuel shortage. The absence of ice cleats was particularly noticeable when the roads were ice-covered.

signed Harwich

Oberleutnant and Company Commander

—

The *28. Jäger-Division* fought in the Goldap-Angerapp area in 1944/1945, after which it withdrew via Braunsberg to Elbing. *Gepanzerte Munitionsschlepper-Kompanie 803* was disbanded before the division was lost in the Heiligenbeil pocket.

gepanzerte Munitionsschlepper-Kompanie 804

Formed in October 1944, *gepanzerte Munitionsschlepper-Kompanie 804* was incorporated into the *198. Infanterie-Division* which was then located in Alsace.

It is interesting to note that deliveries to the division included a prototype of a converted *Borgward BIV, Ausführung A* equipped with a gear transmission.

The division tested the vehicle and filed the following report: (Source: Bundesarchiv/Militärarchiv Freiburg)

—

Report on Test Run on 14 March 1945

The following was determined after today's test drive in terrain west of Oak Mountain with two *BIV A* demolition-charge carriers converted into ammunition carriers:

a.) Chassis number 360261 equipped with Trilok transmission A II 260

b.) Chassis number 360504 equipped with 5-speed gear transmission

Concerning a.) The run was made with a payload of 820 kilograms (including crew) on and off the road (normal payload 620 kilograms including crew). A speed of 30 kilometers an hour was reached on the road. Maneuverability, achieved by braking the steering mechanism, is very poor. Turning radius is 7 meters. Fuel consumption is very high since the vehicle must nearly always be driven at full throttle. During the test run the vehicle consumed 10 liters of fuel while covering 8 kilometers.

Concerning b.) The vehicle with the gear transmission was driven in the same terrain with the same load (including crew) of 820 kilograms. Road speed was 40 to 42 kilometers an hour. A climb gradient of 30 to 35 degrees was achieved. Every incline was negotiated 30% faster than with the above-described vehicle. The rigid connection between engine and transmission produces greater maneuverability (turning radius 5 meters). The ability to select gears results in lower fuel consumption. During the test run the vehicle consumed 10 liters of fuel while covering 11 kilometers.

Summary of Comparison between Trilok and Gear Transmissions

a.)
1. Slow in movement on road and cross-country
2. Poor maneuverability
3. Higher fuel consumption (limited radius of action)

b.)
1. Faster in movement on road and cross-country
2. Greater maneuverability
3. Lower fuel consumption (greater radius of action)
4. Little wear on the clutch
5. No requirement for Trilok oil

(signature illegible)

Oberleutnant and Company Commander

Infanterie-Schule Döberitz

—

The *198. Infanterie-Division* was part of the *LXXX. Armee-Korps* in the *19. Armee* and, after fighting in Alsace, it withdrew through the Pfalz into the Rhine-Neckar region. After retreating through the line Sinsheim — Heilbronn, the division surrendered to the Americans., By then it was no more than a *Kampfgruppe*.

gepanzerte Munitionsschlepper-Kompanie 805

Formed at *Infanterie-Schule Döberitz* in mid-October 1944, *gepanzerte Munitionsschlepper-Kompanie 805* was assigned to the *304. Infanterie-Division*, which was stationed in Upper Silesia, on 7 November 1944.

gepanzerte Munitionsschlepper-Kompanie 806

The last unit to be formed was *gepanzerte Munitionsschlepper-Kompanie 806*.

On 11 March 1945 the General Staff of the Army decided not to proceed with formation of the unit. This order was then apparently rescinded at some point. On 23 March 1945 the acting command of the *III. Armee-Korps* (*WK-Kommando III*) advised the *OKH* that the company had departed by rail for the west on 21 March. There the armored ammunition carrier company was intended to be deployed with the *180. Infanterie-Division*. The only operational report concerning the company is an order for the issuance of additional ammunition carriers. (Source: Bundesarchiv/Militärarchiv Freiburg)

—

Telex

To *OB West*

Two additional armored ammunition carriers with personnel in trucks are to be dispatched to *gepanzerte Munitionsschlepper-Kompanie 806*, which is to be attached to the *180. Infanterie-Division*. Request you advise *Inf.Schule/Lehrstab 3*/Döberitz-Elsgrund of where to deliver the vehicles.

OKH/ObdE/AHA/In 2 Nr. 1655 dated 26 March 1945

For the Command:

Freytag

Oberst

—

In the second half of March 1945 the *180. Infanterie-Division* was attached to the *1. Fallschirmarmee* and fought south of Wesel after the Allied crossing of the lower Rhine in the Walsum-Voerde area starting 24 March. What was left of the unit was captured by the American in the Ruhr pocket.

Regrettably, there are almost no details concerning the *Munitionsschlepper-Kompanien 801* to *806*. Apart from *gepanzerte Munitionsschlepper-Kompanie 803* little can be said about their operations or losses.

On 22 March 1945 *Infanterie-Schule Döberitz, Abteilung V (K)*, advised the *ObdE/In 2* of the school's vehicle inventory. Listed therein were 42 Borgward BIV's in the school's replacement and training platoon. These vehicles were scheduled to be converted as replacement vehicles or stripped for parts.

Gliederung der gepanzerten Munitionsschlepper - Kompanie 801 (Stand: Herbst 1944)

M.Jaugitz / Trojca 2000.

Factory model of the armored ammunition carrier; this one still has the early-style roadwheels (Spielberger)

Right, below and top opposite page: Experimental version of the armored ammunition carrier with a 5 cm *Pak 38* (*Pz.Sfl. Ia*). (Bundesarchiv 79/94/19)

Above and right: Interior shots of the fighting compartment. (Spielberger)

Opposite page: The serial production models were tested at the Borgward facilities. These have already had the new drive sprocket and dead track mounted. (Franzen)

Above and right: The ability of a crew of the armored ammunition carrier to work with a forward infantry post is tested during an exercise. (Leon)

The armored ammunition carriers were first employed with the *1. Infanterie-Division* in the summer of 1943. (Bundesarchiv 461/217/18)

Opposite page bottom and above: *Gepanzerte Munitionsschlepper-Kompanien 801-806* were equipped with modified *BIV*, *Ausführung A* carriers starting in the fall of 1944. Both of these pictures were taken shortly after the end of the war.

2.4.2: Panzer-Vernichter-Abteilung 1

In January 1945 *Panzer-Versuchs-und Ersatz-Abteilung 300* began searching for trained *BIV* drivers and volunteers for a new special unit. Those who volunteered were transported to the Grafenwöhr training area for training on the *Panzerschreck*. After three days of training and test firing, the special unit was transferred to Neuruppin to the *Panzer-Ersatz-Abteilung 5*. At that point the unit was designated as *"Einheit Langsdorf"*.

At the beginning of February 1945 the *IN 6* ordered the radio equipment removed from 56 *BIV's* being held by the Magdeburg-Gommern armor ordnance office. The maintenance and ordnance facilities of the replacement battalion were ordered to mount six *Panzerschrecks* on the vehicles and augment their armor. Thanks to twelve-hour shift work, the replacement battalion succeeded in converting the vehicles as requested.

The vehicles, designated as the *BIV-Panzerjäger*, were then issued to the personnel of the unit, which had been redesignated as *Panzer-Vernichter-Abteilung 1*.

Johannes Hentschel, a former member of the *3./ Panzer-Vernichter-Abteilung 1*, wrote:

We received new *BIV's*. A second seat, consisting of a wooden board, had been installed next to the driver's seat. This was mounted higher than the driver's seat. In front of it was a mounting for six "Stovepipes". A protective shield was installed in front of the operator. Mounted on the shield were the aiming mechanism and the firing buttons. In addition to the rounds in the tubes, 18 reserve rounds were housed around the gunner's seat on the left side.

Difficulties were encountered during practice firing as glass shrapnel hit the gunner after firing. This was due to the fact that the propellant charge of the *Panzerschreck* projectile was housed in a glass container.

The newly-formed unit consisted of three companies. Battalion commander was Hauptmann Groß.

At the end of February the unit was moved from Altruppin to Berlin, where it was billeted in the *Landespioneer* School in Berlin-Karlshorst. Some of the vehicles were parked on the race track there, while one section of six *BIV's* was assigned to guard the *OKH* in Fehrberliner Square.

After American forces began marching on Berlin, *Panzer-Vernichter-Abteilung 1* was moved on the *Autobahn* to Ferch south of Lake Schwielow.

After the start of the Soviet offensive on the Oder Front, the unit was withdrawn and, during the night of 19-20 April, was rushed in the direction of Straußberg. Use of the *BIV* tank destroyers against armored vehicles was not possible on account of the large numbers of Russian infantry.

Panzer-Vernichter-Abteilung 1 subsequently withdrew into the Buch-Buchholz area northeast of Berlin. There some of the *BIV's* had to be abandoned and blown up. The crews then engaged the advancing Soviet tanks with *Panzerfäuste* and hand-grenade bundles.

The unit became further dispersed during these fighting withdrawals. Parts of the battalion engaged Soviet armored forces with the *BIV* tank destroyers in the "Unter den Linden" area, at the Brandenburg Gate, on the Frankfurter Allee and the Landsberger Allee and in the area around the police headquarters.

Many personnel of *Panzer-Vernichter-Abteilung 1* were seriously wounded or killed in this fighting. When the survivors were taken prisoner on 2 May 1945, they saw soldiers of the Red Army driving the captured *BIV* tank destroyers through the city.

The majority of the *Panzerjäger BIV's* were lost at the end of April 1945 in the vicinity of the Brandenburg Gate. (Leon)

The *Borgward BIV, Ausführung C* served as the basis of the *Panzerjäger BIV*. The six-barreled weapon was mounted on the right side. (Leon)

Above and top of next page: The *BIV, Ausführungen A* and *B* had the weapons mounted on the left-hand side.

The wrecks of *BIV's* in front of the Brandenburg Gate.

There were also variations where the *Panzerschreck* tubes were mounted above the engine compartment. (Leon)

After the fighting was over, Russian soldiers moved through the streets of Berlin on captured *Panzerjäger BIV's*. Note the smoke-grenade launchers and the battalion insignia on the demolition container. (Fleischer)

Chapter 3:
Color Section

Note: The color illustrations are not presented in a constant scale. They have been enlarged wherever possible to show details of the paint scheme employed.

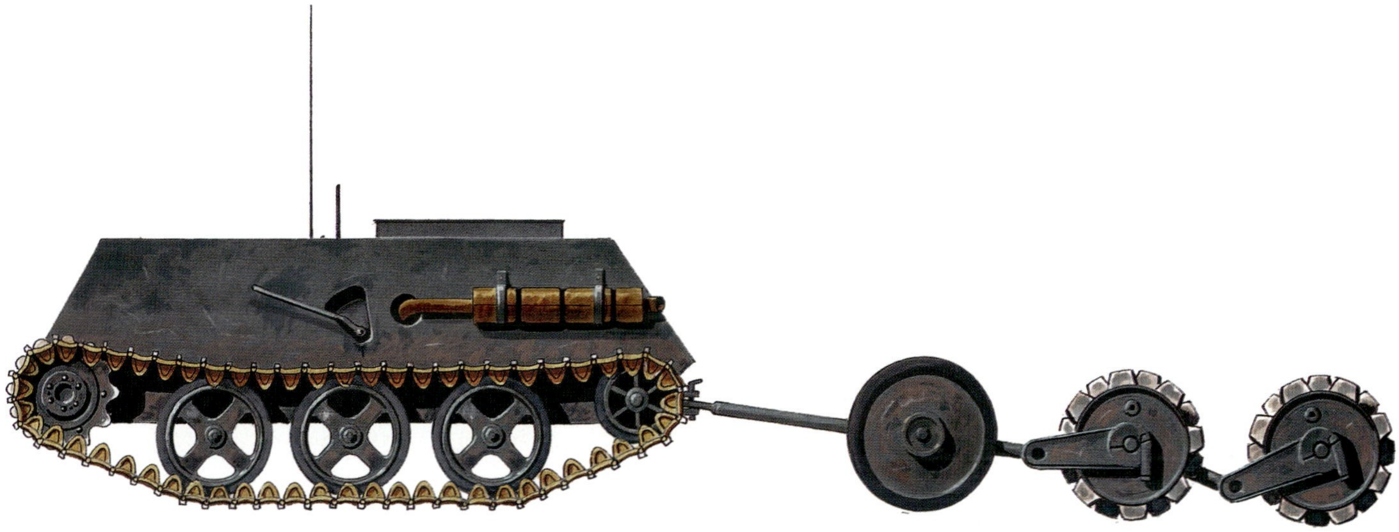

A *Borgward BI Minenräumwagen* of *Minenräum-Kompanie 1*. Blieskastel, summer of 1940.

A *Borgward BIV Minenräumwagen* of *Minenräum-Abteilung 1*. Russia, July 1941.

Kleiner Panzerbefehlswagen I (Sd.Kfz. 265) of *Minenräum-Abteilung 1*. Russia, July 1941.

A *Panzer III* of the *1./Panzer-Abteilung 300 (Funklenk)*. Sevastopol, June/July 2001.

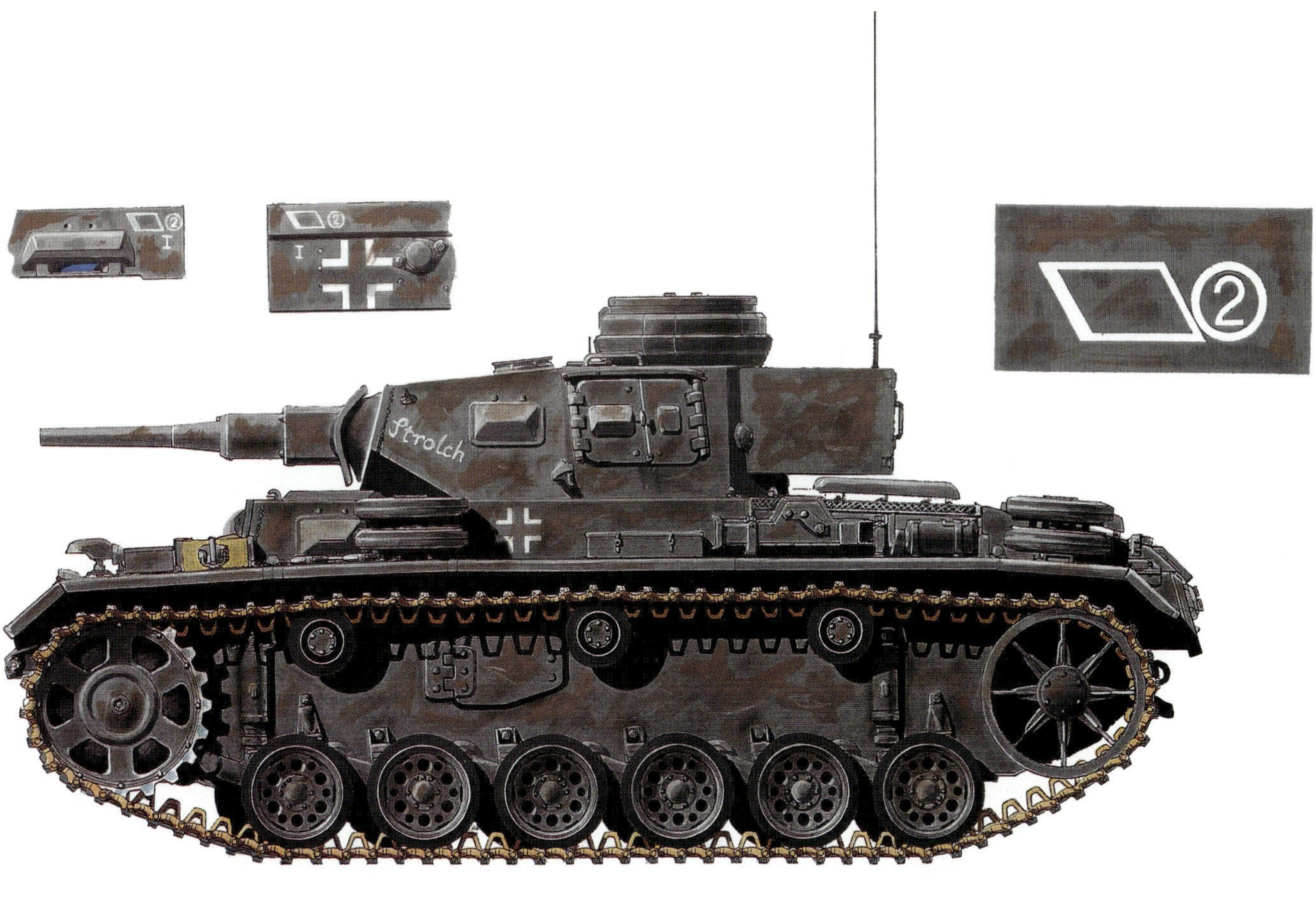

A *Panzer III* of the *2./Panzer-Abteilung 300 (Funklenk)*. The Crimea and Southern Russia, summer of 1943.

A *Borgward BIV*, *Ausführung A* of the *1./Panzer-Abteilung 300 (FL)*. The Crimea, June/July 1942.

A wire-guided Bren Carrier of the *3./Panzer-Abteilung 300 (FL)*. The Crimea, June 1942.

A Utility Tractor of the *3./Panzer-Abteilung 300 (Funklenk)*. The Crimea, June 1942.

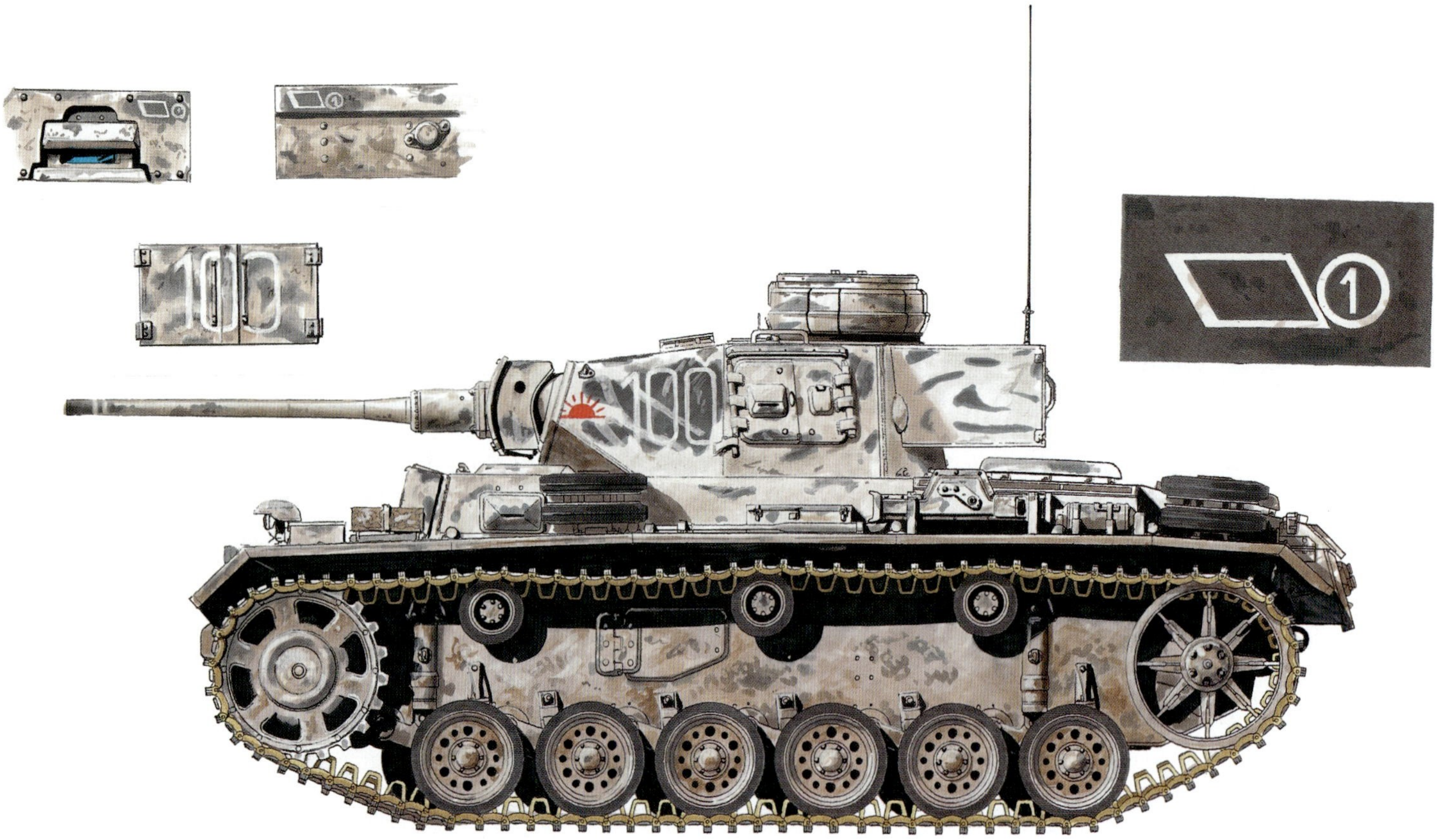

A *Panzer III, Ausführung L* of the *1./Panzer-Abteilung 301 (Funklenk) (Kompanie von Abendroth)*. The Southern Front of Russia, winter of 1942/1943.

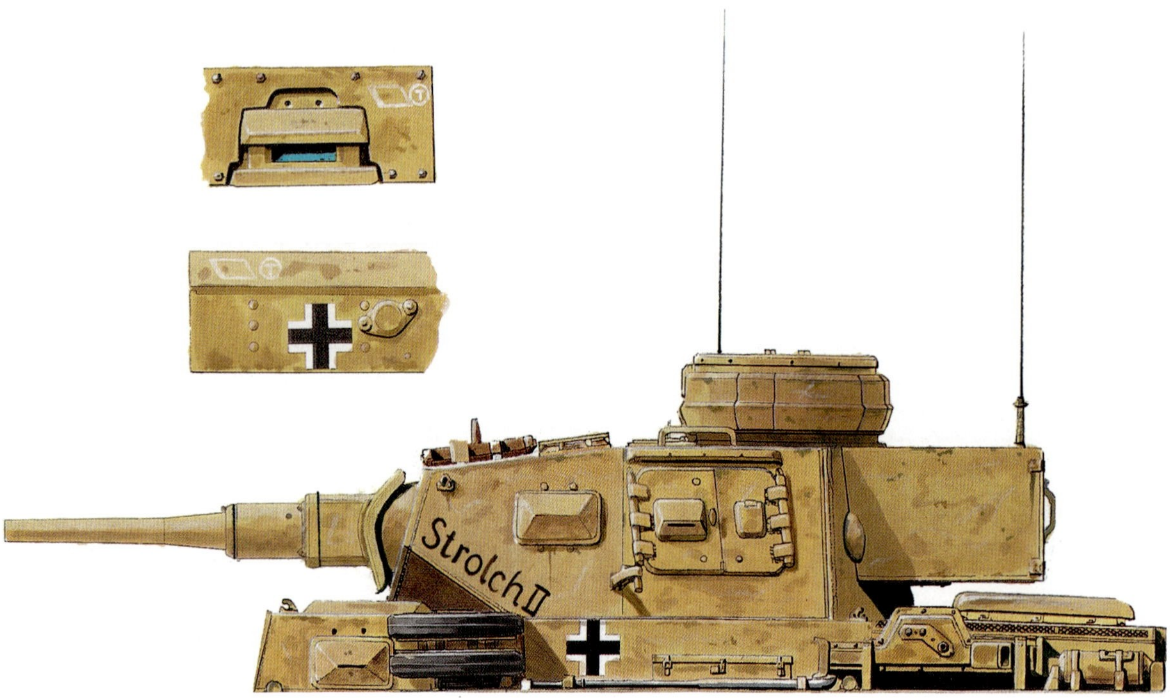

A *Panzer III* of the *Versuchskommando (FL) Tropen*. El Daba (North Africa), October/November 1942.

A *Panzer III, Ausführung J* command and control tank of *Versuchskommando (FL) Tropen*. El Daba area, October/November 1942.

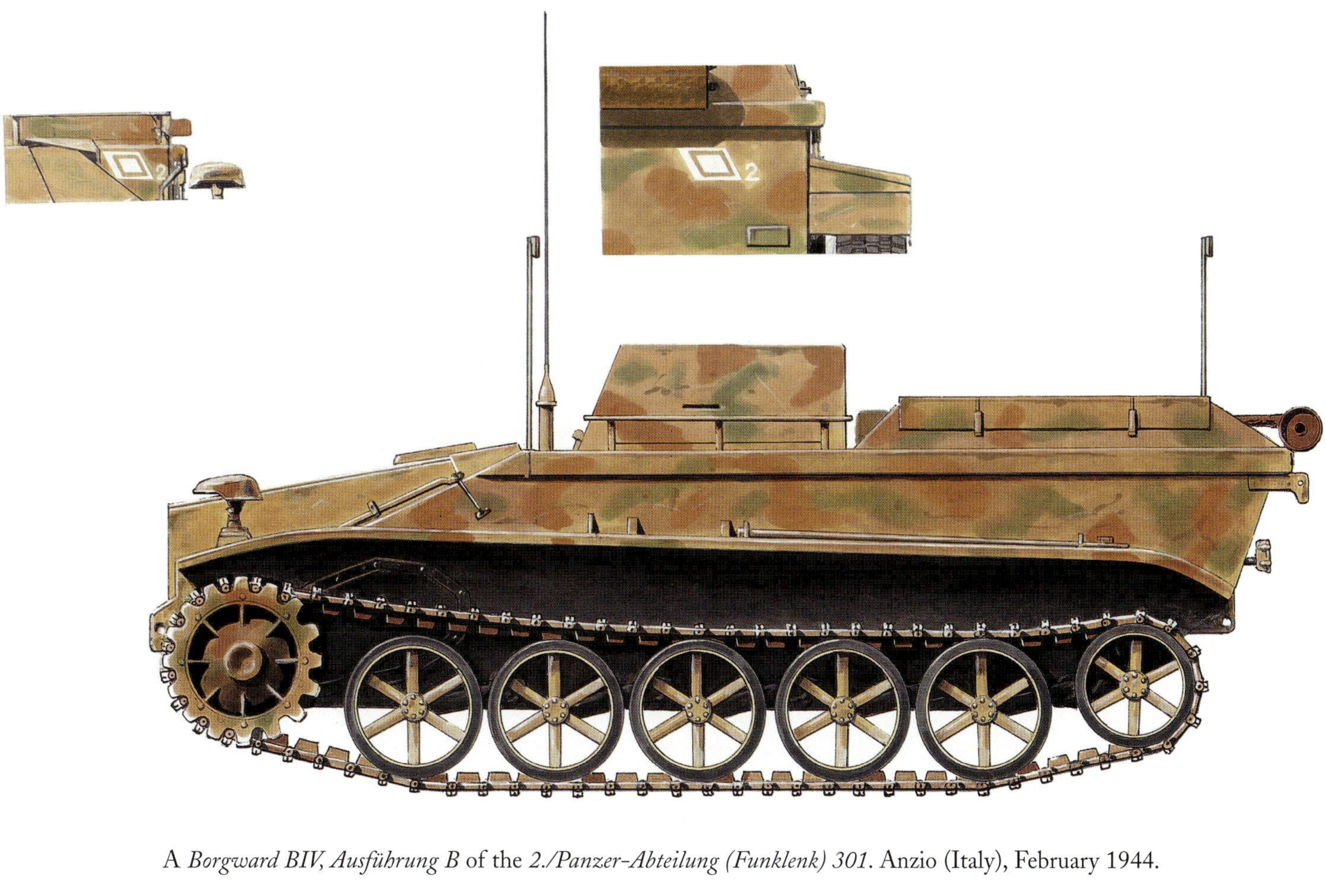

A *Borgward BIV, Ausführung B* of the *2./Panzer-Abteilung (Funklenk) 301*. Anzio (Italy), February 1944.

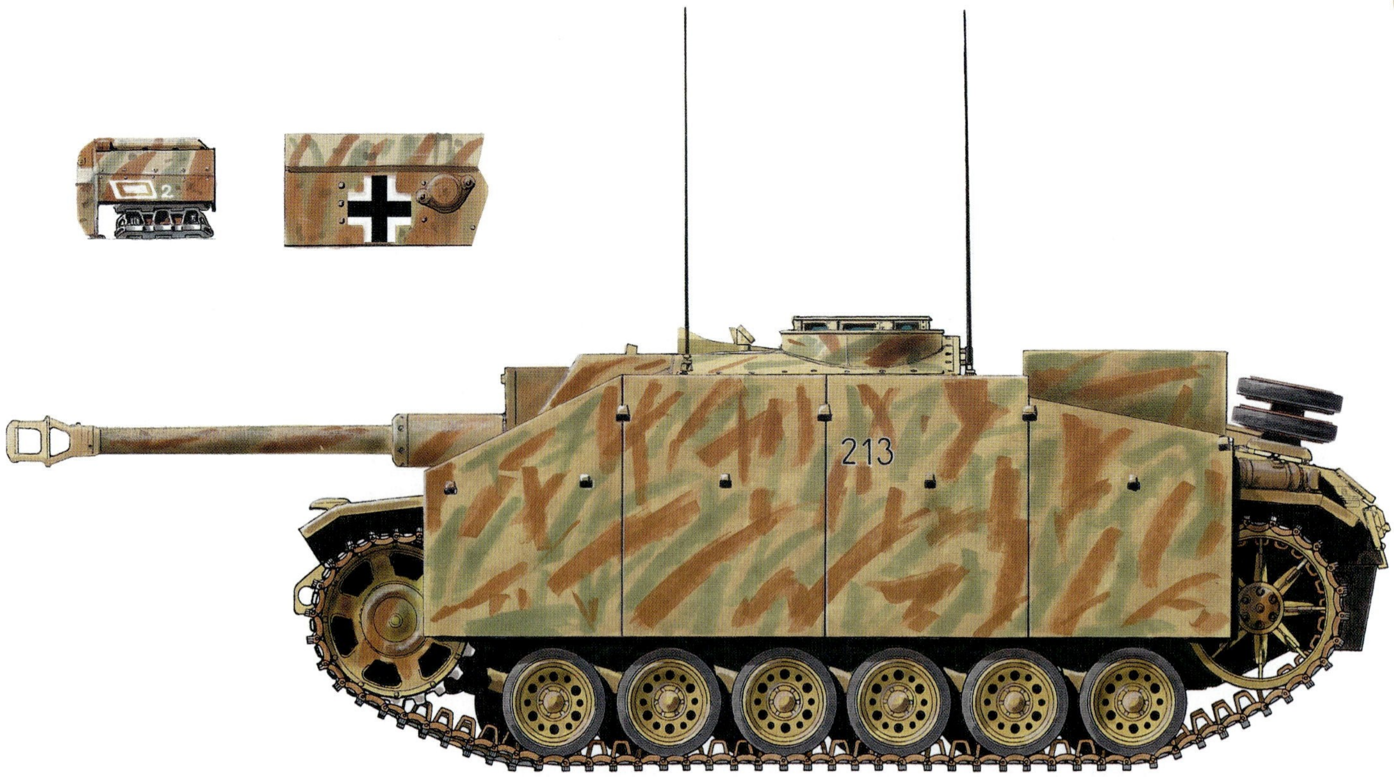

A *Sturmgeschütz III* command and control vehicle of the *2./Panzer-Abteilung (Funklenk) 301*. Galicia, summer of 1944.

A *Sturmgeschütz III* command and control vehicle of *Panzer-Kompanie (Funklenk) 311*. Poltawa, summer of 1943.

A *Borgward BIV*, *Ausführung A* of *Panzer-Kompanie 311 (Funklenk)*. Poltava (Russia), summer of 1943.

Leichter Zugkraftwagen (Sd.Kfz. 11) of *Panzer-Kompanie (Funklenk) 311*.

This page and next: Front, rear and side views of a *Borgward BIV* of *Panzer-Kompanie (Funklenk) 312*. Kursk/Orel, 1943.

A *Sturmgeschütz III* command and control vehicle of *Panzer-Kompanie (Funklenk) 312*. Kursk/Orel, July 1943.

A *Panzer III, Ausführung L* command and control tank of *Panzer-Kompanie (Funklenk) 312*. Kursk, July 1943.

A *Panzer III, Ausführung N* command and control tank of *Panzer-Kompanie (Funklenk) 313*. Kursk, July 1943.

A *Borgward BIV, Ausführung A* (modified) of *Panzer-Kompanie (Funklenk) 313*. Kursk, July 1943.

Borgward BIV, Ausführung A (modified) of *Panzer-Kompanie (Funklenk) 313*. Kursk, July 1943. The left-hand view illustrates the placement of the tactical insignia on the front of the vehicle.

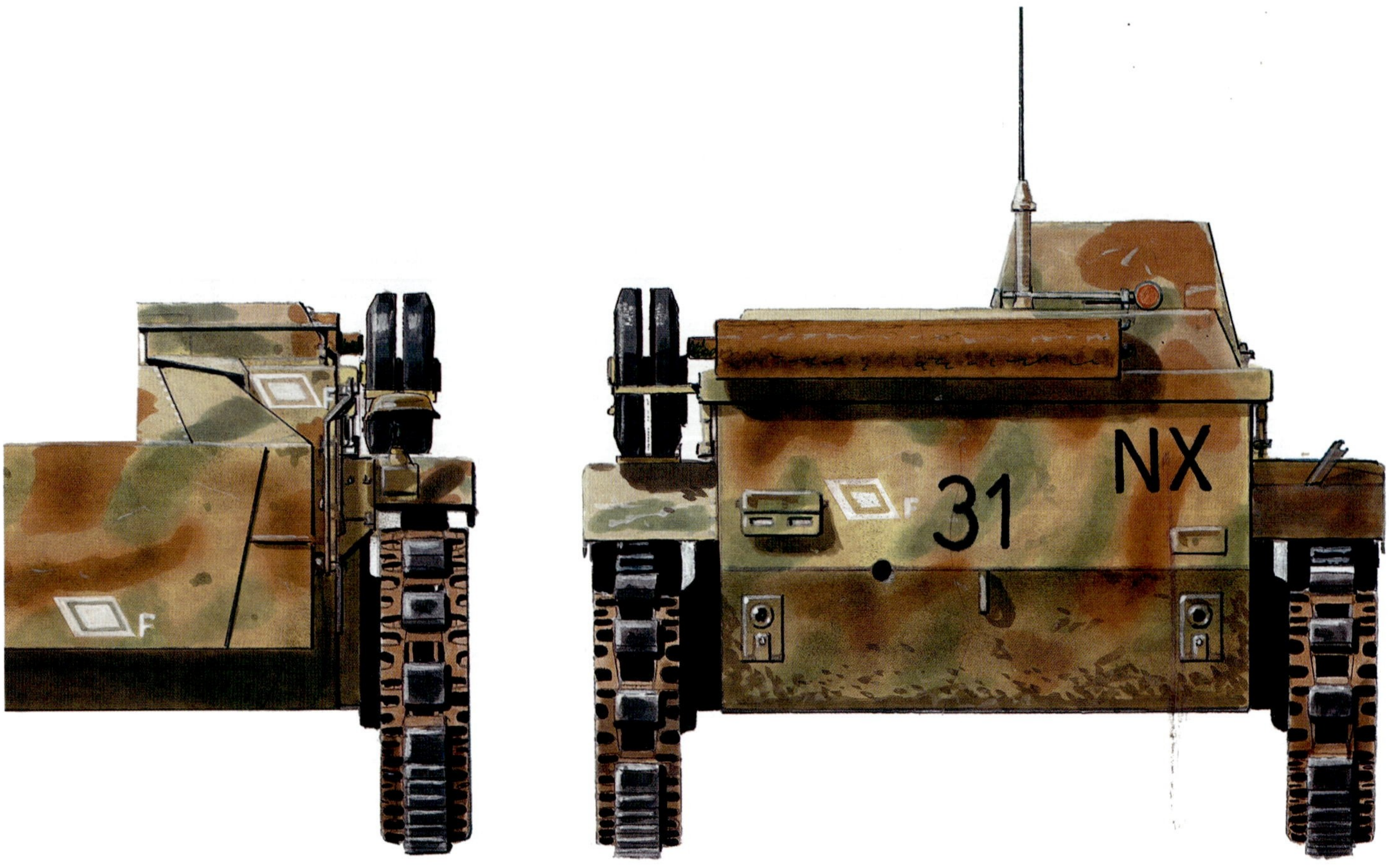

Tiger I command and control tank of the *3./schwere Panzer-Abteilung 508*. Italy, 3 June 1944.

Borgward BIV of the *3./schwere Panzer-Abteilung 508*. Italy, 3 June 1944.

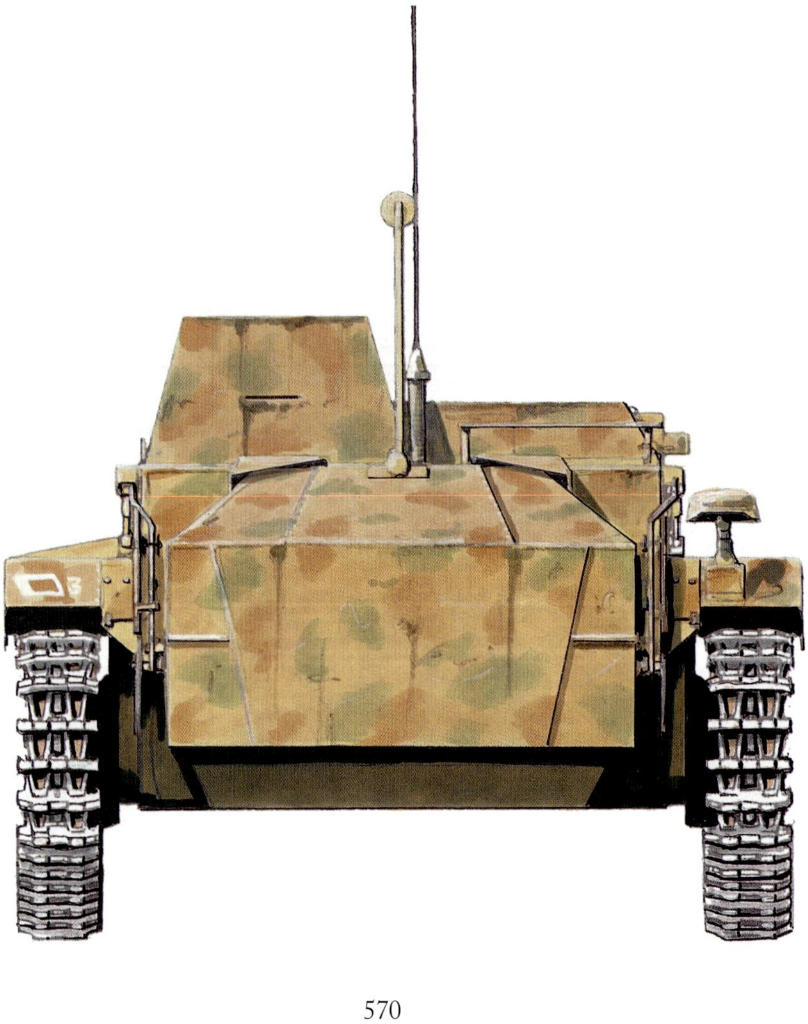

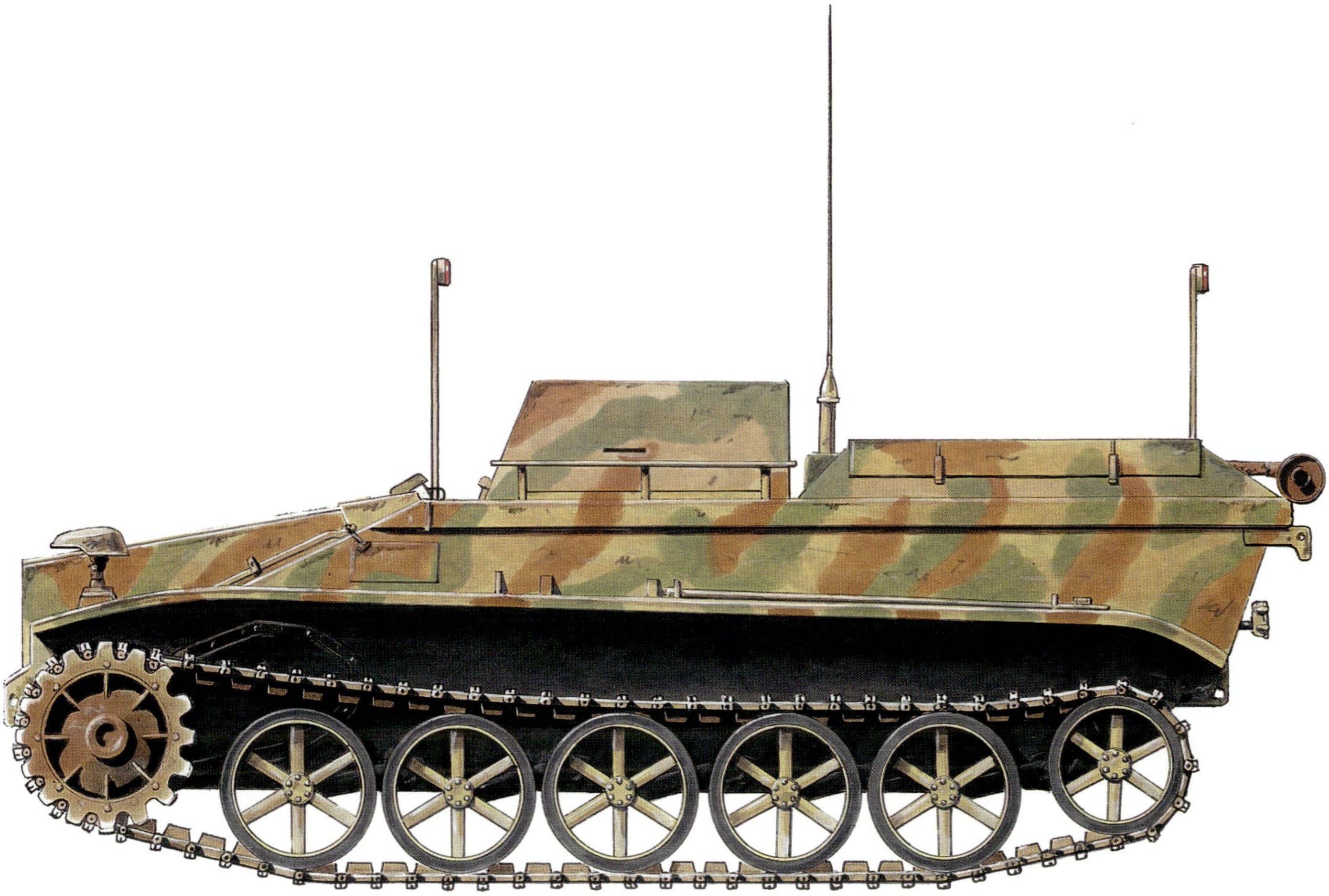

Borgward BIV of *Panzer-Kompanie (Funklenk) 314*. Kursk, July 1943.

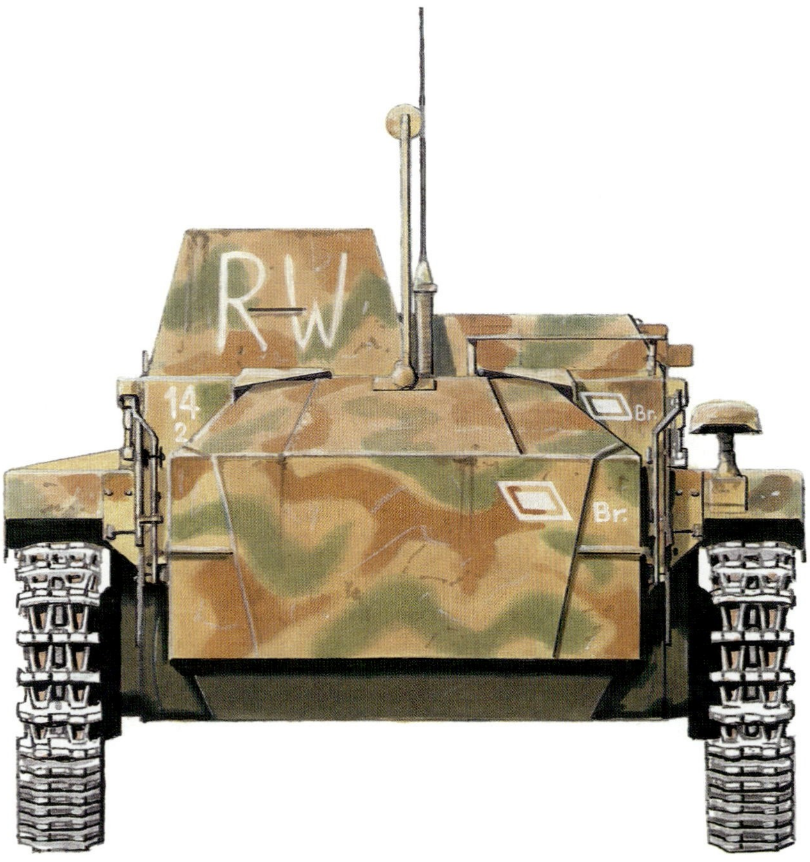

A *Sturmgeschütz III* command and control vehicle of *Panzer-Kompanie (Funklenk) 314*. Kursk, 1943.

Tiger I command and control tank of the *3./schwere Panzer-Abteilung 504*. Italy, summer of 1944.

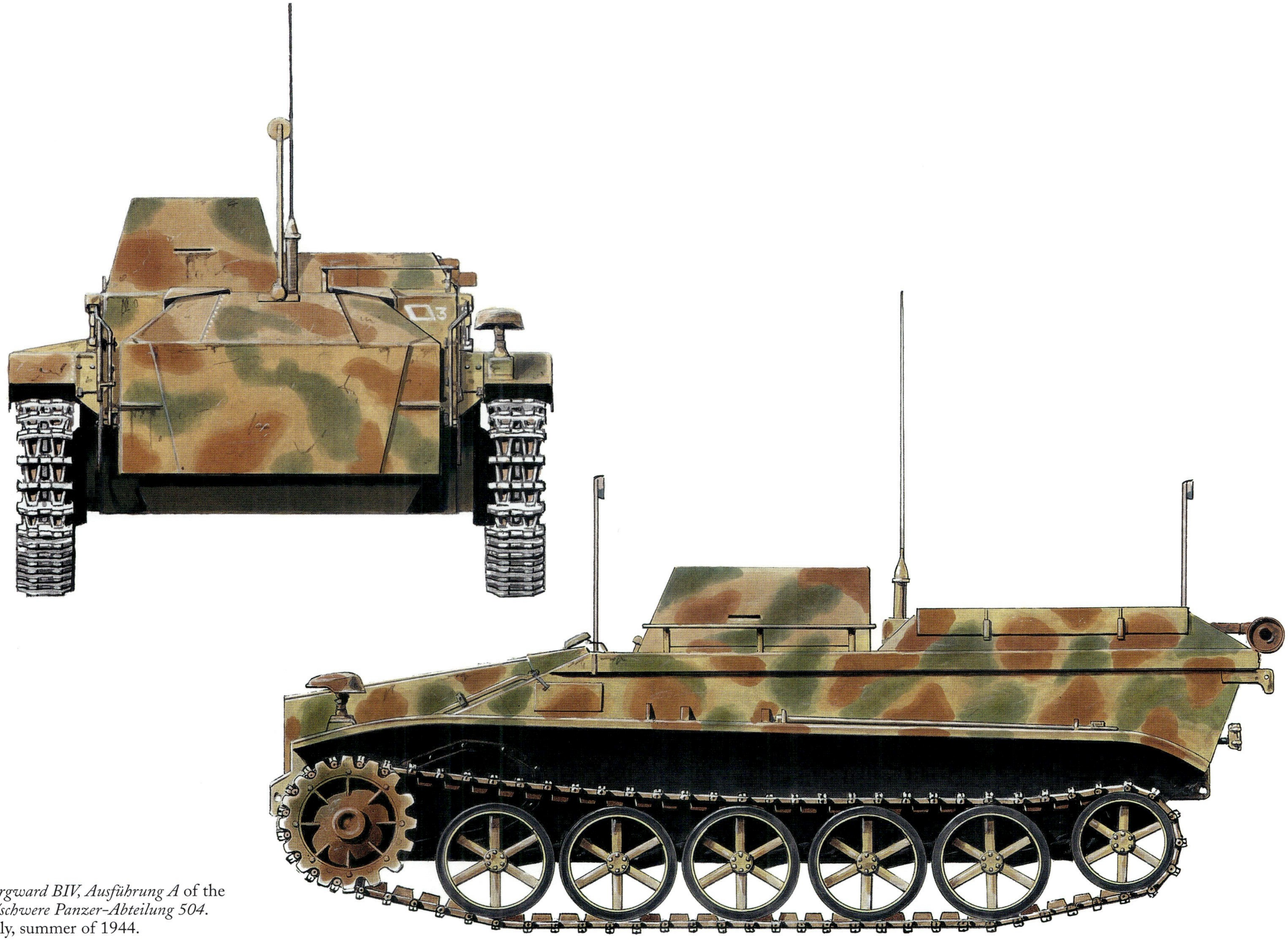

Borgward BIV, Ausführung A of the
3./schwere Panzer-Abteilung 504.
Italy, summer of 1944.

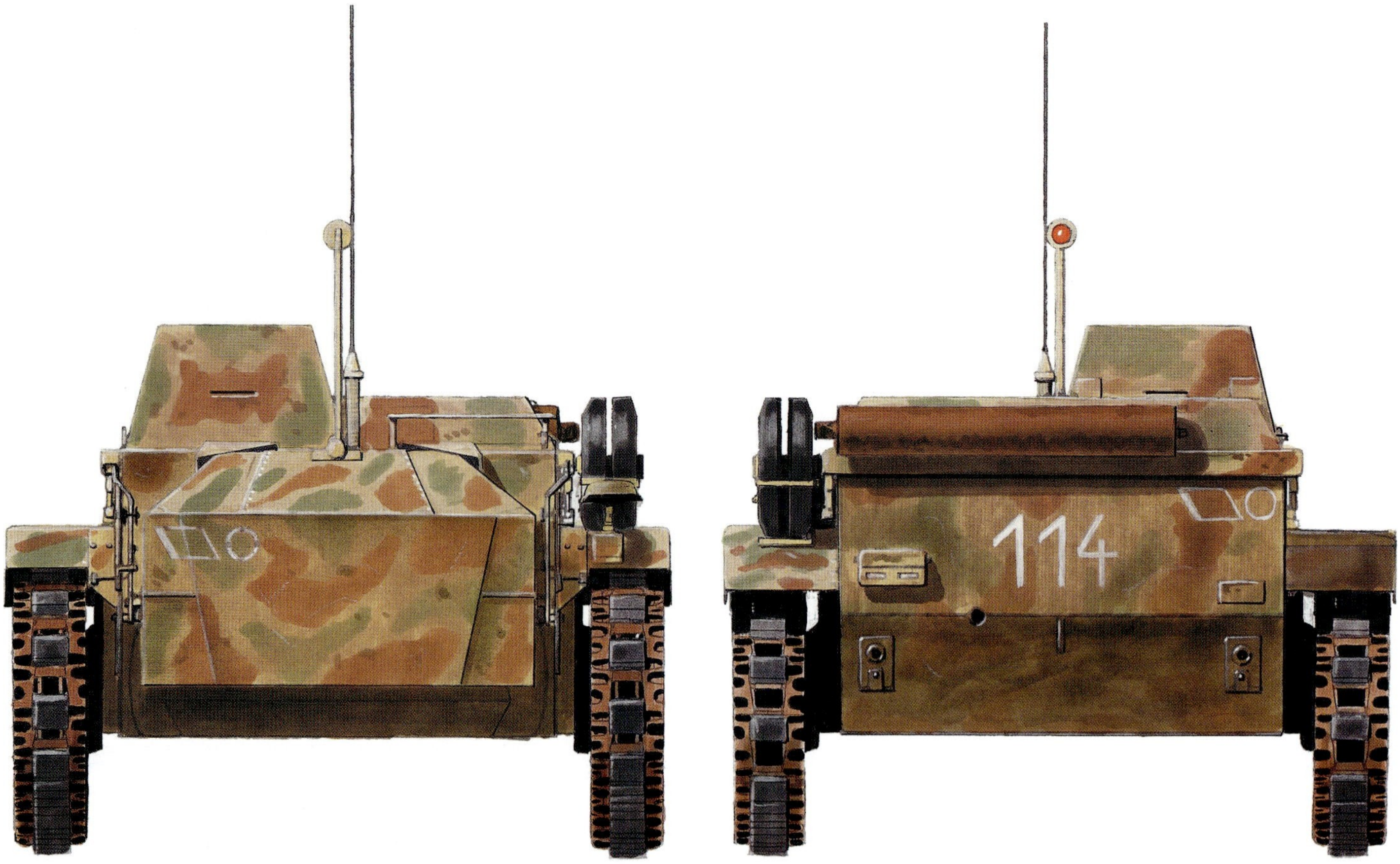

A *Borgward BIV*, *Ausführung A* (modified) of *Panzer-Kompanie 315 (Funklenk)*. Normandy (Caen), June 1944.

A *Sturmgeschütz III* command and control vehicle of *Panzer-Kompanie (Funklenk) 315*. Normandy, June 1944.

A *Sturmgeschütz III* command and control vehicle of *Panzer-Kompanie (Funklenk) 316*. France, May/June 1944.

Tiger II command and control tank of the *1./schwere Panzer-Kompanie (Funklenk)* (later: *Panzer-Kompanie 316 (Funklenk)*). France, May – June 1944

A *Borgward BIV, Ausführung B* of *Panzer-Kompanie (Funklenk) 316*. Normandy, July 1944.

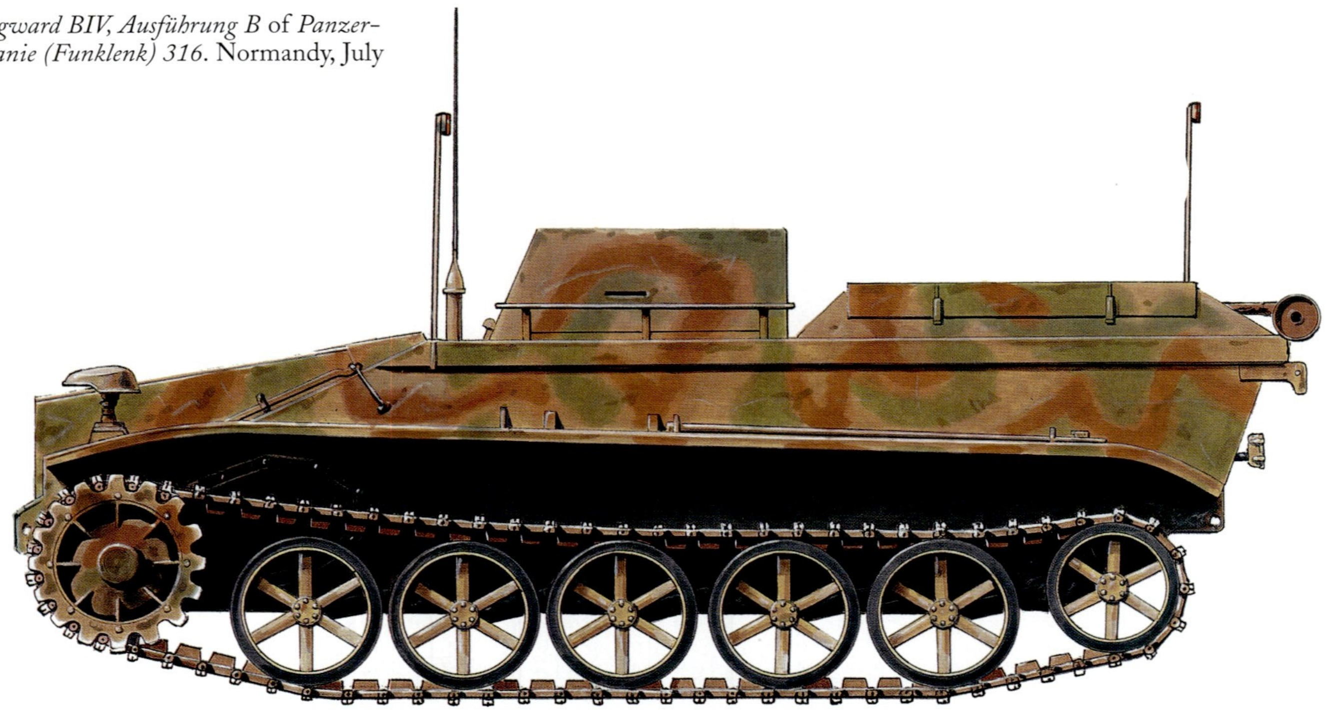

A *Sturmgeschütz III* command and control vehicle of the *3./Panzer-Abteilung (Funklenk) 302*. Warsaw, August 1944.

A *Sturmgeschütz III* command and control vehicle of the *2./Panzer-Abteilung 302 (Funklenk)*. East Prussia, winter of 1944/1945.

A *Borgward BIV*, *Ausführung C* of *Panzer-Abteilung (Funklenk) 302*. Warsaw, August 1944.

A *Sturmgeschütz III* command and control vehicle of the *4./Panzer-Abteilung (Funklenk) 301*. Normandy, June/July 1944.

Rear, side and detailed views of a *Borgward BIV* of the *4./Panzer-Abteilung (Tiger/Funklenk) 301*. Normandy, June and July of 1944.

A *Sturmgeschütz III, Ausführung G* command and control vehicle of *Panzer-Kompanie (Funklenk) 319*. Belgium, September 1944.

Tiger I command and control tank of the *1./Panzer-Abteilung (Tiger/Funklenk) 301*. The Ruhr Front (Germany), November and December 1944.

A *Borgward BIV* of the *1./Panzer-Abteilung (Tiger/Funklenk) 301*.
Germany, January/February 1945.

A *Borgward BIV*, *Ausführung C* of *Panzer-Zug (Funklenk) 303* employed with the *25. Panzergrenadier-Division*. Germany (Oder front), April 1945.

A *Sturmgeschütz III* of *Panzer-Versuchs- und Ersatz-Abteilung 300 (Funklenk)*. Eisenach and Berka Training Area (Germany), spring of 1943.

A *Sturmgeschütz III, Ausführung F/8* command and control vehicle of the *Panzer-Versuchs- und Ersatz-Abteilung 300* (Funklenk). Eisenach (Germany), autumn 1943 to spring 1944.

A *Borgward BIV, Ausführung A* of *Panzer-Versuchs- und Ersatz-Abteilung (Funklenk) 300*. Eisenach/Berka Training Area (Germany), summer of 1943.

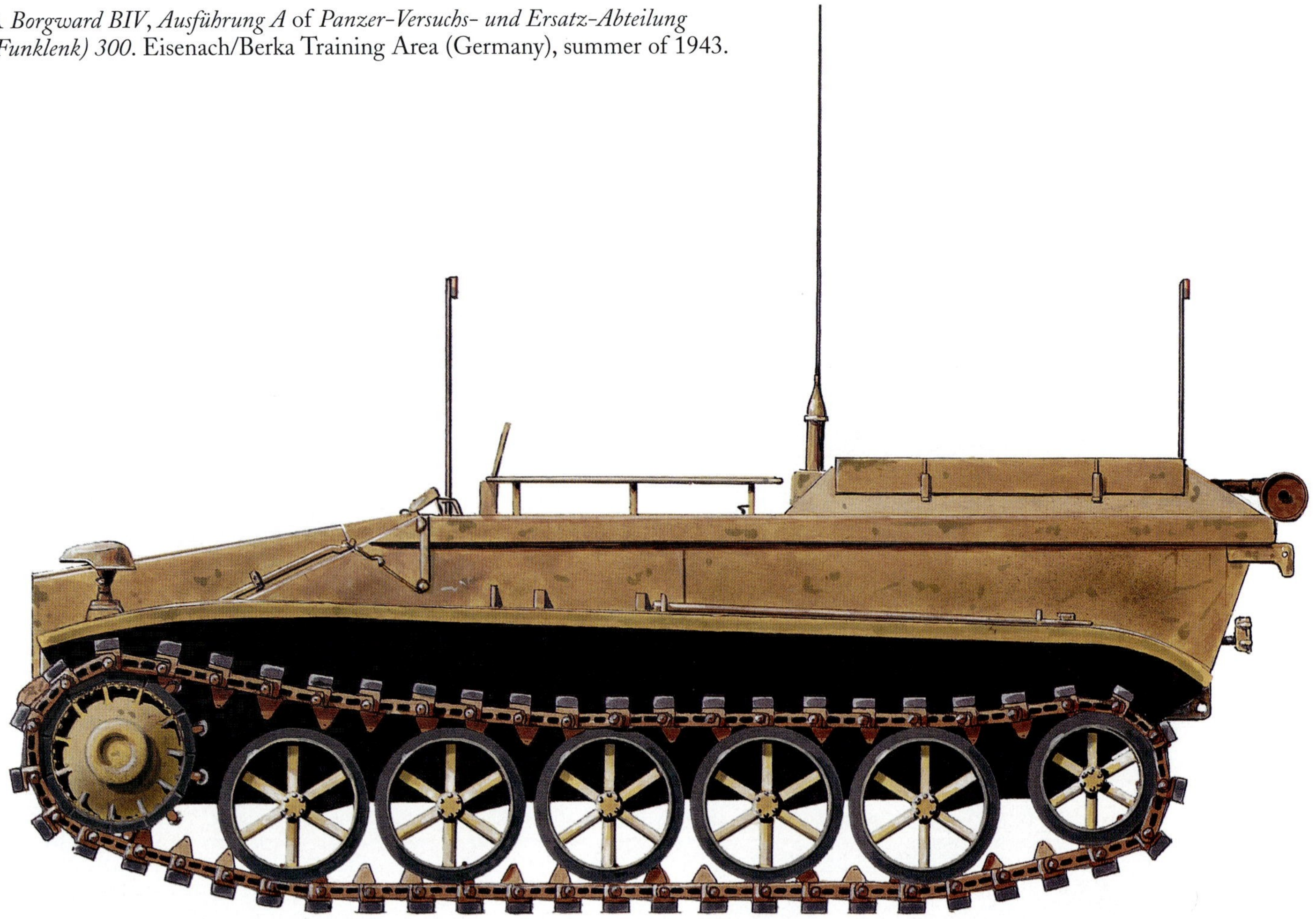

A *Borgward BIV, Ausführung A* with a television camera of the *1./Panzer-Versuchs- und Ersatz-Abteilung (Funklenk) 300*. Eisenach (Germany), spring of 1943.

A *Springer* of *Panzer-Versuchs- und Ersatz-Abteilung (Funklenk) 300*. Eisenach (Germany), the spring of 1945.

A *Borgward BIV* of the *gepanzerte Munitionsschlepper-Kompanie 806*. Germany, April 1945.

A modified *Borgward BIV* with the 6-barelled *Panzerschreck* mounting. These were used by *Panzer-Vernichter-Abteilung 1*. Berlin, April 1945. The detailed view above shows the three smoke-grenade launchers mounted to the front of the vehicle.

Chapter 4: Biography and Doctrinal Extracts

4.1: Highly-Decorated Members

Ferdinand von Abendroth

Since the attachment of remote- and radio-controlled units to larger formations (regiments, divisions, corps, etc.) lasted only a relatively short time, the commanders of the higher formations knew little about the successes achieved by wire- or radio-controlled demolition-charge carriers in most cases. Consequently, it was the duty of the unit commander (company commander, etc.) to apply for decorations for soldiers in his command who had distinguished themselves in action.

During the war only one member of the remote-controlled force, *Hauptmann* Ferdinand von Abendroth, was awarded a high-level decoration — the German Cross in Gold. It was awarded for von Abendroth's actions in the taking of Fortress Sevastopol and during the winter of 1942-43.

Ferdinand von Abendroth was born in Kössern, Saxony on 4 January 1916 and joined the *Wehrmacht* as a young man. The following is a summary of Ferdinand von Abendroth's career during the war:

1 September 1939 – 20 March 1942: Assigned to the *3./Panzer-Regiment 31*
20 March 1942 – 6 July 1943: Company commander of the *1./Panzer-Abteilung 300/301*
6 July 1943 – 20 May 1944: Company commander of *Panzerkompanie 315 (Fkl)*
20 May 1944 – 15 July 1944: Battalion commander of *Panzer-Abteilung (Fkl) 301*

Ferdinand von Abendroth was seriously wounded on the afternoon of 15 July 1944. He was transferred to the military hospital at Lemberg, where he died as a result of his wounds on 16 July. He was posthumously promoted to the rank of *Major* retroactive to 1 July 1944.

Ferdinand von Abendroth was considered a caring, circumspect and experienced officer. He was well thought of by his superiors and loved by his men. This was not only stressed by former members of his company but also in official evaluations.

He was awarded the following decorations:

22 September 1939: Iron Cross, First Class

17 October 1939: Iron Cross, Second Class

27 February 1940: Wound Badge in Black

18 August 1941: Tank Combat Badge in Silver

22 December 1942: Crimea Shield

3 August 1943: German Cross in Gold

4.2: Documents

4.2.1: Instructions on the Employment of Mine-Clearing Vehicles (6 June 1941)

H.Qu.O.K.H., 6 June 1941

Oberkommando des Heeres

Gen.St.d.H./Ausb.Abteilung(II)/Gen.d.Schn.Truppen

Nr. 600/41 g Kdos

Secret

I authorize the distribution of
"Instruction on the Employment of
Mine-Clearing Vehicles
(*M.R. Wagen*) dated 6 June1941."

For the Command

Halder

- 1 -

Table of Contents

- 2 -

Introduction

These instructions pertain to the current state of technical development (May 1941). The equipment is usable in the field, however, development is not yet concluded.

- 3 -

A.) Nature and Mode of Operation of the Weapon

1.) Mine-clearing vehicles are used to create lanes in minefields (used as mine clearers) and to engage enemy field positions or to eliminate obstacles (used as demolition-charge carriers) (see section 6.)

Mine-clearing vehicles are committed unmanned and guided to the target by remote control, destroying it through detonation.

2.) The advantages of the weapon lie in its great explosive effect and its effect on enemy morale. Remote control gives the weapon great accuracy and avoids the use of personnel. Because of its small size it offers the enemy only a small target.

At the present time its disadvantages are its slow speed and poor performance. Mine-clearing vehicles cannot negotiate large trenches and craters with walls.

3.) The mine-clearing vehicle is capable of the following tasks:

a) Destruction or damaging of concrete bunkers and heavily fortified field installations in attacks against a stationary front or against fortified field positions.

b) Destruction of small minefields or creation of lanes in minefields (1 mine-clearing vehicle with a 300 kilogram high-explosive charge takes out mines in a circle with a 40-meter diameter.

c) Elimination of barricades and obstacles in open terrain and on roads. (Effectiveness against dug-in railroad rails is limited; use of 3 vehicles is necessary to create a lane through iron, concrete and dragons teeth obstacles.)

d) Destruction of enemy pockets of resistance, especially in towns and on wood lines.

e) Drawing of enemy fire, especially from antitank weapons. This is always a secondary role for the vehicles.

f) Laying of smoke or gas, offensively and defensively*)

g) Movement through areas under heavy enemy fire*)

Because of their low speed and weight, mine-clearing vehicles are not suited to overrunning stake and tree barricades or for dragging away wire obstacles.

4. The mine-clearing vehicle is a small box-shaped fully-tracked vehicle with a radio-receiver system. The hull consists of concrete, with an additional armor plate on the front. The vehicle can carry a demolition charge weighing up to 515 kilograms. The size of the charge depends on the mission and the nature of the target. The vehicle can accommodate other material in place of, or in addition to, the demolition charge.

5. The remote-control process takes place from a thinly-armored *kleiner Panzerbefehlswagen* equipped with a transmitter system. Two mine-clearing vehicles can be employed at the same time from a single *Panzerbefehlswagen*. For a successful remote-control operation the controller must be able to observe the mine-clearing vehicle(s) all the way to the target. To achieve this, it may be necessary for the armored command vehicle to follow the mine-clearing vehicle(s) in stages, moving from cover to cover.

While under remote-control, moving mine-clearing vehicles cannot alter their speed and, if stopped, cannot resume movement.

6. Ignition of a mine-clearing vehicle's charge occurs

When used as a mine clearer: By detonation when the vehicle moves over a mine,

When used as a demolition-charge carrier: By radio command as soon as the controller recognizes that the vehicle has reached its target (see subparagraph 5).

*) Note: In cases f) and g) the vehicles carry smoke grenades, chemical-warfare agents and ammunition (see subparagraph 4)

As a rule, the vehicle is completely destroyed when the charge is detonated. In normal terrain (normal vegetation) detonation of the charge results in a crater approximately 1.5 meters deep and 5 meters in diameter.

If the detonation of an enemy mine does not ignite the demolition charge, proceed as per subparagraph 23.

7. Because of the low march speed (see subparagraph 8) and to reduce wear, the mine-clearing vehicles are usually loaded aboard commercial-type heavy trucks for longer marches. After leaving the trucks, the mine-clearing vehicles are hand-driven to friendly front lines, then remotely controlled in action.

8. Mine-Clearing Vehicle Technical Data:

Height: 1140 mm
Width: 1700 mm
Operational Weight: 2.8 tons
Speed in 1st gear: 2 kilometers an hour
Speed in 2nd gear: 5 kilometers an hour
(Choice of gears depends upon the terrain)
Range: 30 kilometers
Maximum possible distance between the mine-clearing vehicle and the command and command vehicle during remote-control operation: 2 kilometers
Elapsed time from time of ignition command to actual ignition: 12 seconds
Armor protection against rifle fire and small-caliber fragments

B.) Organization, Equipment, March Lengths

9. Mine-clearing units are army-level units.

The smallest fighting unit is the single remotely-controlled mine-clearing vehicle (see subparagraph 5).

A section consists of an armored command and control vehicle and three mine-clearing vehicles. Three sections form a platoon; three platoons a company. A battalion (mine-clearing) consists of two companies plus a headquarters and a signals and vehicle maintenance section.

10. A heavy truck is assigned to each mine-clearing vehicle (see subparagraph 7).

Each squad has a portable loading ramp and other loading equipment.

Each company is equipped with light machine-guns for self-defense and antiaircraft defense. Each also has a signals and vehicle maintenance section as well as men trained in combat-engineer techniques.

11. When halted the company road-march length is 2,400 meters, with an interval of 20 meters between vehicles. Passage time for the company (mine-clearing vehicles loaded) is approximately 12 minutes at a speed of 30 kilometers an hour.

C.) Operations

a) General:

12. Because of the high value of the equipment, the relatively small number of mine-clearing vehicles available for use and the unique nature of the weapon, mine-clearing vehicles can only be used at the *Schwerpunkt* of combat operations.

13. Surprise of the enemy and very careful preparation before operations are conditions for success. Mine-clearing vehicles are therefore only to be committed to the attack after careful planning.

It is advisable to exploit the cover of darkness to transport and prepare the mine-clearing vehicles. Using artillery fire or aircraft to mask noise can be advantageous. The operational preparation of mine-clearing vehicles must be safeguarded by other combat arms, especially infantry and antitank forces.

14. Mine-clearing vehicles can be committed in company, platoon or section strength or even individually. Platoon-size commitment is preferred.

The number of mine-clearing vehicles committed, and the way they are used, depends on the mission and target as well as the nature of the terrain and the available cover.

Flat terrain without obstacles offering a good view makes it possible to commit a larger number of mine-clearing vehicles simultaneously, however, it also favors the enemy's defensive measures.

Overgrown, rolling terrain usually restricts operations to single mine-clearing vehicles. The same applies to missions to clear obstacles from roads.

Deploying the entire company at once is not possible because of the limited number of available transmitter frequencies. If an especially powerful effect is required, the platoons can be committed one after the other with an interval of ten minutes between platoons.

For technical reasons a section's mine-clearing vehicles can only be controlled by their respective command and control vehicle. An exchange of vehicles within the mine-clearing unit* is therefore not possible during operations.

15. Mine-clearing units fight as part of combined-arms operations. (See subparagraph 24.)

During operations, the mine-clearing unit is generally attached to the attack force in order to assure proper coordination.

16. The operations order to the leader of the mine-clearing unit must contain:

1.) Enemy situation; friendly intentions
2.) Mission of the mine-clearing unit
3.) Coordination with other arms
4.) Operational security
5.) Timing of the attack

b. Scouting

17. Every mission must be preceded by thorough scouting of the terrain and target during daylight. Contact is to be established early on with combat units in the sector. The scouting includes:

a) Approach routes (suitability for heavy trucks)
b) Unloading area (for unloading the mine-clearing vehicles from the trucks)
c) Assembly area (starting point for the attack)
d) Direction to the target
e) Negotiability of the terrain to the target

*) Note: Within this set of employment instructions, mine-clearing unit means whatever number of mine-clearing vehicles is available, from a section upwards.

Pre-planned aerial photo reconnaissance can provide valuable information for the use of mine-clearing vehicles, especially concerning the suitability of the terrain for employment of the vehicles.

c. Moving into Position:

18. Prior to employment the vehicles must be made operational at least 6 kilometers behind the front. This includes fueling, installation of demolition charges or special munitions (see subparagraph 4, note) and checking and tuning of the transmitter and receiver.

19. After the terrain reconnaissance is complete, the vehicles are moved forward into the unloading area, where the mine-clearing vehicles are unloaded from the trucks. The unloading area is best located on or near a road. The trucks remain in the unloading area at the disposal of the leader of the mine-clearing unit. Unloading a mine-clearing squad takes approximately 20 minutes.

20. After unloading the trucks, the vehicles, steered by hand, move up into the assembly area and position themselves in the specified direction of departure. The mine-clearing vehicles' demolition charges are armed; the remote-controlled equipment switched on. Radio silence is in effect.

21. The unloading and assembly areas must be concealed from ground and air observation by the enemy and must be protected against infantry fire. To the extent that the enemy situation and the terrain permit, the two areas are to be collocated. If this is not possible, an effort must be made to minimize the distance between them in order to simplify moving into position and preserve equipment and time.

The assembly area (line of departure) must be located as near as possible to friendly lines to minimize the distance the mine-clearing vehicle has to travel to the target.

Terrain obstacles between the unloading and assembly areas and even in friendly front lines that might hamper the movement of the mine-clearing vehicles must be eliminated on a timely basis; if necessary, with the support of other units.

22. When the order to attack is issued the drivers start the mine-clearing vehicles for the section leader. The latter then guides them to the target using the command transmitter.

23. Remote-controlled equipment must not be allowed to fall into enemy hands. After the operation any mine-clearing vehicles still operational or parts of vehicles must be recovered. If this is not possible, they must be fired on and destroyed.

D. Cooperation with Other Units

24. As long as an active enemy defense can be expected, mine-clearing vehicles may only be committed with other units and never alone.

25. As a rule, mine-clearing vehicles are committed with armored formations in order to create lanes in minefields or eliminate antitank obstacles prior to an armored attack. When used with infantry or motorized units they will usually be employed as demolition-charge carriers for the engagement of combat positions or elimination of obstacles.

26. Success depends on the responsible commander establishing the timing and location of the coordinated effort (attack plan). This applies to individual mine-clearing vehicles as well as for the mine-clearing section, platoon, etc.

27. If the attack is to be preceded by an artillery barrage, all efforts should be made to avoid firing on the sector allocated to the mine-clearing vehicles so as to avoid creating shell craters which might hamper their movement.

28. Mine-clearing vehicles are always committed ahead of the attack force. The attack force must complete attack preparations before the mine-clearing vehicles move off from the departure point. The departure time of the vehicles must be known to the attack force.

The timing of the attack force's departure from its assembly area and the manner in which it is to follow the vehicles depends on the enemy situation, terrain, distance to the target and safety requirements (see subparagraph 31).

The closer the attack force stays to the mine-clearing vehicles, the less time the enemy will have to defend and the better it will be able to exploit the success of the mine-clearing vehicles.

29. The mine-clearing vehicles will often be tasked to support assault teams. Depending on the mission and terrain, one to three vehicles are attached to each assault team. Furthermore, it may be advisable to assign tanks as part of the attack team. The role of the vehicles in the assault team is based on subparagraph 28.

30. In order to ensure a trouble-free advance, the mine-clearing vehicles require sufficient covering fire from artillery, heavy infantry weapons, tanks or antitank guns from the time of departure until arrival at the target. The purpose of these weapons is to silence the targets against which the mine-clearing vehicles are committed and take out enemy antitank defenses and observation posts. The use of artificial smoke is usually advantageous, however, the mine-clearing vehicle must remain visible to the controller.

Placing smoke on the target briefly can assist the controller in locating the target and guiding the vehicle to it.

31. Because of the charge's great explosive force, the following safety measures are required to avoid casualties:

In the attack riflemen and other troops not under cover must be 150 to 200 meters from the mine-clearing vehicle at the moment of detonation. If there is good cover available, the range can be less. Even with hatches closed, tanks may only approach to within 100 meters of the detonation site.

If, during pre-attack preparations, enemy action creates the possibility of a premature detonation, the above safety measures also apply for the assembly area.

The commanders of the mine-clearing units shall also inform other units of applicable safety measures as required.

4.2.2: Instructions for the Use of Remotely-Guided Demolition-Charge Carriers (Panzer-Abteilung 300)

A. Nature and Mode of Operation of the Weapon

1.) The remotely-controlled demolition-charge carrier is used to engage field positions, eliminate obstacles, engage super-heavy enemy tanks and to detect mines. The demolition-charge carrier is employed unmanned, guided to its target by remote control and destroys the target through the detonation of the charge it carries.

2.) The weapon's advantages are its great explosive effect and its effect on morale. Remote-control gives the vehicle great accuracy and places no personnel at risk. Because of its diminutive size it presents the enemy with a small target.

The demolition-charge carrier cannot negotiate deep ditches or craters with steep walls.

3.) The demolition-charge carrier is especially well suited to the following tasks:

a) Engaging and destroying antitank obstacles, iron and concrete hedgehogs, angle iron obstacles and wire-cable obstacles,
b) Destroying roadblocks of all types including hidden charges,
c) Destroying bunkers, dugouts and nests of resistance,
d) Destroying all kinds of wire obstacles,
e) Defending against massed infantry attacks through psychological and actual effects,
f) Delivering smoke or chemical warfare agents,
g) Blowing up targets to great effect in built-up areas,
h) Detecting mines and creating lanes in minefields approximately 40 meters in diameter,
i.) Drawing enemy fire.

The last two tasks are usually secondary tasks which are completed simultaneously with other tasks.

B. Organization, Equipment and March Lengths

4.) In terms of terrain-crossing ability and march speed the companies are rated the same as armor units. They are capable of defending themselves against air and ground attacks. Their equipment with unit trains, motor vehicles, maintenance sections and signals equipment is similar to that of tank companies.

C. Operations

5.) Because of the high value of the equipment, the relatively small number of demolition-charge carriers available, and the unique nature of the weapon, mine-clearing vehicles can only be used at the *Schwerpunkt* of combat operations.

6.) Demolition-charge carriers can be committed singly or in company, platoon or section strength. Platoon-size commitment is the rule.

The number of demolition-charge carriers committed and the way they are used depends on the mission and target as well as the nature of the terrain and the available cover.

Flat terrain without obstacles offering a good line of sight makes it possible to commit a larger number of demolition-charge carriers simultaneously, however, it also favors the enemy's defensive measures.

Overgrown, cut-up terrain usually restricts operations to a single demolition-charge carrier. The same applies to missions to clear obstacles from roads.

7. *Panzer-Abteilung 300* fights as part of a combined-arms team. In order to assure proper coordination between them and *Panzer-Abteilung 300*, the battalion is generally attached to the attack force.

8. The operations order to the leader of *Panzer-Abteilung 300* must contain:

a) Enemy situation; friendly intent
b) Mission of *Panzer-Abteilung 300*
c) Coordination instructions for other units
d) Security of the assembly area
e) Time of attack

9. The medium company is only suited for operations against stationary fronts or fortified lines of resistance, because its remotely-controlled radius of action is only about 1,200 meters and the wire must be rewound after a successful operation (duration about 20 minutes). In addition, the wire-controlled vehicle is especially suited to the defense against super-heavy tanks and massed infantry attacks. The explosive effect of the Bren Carrier is approximately equal to that of the heaviest bombs.

D. Combined Arms Operations

10.) Demolition-charge carriers must be committed in conjunction with other units, never independently, since the immediate exploitation of the explosive effect offers the basis for success.

11.) The use of remotely-controlled demolition-charge carriers is to be included in the plan of attack. One exception is employment as part of armored spearheads.

12.) If the attack is to be preceded by an artillery barrage, all efforts should be made to avoid firing on the sector allocated to the demolition-charge carriers so as to avoid creating shell craters which might hamper their movement.

13.) Demolition-charge carriers are always deployed in front of the attack force.

The attack force must complete attack preparations before the demolition-charge carriers move off from the line of departure. The attack force must know the departure time of the vehicles. The timing of the departure of the attack force from its assembly area and the manner in which it is to follow the vehicles depends on the enemy situation, terrain, distance to the target and safety requirements (see subparagraph 16).

The closer the attack force stays to the demolition-charge carriers, the less time the enemy will have to defend. It will also be better able to exploit the success of the demolition-charge carriers.

14.) The demolition-charge carriers may be tasked to support assault teams. Depending on the mission and terrain, one to three vehicles will generally be assigned to each assault team. Furthermore, it may be advisable to assign tanks and flamethrowers as part of the attack team. The role of the vehicles in the assault team is based on subparagraph 13.

15.) In order to ensure a smooth advance, the demolition-charge carriers require sufficient covering fire from artillery, heavy infantry weapons, tanks or antitank guns from the time of departure until arrival at the target. The purpose of these weapons is to pin down the targets against which the demolition-charge carriers are committed and take out enemy antitank defenses and observation posts. The use of artificial smoke is usually advantageous, however, the demolition-charge carrier must remain visible to the controller.

Placing smoke on the target briefly can assist the controller in locating the target and guiding the vehicle to it.

16. Because of the charge's great explosive force, the following safety measures are required to avoid casualties:

In the attack, exposed riflemen and other troops must be 100 to 150 meters from the demolition-charge carrier at the moment of detonation and in depressions in the ground. If there is good cover available, the range can be less. Even with hatches closed, tanks may only approach to within 50 meters of the detonation site.

If, during pre-attack preparations, enemy action creates the possibility of a premature detonation, the above safety measures also apply for the assembly area.

The officers of *Panzer-Abteilung 300* shall also inform other units of applicable safety measures as required.

4.2.3: Interim Guidelines for the Employment of Radio-Controlled Armored Vehicles (2 April 1943)

—

Secret!

Interim Guidelines
For the Employment of Radio-Controlled Armored Vehicles
(Radio-Controlled Armored Vehicles)
2 April 1943

—

Inspector General of the Armored Forces
Berlin, 2 April 1943

I hereby authorize the following manual:

"Interim Guidelines for the Employment of
Radio-Controlled Armored Vehicles"

Guderian

This is a secret document as defined in Section 88 of the German Criminal Code (24 April 1934 edition). Misuse will be punished according to the provisions of this law, provided other penal clauses do not apply.

—

Introduction

The manual "Guidelines for the Employment of Remotely-Controlled Armored Vehicles" is applicable to the current state of technical development (March 1943). The remote-controlled tank is deemed operational for use in the field, however, development of the weapon has not yet been completed. To date no experience has been gained in winter warfare.

For these reasons, therefore, it is the responsibility of the unit commander to whom the remote-controlled tanks are attached to inform himself about this special weapon's characteristics, capabilities and operational principles before employing it and committing it only to those missions which are justifiable given the nature and value of the weapon.

—

Table of Contents

—

A. Nature and Missions of the Weapon

1. The remote-controlled tank is a special weapon of the armored forces which is organized in separate companies within the *Panzer-Regiment* or in separate battalions (army-level unit).

The largest unit which can be committed as a body is the company.

The simultaneous employment of all of the companies in the battalion formation is not recommended for tactical and technical reasons. In general, the battalion is responsible for training, development of the equipment and supply.

2. The remote-controlled tank is driven by a crewman while on the march. When employed, it is remotely controlled by radio from a command and control vehicle (*Panzer III*, *Panzer VI* or *Sturmgeschütz*). The otherwise unarmed vehicle is equipped with a 450-kilogram jettisonable charge and smoke discharging equipment.

3. The driving and control characteristics of the remote-controlled tank exclude missions when the terrain between the controller and the target has been torn up by heavy bombs or artillery fire, is crossed by multiple lines of trenches or positions, or allows no line of sight. Missions are also excluded in marshy terrain or outside cleared lanes in heavily wooded or vegetated areas.

4. The remote-controlled tank is both a reconnaissance and combat vehicle. Its employment avoids losses in men and equipment. When employed as part of an armored force, its primary missions are reconnaissance and the engagement of targets which might threaten the armored attack and which cannot successfully be engaged by other weapons.

5. Within radio range, which can be up to 2,000 meters depending on the line of sight, it is primarily suited for the fol-

lowing missions:

Missions as a reconnaissance vehicle:

a) Combat reconnaissance of enemy defenses in front of the armored spearhead and in the attack zone of an armored unit by drawing defensive fire and moving over mines.

b) Terrain reconnaissance of the attack sector for drivability (marsh land, steep banks, narrows, antitank ditches, hard-to-see defiles running across the direction of the attack, etc.).

Missions as a demolition-charge carrier:

a) Elimination of field and road obstacles and barriers.

b) Destruction of permanent and heavily-fortified or field-type defensive positions.

c) Destruction of personnel in enemy positions (lethal effects within a radius of 30 – 40 meters, temporary incapacitation up to 80 meters).

d) Destruction of the heaviest enemy tanks when they cannot be destroyed by other weapons (running gear of heavy tanks destroyed, crew killed by explosion).

e) Destruction of bridges and other manmade structures when enemy action makes demolition by combat engineers impossible.

Smoke Operations:

f) Laying of smoke to conceal the vehicle itself, blinding of individual targets with smoke, laying down smoke over entire sectors.

g) Creating decontaminated areas in contaminated terrain. Supplying follow-on decontamination teams with decontaminants in combat (for technical details relating to f) and g) see subparagraph 11).

6. Moving over mines results in the detonation of the 450-kilogram jettisonable charge, destroying the vehicle and setting off other mines in a radius of 15 meters.

The remote-controlled tank is therefore only capable of indicating the presence of mines through its detonation and rendering them harmless within a limited radius. It is not, however, capable of deliberate mine-clearing operations.

B. Technical Details, Organization and Equipment

I.

7. The *BIV* remote-controlled tank is a fully-tracked vehicle.

Height: 1.25 meters
Width: 1.80 meters
Length: 3.35 meters
Weight (operational): 3.6 tons
Range:
Driven by a crewman: 150 kilometers
Radio controlled (maximum distance of the remote-controlled tank from the control vehicle): up to 2 000 meters
Off-road capability: Similar to that of the *Panzer III*
Ground clearance: 0.20 meters
Trench-crossing ability: 1.35 meters
Climbing ability: 0.45 meters
Maximum incline: 35°
Fording ability: 0.80 meters
Armor: frontal only, proof against small arms

8. The control tank is a *Panzer III*, *Panzer IV* or *Sturmgeschütz*. It houses the transmitter equipment internally but retains its armament (main gun and machine gun). The command and control vehicle is only capable of controlling one remote-controlled tank at a time.

9. The command and control vehicle's transmitter equipment and the remote-controlled tank's receiver allow the following commands to be transmitted, received and executed: "Start, stop, right, left, faster, slower, forwards, backwards, ignite, jettison, lay smoke."

10. The jettisonable charge is safe to handle, provided the fuse capsule is removed during the march and when secured. It can then only be detonated by a direct bomb or shell hit. Smaller-caliber hits merely penetrate the explosives. If the remote-controlled tank catches fire, the jettisonable charge merely burns slowly.

Nevertheless, a safety distance of 20 meters from the remote-controlled tank is to be maintained at all times while on the march, in a billeting area and in the assembly area.

11. A smoke discharger can also be installed on the remote-controlled tank. It accommodates eight *Langnebelkerzen 42* smoke grenades (smoke duration 25 to 30 minutes). On command, each smoke candle can be discharged from the tank and ignited or ignited while still on the moving remote-controlled tank.

12. In lieu of the smoke grenades the vehicle can accommodate and fire/deposit:

eight decontamination grenades or
eight containers of decontaminant.

The decontamination grenades explode after being fired, spreading decontaminant and creating decontaminated islands for two to three men in contaminated terrain. The containers hold 25 – 30 kilograms of decontaminant and are used only to supply decontamination teams in action.

II.

13. The smallest combat unit is the platoon (see section C subparagraph 18).

The platoon is organized as follows:

Platoon headquarters
3 sections, each with
1 command and control vehicle

4 remote-controlled tanks

1 prime mover with reserve jettisonable charges and smoke grenades
Total: 4 command and control vehicles and 12 remote-controlled tanks.

The company is organized as follows:

Company headquarters
2 platoons
Combat trains
Maintenance section
Total: 10 command and control vehicles and 36 remote-controlled tanks (includes 12 in reserve with the combat trains)

The battalion is organized as follows:

Battalion headquarters
1 Headquarters company
3 Remote-controlled tank companies
1 Maintenance platoon
Total: 32 command and control vehicles and 108 remote-controlled tanks

14. Length of the entire company in march column:

20 meters interval between vehicles, including trains: 2,200 meters

Without trains: 1,500 meters

Passage time for the entire company:

At 20 kilometers an hour (with trains): 6.5 minutes

At 20 kilometers an hour (without trains): 4.5 minutes

15. Fuel consumption per 100 kilometers for a remote-controlled tank company is 7,000 liters; for a radio-controlled tank battalion 24,000 liters.

16. Battalion radio equipment includes the *Fu 5* sets in the command and control vehicles plus six 30-Watt sets, two of which are installed in command and control vehicles.

C. Operational Guidelines

a) General

17. Remote-controlled tank units generally operate as part of *Panzer-* or *Panzer-Grenadier-Divisionen.* As a rule, the remote-controlled tank company is attached to a *Panzer-Regiment* for operations, the remote-controlled tank platoon to a *Panzer-Abteilung.*

18. The smallest combat unit is the platoon. The attachment of single sections is out of the question for tactical reasons, as each section can only control one remote-controlled tank at a time. Furthermore, if the platoon is split up neither adequate command nor supply is assured.

19. The negotiability of the terrain and the cover offered by it are decisive factors in the use of remote-controlled tanks.

Flat terrain without obstacles and offering good fields of view makes it possible to commit a larger number of remote-controlled tanks simultaneously, however, it also favors the enemy's defensive measures.

Densely vegetated and broken terrain usually restricts operations to single remote-controlled tanks. The same applies to missions to clear obstacles from roads.

20. Close cooperation between the commander of the remote-controlled tank units and the force commander are of vital importance to the success of an operation by remote-controlled tanks.

Before an operation the commander of the remote-controlled tank unit is to be listened to; he makes recommendations for the employment of his unit.

During an operation close signals communications must exist between him and the force commander to ensure that results achieved by the remote-controlled tanks are exploited at once.

21. The employment of even the smallest combat unit of remote-controlled tanks must be prepared according to a plan and supported by other units.

In addition to providing cover for the control site, from which the command and control vehicle conducts the radio-control process, support includes providing strong fire support for the advance of the remote-control tank in all types of missions.

The units assigned to this purpose must follow the movements of the remote-controlled tanks closely so they can immediately pin down enemy defensive weapons as they are recognized.

With its armament, the command and control vehicle is capable of assuming this task on its own.

22. The element of surprise is a significant element in the successful use of this special weapon. Not only does it limit the enemy's defensive readiness, it also increases the demoralizing effect which the unmanned yet controlled vehicle and its tremendous destructive force have on the enemy.

23. The use of artificial smoke by the units providing covering fire for the purpose of blinding the enemy's defensive weapons and observation posts can be beneficial in all types of mission. The remote-controlled tank and the target must remain visible to the controller, however.

24. To the extent possible, friendly artillery must avoid firing on sectors intended for use by remote-controlled tanks so as to avoid cratering the terrain and making it impassable.

25. Before a mission, remote-controlled tanks must usually make a brief halt for servicing and technical preparations.

Like other armored units, remote-controlled tank units should be given the opportunity to assemble and service their vehicles after an operation. Unlike armor units, they are not capable of exploiting success and holding ground which has been won.

26. The use of remote-controlled tanks must not always result in their destruction (blowing up). Remote-controlled tanks used for combat or terrain reconnaissance which cannot be guided back by remote control are to be recovered and, if possible, repaired.

Remote-controlled tanks, even damaged ones, must not be allowed to fall into enemy hands. When capture is imminent they are to be destroyed by self-demolition or gunfire.

For this purpose, support from the unit to which the remote-controlled tank unit is attached must be assured.

b.) Details

27. Remote-controlled tanks are to be used for zone reconnaissance in front of the armored spearhead only when distance and terrain make imminent enemy contact likely.

A lengthy remote-controlled procedure, such as on a winding road with many turns, is difficult and time-consuming.

For the purpose of reconnoitering the attack zone of an armored unit prepared to launch a deliberate attack, several remote-controlled tanks are to be committed so that the entire breadth of the attack zone can be reconnoitered. If there is a flanking threat, remote-controlled tanks are to be committed to reconnoiter.

28. If remote-controlled tanks are committed to reconnoiter or destroy targets before a deliberate attack, their use is to be incorporated as part of the overall plan of attack.

In addition to safety measures and providing cover, observe the following guidelines:

a) As with any other weapon, thorough scouting of the terrain and target are necessary. Approach route, assembly area, control site and direction of departure to the target must be determined.

b) The remote-controlled tanks are to move into the assembly area shortly before the start of the attack so as not to expose them to artillery fire prematurely, to which they are extremely vulnerable.

The control site is to be located close to friendly lines in order to take full advantage of the limited radio-control range. On leaving the attack position, which is usually reached using a crewman in the vehicle, the remote-controlled tanks then pass the control site without stopping. Only the command and control vehicles stop there.

c) The attack force must be in the attack position before the remote-control tanks move off, allowing it to follow up immediately and exploit the effect of the remote-control mission.

The timing of the departure of the attack force behind the remote-control tanks depends on the terrain, the radio-control range and safety rules. Tanks can approach to within 75 meters of a remote-controlled tank and *Panzergrenadiere* in good cover to within 150 meters without being threatened by the detonation.

29. In attacks against fortified positions, multiple rows of antitank obstacles and built-up areas which have been converted to strongpoints, decisive success can only be assured through the combined use of several remote-control tanks against a few targets.

30. Remote-control tanks can be used to support assault teams in towns to good effect. Especially close coordination with the assault teams is necessary and must be worked out in detail. The explosive effect of the jettisonable 450-kilogram charge is especially great in street fighting.

31. The employment of remote-control tanks against enemy tanks is to be limited to engaging super-heavy tanks, because the speed and mobility of lighter tanks makes them difficult remote-control targets. The integrity of the platoon is also to be maintained when engaging enemy tanks, since stationary tanks or those approaching head-on should be attacked with two remote-control tanks in order to divide the enemy's observation and defensive fire. For defense against an enemy tank attack, remote-control tanks can occupy reverse-slope positions. Against enemy tanks which have broken through, remote-control tanks may be brought forward or employed against the enemy's rear.

32. The general rules for the use of smoke also apply to the use of smoke by remote-control tanks, however special advantage may be gained in the following circumstances:

a) Laying down smoke from moving remote-control tanks,

b) Moving the smoke containers to an object which is to be covered with smoke and dropping them at a precise location,

c) Forming precise lanes or zones of smoke by the successive dropping of smoke containers.

The variety of possibilities for laying down smoke screens by remote-control tanks is especially valuable in blinding defensive weapons which cannot otherwise be eliminated — especially on the flanks — and while withdrawing from the enemy.

33. Remote-control tanks are to be used to carry decontamination grenades and containers of decontaminating agents when enemy action prevents decontamination teams from decontaminating an area or the open delivery of decontamination agents to the battlefield.

—

4.2.4: Interim Guidelines for the Employment of the Heavy Remote-Control Tank Company (15 April 1944)

Secret!

Interim Guidelines

For the Use of the Heavy Remote-Control Tank Company

15 April 1944

—

Gen.Insp.d.Pz.Tr.
Nr. 3640/44 geh.

OKH Headquarters, 15 April 1944

I hereby authorize the following manual:

"Interim Guidelines for the Employment of
Radio-Controlled Armored Vehicles" (15 April 1944)

Guderian

This is a secret document as defined in Section 88 of the German Criminal Code (24 April 1934 edition). Misuse will be punished according to the provisions of this law, provided other penal clauses do not apply.

—

Table of Contents

—

A. Nature and Missions of the Weapon

1. The heavy remote-controlled tank company is a special company which is incorporated into heavy tank battalions. The battalion commander determines its use.

2. The heavy remote-control tank company is organized as follows:

a) Combat Elements:

Company headquarters: Two *Panzerkampfwagen VI's*
3 Platoons, each with: 4 *Panzerkampfwagen VI's* and 9 remote-control tanks
1 medium SPW.
(Per platoon three sections each with 1 *Panzerkampfwagen VI* and 3 remote-control tanks.)

b) Wheeled echelon:

Company headquarters,
Company maintenance section,
Combat trains,
Baggage trains.

Total: 14 command and control tanks and 36 remote-control tanks (including 9 in reserve with the combat trains).

3. The remote-control tank is driven by a crewman while on the march. In action it is remotely controlled by radio from a command and control tank (*Panzerkampfwagen VI*).

The unarmed vehicle is equipped with a 450-kilogram jettisonable charge and smoke generation equipment.

4. The driving and control characteristics of the remote-control tank exclude missions when the terrain between the controller and the target has been torn up by heavy bombs or artillery fire, is crossed by multiple lines of trenches or positions, or is broken in nature, reducing visibility. Missions are also excluded in marshy terrain or outside cleared lanes in broken wooded or brush-covered areas. Guiding a tank on narrow roads is not possible and cannot be successful.

5. The remote-control tank is both a reconnaissance and combat vehicle. Its use avoids losses in men and equipment. When employed as part of an armored force, its primary missions are reconnaissance and the engagement of targets which might threaten the armored attack and which cannot successfully be engaged by other weapons.

6. Within radio range, which can be up to 2,000 meters depending on observation possibilities, it is primarily suited for

the following missions:

Missions as a reconnaissance vehicle:

a) Zone reconnaissance of enemy defenses in front of the armored spearhead and in the attack zone of an armored unit by drawing defensive fire and moving over mines.

b) Terrain reconnaissance of the attack area for drivability (marsh, steep banks, defiles, antitank ditches, hard-to-see ravines running across the direction of the attack, etc.). Results of scouting of marshy areas are unreliable as remote-control tanks exert less ground pressure than normal tanks.

Missions as a demolition-charge carrier:

c) Elimination of terrain and road obstacles and barricades, including the elimination of small mine obstacles (see also subparagraph 7).

d) Destruction of permanent and heavily-fortified or field-type defensive positions.

e) Destruction of personnel in enemy positions.

f) Destruction of the heavy enemy tanks when other weapons are insufficient (running gear of heavy tanks destroyed, crew killed by explosion).

g) Destruction of bridges and other manmade structures when enemy action makes demolition by combat engineers impossible.

Smoke Missions:

h) Laying of smoke to conceal the vehicle itself, blinding of individual targets with smoke, laying down smoke over entire sectors.

7. Driving over mines results in the detonation of the 450-kilogram jettisonable charge, destroying the vehicle and setting off other mines within a radius of 15 meters. It is capable of indicating the presence of mines through its detonation and rendering them harmless within a limited radius. It is not, however, capable of performing deliberate mine-clearing operations.

8. The command and control tank is a *Panzer VI* with transmitter equipment for remote-control tanks and full weaponry (main gun and machine guns). It is capable of controlling two remote-control tanks simultaneously.

9. The command and control tank's transmitter equipment and the remote-control tank's receiver allow the following commands to be transmitted, received and executed: "Start, stop, right, left, faster, slower, forwards, backwards, ignite, jettison, lay smoke."

10. The jettisonable charge is safe to handle provided the blasting cap is removed during the march and in billeting areas. It can then only be detonated by a direct bomb or shell hit. Smaller-caliber hits merely penetrate the explosives.

A safety distance of 25 meters from the remote-control tank is to be maintained at all times while on the march, in a billeting area and in the assembly area.

11. A smoke discharger can also be installed on the remote-control tank. It accommodates four *Nebelkerzen 39* smoke grenades. On command, the smoke grenades are ignited and ejected by a propellant charge. Smoke duration is from 10 to 15 minutes.

12. The heavy remote-control tank company is equipped with the same signals equipment as a conventional heavy tank company.

B. Operational Guidelines

a) General

1. The heavy remote-control tank company bolsters the fighting strength of the heavy tank battalion. Without remote-control tanks it can carry out the same missions as any heavy tank company.

2. The heavy remote-control tank company is to be committed only as part of the battalion. The high value of the demolition-charge carrier limits its use to decisive targets.

The smallest combat unit for remote-control tanks is the platoon.

3. The negotiability of the terrain and the cover it offers are decisive factors in the employment of remote-control tanks. Flat terrain without obstacles and offering good visibility makes it possible to commit a larger number of remote-control tanks simultaneously, however, it also favors the enemy's defensive measures. Unfavorable terrain makes it necessary to restrict the number of remote-control tanks committed.

4. Coordination within the battalion is of vital importance to the success of the remote-control tanks. Before an operation, the commander of the remote-control tank unit is to have input.

5. Pre-planning and support by other units is necessary.

In addition to providing cover for the control site, from which the control tank conducts the radio control process, support includes providing strong fire support for the advance of the remote-control operation.

The units assigned to this purpose must follow the movements of the remote-control tanks closely so they can immediately pin down enemy defensive weapons as they are recognized. The command and control tank is not capable of performing this mission alone.

6. The element of surprise is a significant element in the successful use of this special weapon in the offense and while conducting reconnaissance. Not only does it limit the enemy's defensive readiness, it also increases the demoralizing effect which the unmanned and yet controlled vehicle and its tremendous destructive force have on the enemy.

7. The use of artificial smoke by the units providing covering fire for the purpose of blinding the enemy's defensive weapons and observation posts can be beneficial in all types of missions. The remote-control tank and the target must remain visible to the controller, however.

8. To the extent possible, friendly artillery must avoid firing on sectors intended for use by remote-control tanks so as to avoid cratering the terrain and making it impassable.

9. After the command and control tank has taken control of the remote-control tank, the latter's driver is to be picked up by the medium *SPW*.

The medium *SPW* follows the attack with the remaining remote-control tanks at a distance outside the concentrated defensive fire which falls on the *Tiger* unit. Visual contact with the battalion must be maintained.

10. The use of remote-control tanks must not always result in their destruction (detonation). Remote-control tanks used for terrain or combat reconnaissance which cannot be guided back by remote control are to be recovered and, if possible, repaired.

11. Remote-control tanks, even damaged ones, must not be allowed to fall into enemy hands. When capture is imminent they are to be destroyed by self-demolition or gunfire.

12. Every period of inactivity is to be used to service and repair the remote-control tanks.

13. Because of the danger of interception, checking of the radio-control equipment is to be done only with test equipment or tuning antennas.

<u>b.) Details</u>

1. For the purpose of reconnoitering the attack zone of an armored unit about to conduct a deliberate attack, several remote-control tanks are to be committed so that the entire width of the attack zone can be reconnoitered. (Do not spare remote-control tanks!)

Remote-control tanks can eliminate surprises on the flanks.

2. In continuous antitank barricades and all known antitank obstacles, the remote-control tanks create lanes with their explosive effect. In-depth antitank obstacles require the use of several remote-control tanks in succession.

Antitank ditches require the use of two remote-control tanks with approximately 5 meters between them. Effectiveness against antitank ditches depends upon ground conditions.

Several remote-controlled tanks must be committed to create lanes through a minefield. All tank crews are to closely observe the route taken by the demolition-charge carriers and their detonation points.

Remote-control tanks cannot detect randomly emplaced mines effectively. In a deliberate attack, the use of remote-control tanks is to be phased in according to the plan of attack. In addition to safety, providing cover and covering fire, observe the following guidelines:

a) Thorough reconnaissance of the terrain and target are necessary for the employment of remote-control tanks (approach route, assembly area, control site and direction to the target must be determined).

b) The remote-control tanks are to move into the assembly area shortly before the start of the attack so as not to expose them to artillery fire prematurely, to which they are extremely vulnerable.

The control site is to be located close to friendly lines in order to take full advantage of radio-control ranges.

c) The attack force must be in position before the remote-control tanks deploy. The rest of the attacking units must follow up immediately and exploit the effect of the remote-control tanks.

d) The remote-control tanks and the control positions will move forward sequentially, from sector to sector, so that the section leader will always have good visibility and guide his remote-control tanks accordingly.

e) The deployment of the attack force behind the remote-control tanks depends on the terrain and the safety distances to be maintained (tanks can approach to within 75 meters of a remote-control tank, *Panzergrenadiere* in good cover to within 150 meters, without being threatened by the detonation).

4. In attacks against fortified positions, multiple rows of antitank obstacles and built-up areas converted to strongpoints, decisive success can only be assured through the combined use of several remote-control tanks against a few targets.

5. Remote-control tanks can be used to support assault teams in towns with good effect. Especially close coordination with the assault teams is necessary and is to be discussed in detail. The explosive effect of the jettisonable charge is especially great in street fighting.

6. The use of remote-control tanks against enemy tanks is to be limited to engaging heavy tanks, because the speed and mobility of lighter tanks makes them difficult remote-control targets. Stationary tanks or those approaching head-on should be attacked with two remote-control tanks in order to divide the enemy's observation and defensive fire.

Against enemy tanks that have broken through, the heavy remote-control tank company may attack from the flank or from behind. (Mobile defense)

Close coordination is to be maintained within the platoon.

7. The general rules for the use of smoke also apply to the use of smoke by remote-control tanks, however, special advantages may be gained through the laying down of smoke by moving remote-control tanks.

(Precise lines, forming lanes or zones of smoke.) The laying of smoke screens by remote-control tanks is especially valuable for blinding defensive weapons which cannot otherwise be eliminated, especially on the flanks and while withdrawing from the enemy.

8. Equipping the remote-controlled tanks with direction lights makes possible their limited use at night in previously scouted terrain. This avoids unexpectedly running into a defense front.

9. The use of remote-controlled tanks in winter is limited by their inability to negotiate snow and the sensitivity of the electrical equipment to severe cold.

4.2.5: Memorandum on the Employment of the Radio-Control Force in "Operation Zitadelle" from 5 to 7 July 1943

(Source: Bundesarchiv/Militärarchiv Freiburg)

Secret

23 July 1943

Reinel

Major

Commander

Panzer-Abteilung (FKL) 301

Panzer-Abteilung (FKL) 301 Operations/Commander *280/43g*

Memorandum on the Continued Employment of the Radio-Control Arm Based on Evaluation of Experience in "Operation *Zitadelle*" from 5 to 8 July 1943

Three separate remote-control tank companies were committed in the attack zone of *Armee-Oberkommando 9* south of Orel on 5 July. Two companies were attached to *schweres Panzerjäger-Regiment 656* and one to *schwere Panzer-Abteilung 505*. The companies were committed as complete units under the command of the company commander who directed the platoons to work with the companies of the larger formations employed at the front. The missions assigned the companies were the same, namely: Conducting reconnaissance in force, detecting minefields and blasting lanes, destroying difficult-to-engage targets such as dug-in antitank weapons and super-heavy tanks.

The reports sent by the units concerning the operations of the remote-control tanks gave the following picture:

1.) Operations by *Panzer-Kompanie (Fkl) 314* with the *I./schweres Panzerjäger-Regiment 656*.

A very dense and deeply echeloned Russian minefield barred access to the Russian main line of resistance, which was also screened by very heavy artillery fire. In accordance with the operations order, the company started to clear three lanes through the minefields. A total of 12 *BIV's* was committed because of the depth of the minefield. The command and control vehicles passed through the resulting lanes without suffering damage. The operations order had called for the combat engineers to mark the lanes, however, because of the extremely heavy artillery fire, they were unable to move forward. This resulted in a halt in the attack. Because of the large number of artillery rounds exploding on the battlefield, the heavy tank-destroyers (*Ferdinands*) were unable to clearly identify the lanes created by the *BIV's*, which had not been marked. Furthermore the *BIV's* left no visible tracks on the grass. As a result, in spite of the lanes that had been created, *Ferdinands* were rendered disabled by mines.

A total of seven *BIV's* were detonated in the subsequent course of the attack. Of these one ran into a trench manned by enemy infantry. The troops in the trench attacked the *BIV* with hand grenades and close-range weapons but were wiped out when the *BIV* exploded. Two *BIV's* were guided against a heavily-occupied patch of woods against which the infantry was unable to advance. They were detonated in place, after which all resistance ceased. A total of four *BIV's* were knocked out by artillery fire. One, which was in an armed condition (blasting caps installed), detonated, while the other three without fuses burned without incident.

2.) Operations by *Panzer-Kompanie (Fkl) 313* with the *II./schweres Panzerjäger-Regiment 656* took place under similar conditions. Four *BIV's* were disabled when one platoon drove into an unmarked German minefield while moving up for a mission. The other platoon created just a single lane in the Russian minefield, using up four *BIV's* in the process. One *BIV* was hit by artillery fire and detonated in the assembly area, whereupon two other *BIV's* caught fire and also exploded. The undersigned was unable to determine the cause of this incident with certainty, since the *BIV* drivers and combat engineers involved were killed. It is suspected that the fuses had already been inserted and were set off by the heat of the burning charge. Another *BIV* was hit by artillery fire while being guided and also exploded.

In the subsequent course of the attack three *BIV's* destroyed three dug-in antitank positions and one bunker, scoring a tactical and psychological success.

3.) Operations by *Panzer-Kompanie (Fkl) 312* with *schwere Panzer-Abteilung 505*.

Employment of the company in the battlefield reconnaissance role in front of the *Tiger* tanks was in keeping with the tactical requirements and achieved good results. One *BIV* was sent against an antitank position containing two or three antitank guns at a range of 800 meters. The charge was detonated in place, destroying the antitank guns and infantry concentrated there. One *BIV* was committed against a T34 at a range of 400 meters. The T34 attempted to ram, but was destroyed when the *BIV* was detonated. Three *BIV's* that were deployed against three heavy gun bunkers at a range of 400-600 meters were detonated in place, destroying all three bunkers. A *BIV* sent against an antitank bunker was set on fire 10 meters from the bunker, but nevertheless reached its target and destroyed the bunker. Two *BIV's* committed against antitank gun crews and infantry gun positions at 800 meters destroyed both. One *BIV* reached a Russian position where it was attacked with gasoline bombs and set alight, detonating its charge. The detonation had a devastating effect on the enemy position. In four cases defensive weapons knocked out *BIV's* during the radio-control process. In two cases the radio equipment was recovered, while the other two burned out. Altogether 20 *BIV's* were committed in four days of combat.

Tactical Evaluation of the Operations

1.) In two cases the remote-control companies — an exceptional offensive weapon — were used with heavy tank destroyers and only once with a proper offensive unit, a *Tiger* battalion. When coupled with the heavy tank destroyers, the clumsiness of the *Ferdinands* had a limiting effect on the offensive impetus of the remote-control units. The latter gained ground quickly, however, the *Ferdinands* were unable to exploit this success quickly enough. After moving forward, the command and control vehicles had to wait too long for the *Ferdinands* and were exposed to fierce defensive fire. On the other hand, the collaborative effort with a pure armored unit, the *Tiger* battalion, produced much more favorable results. As offensive weapons, the *Tigers* were technically and tactically much more capable and better trained, gaining ground in the attack area and retaining their cohesion. This experience proves that the remote-control force can only achieve complete success when coupled with pure tank units.

2.) The operations by *Panzer-Kompanien (Fkl) 313* and *314* with the units of *schweres Panzerjäger-Regiment 656* on either side of the Orel — Kursk rail line south of Glasunowka were against a deeply echeloned defensive position whose approaches were mined. The very heavy enemy artillery fire complicated the attackers' task considerably. Our forces committed to the main effort were inadequate given the enemy's strength. This was especially true of the remote-control units.

Widely deployed in the regiment's attack lanes and each attached to a *Ferdinand* battalion, they quickly used up their strength. The lack of reserves meant that losses suffered by the platoons at the front could not be made good, to say nothing of continuing the drive into the depth of the enemy's main battle area. Indeed there were no mobile reserves available with which a senior remote-control commander could have expanded the penetration into a breakthrough at the identified weak spots. The coordination required with the *Ferdinands*, which was absolutely vital to the overall success, had been well prepared through map exercises and live exercises.

In combat, however, this quickly broke down as a result of the vigorous enemy countermeasures. In spite of energetic efforts by the remote-control officers, close contact could not be restored as they were unable to convince the battalion commanders of their requirements. There, too, a senior remote-control command might have been able to intervene successfully. Of a total of 12 available command and control vehicles, 8 were committed in the front line. The remaining four had to be called up a short time later to cover the losses suffered. Because of the absence of further reserves, the attack soon petered out and was halted in the enemy position.

Based on the reports received, it is the opinion of the undersigned that the attack was a vitally important mission, and it is suggested that it should have been carried out by all of *Panzer-Abteilung (Fkl) 301* under unified command. It had been trained and earmarked for such operations. Suggested employment follows:

Employment of the entire radio-control battalion with two companies in the front line and the positioning of the third company as reserve close to the rear of the forward units.

These companies will be sent forward with 2/3 of their offensive strength — 4 command and control vehicles each on a total width of 2 — 3 kilometers. Given such a width, a total of 4 to 6 sufficiently deep and wide lanes can be created, with losses in command and control vehicles being made good by employing the reserve sections in the second wave. (During the operations at Glasunowka only four lanes were created along a total width of 6 kilometers!)

The first company to arrive at the *Schwerpunkt* is to be relieved immediately by the reserve company as soon as its offensive strength begins to dwindle. After the minefield has been passed, the reserve company will move forward along the entire width of the attack sector and achieve a breakthrough. Meanwhile, the relieved company will assemble its remaining combat-ready elements and serve as reserve company.

Because of its adequate complement of command and supply units (headquarters company, maintenance platoon), the unified command is able to carry out repairs to disabled command and control vehicles and quickly return them to action. The battalion takes advantage of its close contacts and reserve equipment. In this way remote-control companies are not a burden on other units which, because of the unique radio-control equipment, would not be able to assist them anyway. (The commanders of the remote-control units which were employed all complained that nowhere had they found understanding and adequate support in these important matters and, more than anything, they wished that they might operate as part of their own battalion.)

In terms of forces and command and control, such a concentrated employment of remote-control forces would promise success, but only under the following conditions:

3.) In the operation at Glasunowka the failure to achieve total success was mainly attributable to a lack of understanding on the part of the larger formation (*Ferdinands*) and poor coordination. Effective coordination broke down even though it had been practiced in joint exercises and discussed at meetings before the operation. Inter-unit coordination will always be difficult under heavy enemy fire, resulting in the following recommendation:

Structure the remote-control formation (battalion) with sufficient firepower to enable it to achieve initial success on its own and maintain its position until follow-on forces arrive. For this purpose, it should be equipped with fast, heavy tanks (*Tigers*). So equipped, it also serves as a tank unit. The smaller assault gun command and control vehicles drew more enemy fire at

Glasunovwka as they moved up. The enemy recognized them as the command and control vehicles for the demolition-charge carriers. Heavy tanks would have been better suited to such an environment. Furthermore, the assault gun is not suited to such a double purpose (fighting vehicle and command and control vehicle) because of the absence of a rotating turret, its cramped fighting compartment (radio operator is also the loader) and the inadequate view from the provisional periscope cupola.

It seems to go without saying that those who know well the power of the remote-control force and who trust in it, accepting the risk posed to them by enemy action (premature detonation of the demolition charge), will most quickly and forcefully exploit the success achieved by the demolition-charge carriers.

In such operations the secret and very valuable remote-control transmitter equipment should be protected by installing it in the best available tanks.

Within the framework of this memorandum it must be mentioned that in terms of personnel, *Panzer-Abteilung 301* is a proven, battle-tested tank battalion which has been in existence since 1938, something which, regrettably, has become rare in the army (first *Panzer-Abteilung 67*, later the *III./Panzer-Regiment 10*). The battalion has added to its experience by thoroughly learning the role of a remote-control battalion. Because of its composition, the valuable armored equipment (*Tigers*) will be carefully maintained and used successfully in combat.

Technical Assessment of the Operation

Generally speaking, the radio control equipment and the demolition-charge carriers lived up to expectations. Some problems did appear, whose rectification in the short term is possible and necessary.

1.) Radio-Control Equipment

It is absolutely necessary that the radio-control equipment function reliably to at least 2,000 meters. The currently available equipment from the Hell Company in no way meets this requirement. Consequently, because the range is not constant and rarely exceeds 800 to 1,000 meters, there is the danger that the equipment might become uncontrollable and fall into enemy hands.

The radio receiver built by the Braun Company, which is currently under test by *Panzer-Abteilung 301*, meets this requirement.

The command receiver suffers during rail transport. It is not sufficiently vibration proof. Sometimes the tone filters fail, and the relays become deformed. These can be rectified by the units, however, it always requires a great deal of preparation before a remote-control operation.

The section leader's transmitters must be fitted with a frequency selector like those of other leaders so that during operations he can quickly take over from the command and control vehicle of a section leader put out of action and direct the radio-control operation.

2.) Combat Engineer Equipment

It has been shown that charges set on fire by enemy action — which happened a number of times, — burn out without incident if the blasting caps have not been fitted, whereas those with blasting caps in place explode. Consequently, it is clear that the blasting caps should not be fitted until the demolition-charge carriers pass through friendly front lines (start of the radio-control procedure).

In order to screw in the three blasting caps and close the charge cover, the *BIV* driver has to crawl out onto the front of the special vehicle. This takes a certain amount of time, during which he is exposed to extreme danger. This creates the possibility that the driver might be shot or, in his excitement, fail to screw in the detonators properly. It must therefore be recommended that a handgrip be fitted to enable the driver to insert the detonators from his position.

The *detonation relay* does not always work reliably. Approximately 25% of the vehicles which ran over mines failed to detonate when the mine went off. The cause of the problem could not be determined as the *BIV's* in question were destroyed by gunfire.

A simple technical aid is required to mark lanes cleared in minefields (possible use of jettisonable paint packets or laying down of a marking strip).

3.)

Losses of *BIV's* to enemy fire totaled approximately 20% of the vehicles committed. A reduction is possible by increasing speed and mobility when moving cross-country. These requirements and armor protection against heavy machine-gun fire must be required for the new developments (*BIV C, Springer*).

4.) Frequencies

When operating as the main effort, the battalion needs an increase in frequencies from 4 bands to at least 6. This will result in more flexible tactical command.

Summary

Employment of the radio-control force in July 1943 was based on expanded service trials in the previous year in which three separate radio-control companies operated separately. The development of the overall situation and the associated switch from offense to defense limited their use to just a few days. The results achieved were in keeping with expectations. Failure to achieve total success cannot be attributed to a failure of the remote-control force as such. In the opinion of the undersigned, this was due to the commitment of inadequate forces at the decisive points, poor coordination with other units and the inability to quickly and completely exploit the explosive effect.

Based on the operational lessons learned by the undersigned, it is proposed that the following service trial be carried out:

Issuing *Panzer-Abteilung (Fkl) 301* with *Tigers* in the command and control tank role and employment of the entire battalion at the *Schwerpunkt* of an operation as part of a larger armored formation.

Signed
Reinel

—

(Source: Bundesarchiv/Militärarchiv Freiburg)

Secret

Headquarters *OKH*, 8 August 1943

The Inspector-General of Armored Forces
Az. 34a / Ia Ausb. / Ia Org.
Nr. 3019/43 geh.

REFERENCE: Memorandum on the Future Expansion of the Remote-Control Force (*Major* Reinel, commanding officer, *Panzer-Abteilung (Fkl) 301*, dated 25 July 1943)

To:

Gen St d H / Op. (I)

Insp. D. Pz. Tr.

Pz.Ausb. u. Ers.Abteilung 300

Pz.Abteilung (Fkl) 301

The following conclusions have been drawn from the operational use of the remote-control tank companies during the period 5 – 8 July 1943:

1.) In order to achieve success, remote-control units must work closely with armored units within a *Panzer-Division.*

a) Close support of the remote-control tanks by the armor formation. Furthermore, all heavy weapons to be employed in concentration to support the remote-control operation.

b) Immediate exploitation by the armor unit of the success achieved by the remote-control tanks.

2.) Equipping the remote-control units with *Tiger* tanks in the command and control tank role depends on the state of development of the remote-control equipment.

There are two distinct stages:

a) As long as the remote-control equipment is not sufficiently tested, remote-control units must remain special units. Based on availability of equipment, using *Tiger* tanks as command and control tanks cannot be justified at present.

b) As soon as the remote-control equipment has proven itself — roughly after a year — the remote-control units should be equipped with tanks of the type used by the unit with which it will be committed.

At that point in time it would be correct to permanently incorporate the remote-control units into armor formations. This will achieve the desired close coordination in training, combat operations and supply.

3. Organizing the company in two platoons has not proven a success. A third platoon must be requested, however, the platoon must retain its former composition (one platoon-leader tank, three command and control tanks). A corresponding proposal for the organization of a remote-control company with 14 command and control tanks is in development.

4. Employment of the entire remote-control battalion, as recommended by *Major* Reimel, cannot yet be approved. Operations in battalion strength may be necessary for special missions; thorough practice and testing is required.

It is intended that *Panzer-Abteilung (Fkl) 301* be proposed for the first operation by an entire remote-controlled battalion with the following provisos:

a) Close coupling with *Panzer-Divisionen*.

b) Adequate preparation and close coordination with the armor formation.

To summarize, it may be said that the experiences of the remote-control companies are valuable and promising for the future.

The Chief of Staff
Signed: Thomale

4.3: Sources and Bibliography

Published Sources

Armored Fighting Vehicles of Germany in WW II, New York: Arco Publishing Company, 1978

Auerbach, William, **Last of the Panzers , Tanks Illustrated No. 9**, ?: Arms and Armour Press, 1984

Bartelski / Bukowski, **Warszawa 1944**, ?: KAW, 1980

Benamou, Jean-Pierre, **Album Memorial "Bataille de Caen"**, Bayeux (France): Edition Heimdal, 1988

Bender, Roger James, **Uniforms, Organization and History of the Panzertruppe**, San Jose (California, USA): Bender Publishing, 1980

Bernage, Georges, **Album Memorial "Normandie"**, Bayeux (France): Edition Heimdal, 1983

——-, **Normandie: Aout 1944 - La Retraite Allemande**, Bayeux (France): Edition Heimdal, 1988

Bernage / Lannoy / McNair / Baumann, **Album Memorial "Bataille d'Alsace 1944-1945"**, Bayeux (France): Edition Heimdal, 1992

Bernage / Mari / Benamou / Mc Nair, **Album Memorial "Bataille de Normandie"**, Bayeux (France): Edition Heimdal, 1993

Bernage / McNair, **Le Couloir De La Mort**, Bayeux (France): Edition Heimdal, 1994

Buffetaut, Yves, **Au Coeur du Reich**, Paris: ?, 1993

——, **La Bataille du Bocage**, Paris: ?, 1994

——, Yves, **El Alamein**, Paris: ?, 1993

——, **6 Juin, La premier Vague**, Paris: ?, 1993

——, **Bataille pour Moscou**, Paris: ?, 1993

——, **Operation Barbarossa**, Paris: ?,1992

——, **Rhin et Danube**, Paris: ?, 1992

Chamberlain / Doyle / Jentz, **Encyclopedia of German Tanks of WW 11**, ?: Arms and Armour Press, 1978

Carell, Paul: **Verbrannte Erde**, ?: Ullstein-Verlag, 1978

——-, **Der Rußland-Feldzug fotografiert von Soldaten**, ?: Ullstein-Verlag,l 968

Dieckert / Großmann, **Der Kampf um Ostpreußen**, Stuttgart: Motorbuch-Verlag, 1995

Eiermann / Remm, **Kraichgau 1945, Kriegsende und Neubeginn**, Kraichgau (Germany): Verlag Regionalkultur, 1995

Ellis, Chris, **German Infantry and Assault Engineer Equipment 1939-1945**, ?: Bellona Publications, 1976

Engelmann, Joachim / Scheibert, Horst, **Deutsche Artillerie 1939-1945**, ?: C.A. Starke Verlag, 1974

Euler, Helmuth, **Entscheidungsschlacht an Rhein und Ruhr 1945**, Stuttgart: Motorbuch-Verlag, 1980

Feist, Uwe, **Aero Publishing Band 12**, ?: Aero Publishers Inc., 1980

Feist, Uwe / Culver, Bruce, **Tiger**, Ryton Publication, 1992

Fürbringer, Herbert, **Die 9. SS Panzer-Division "Hohenstaufen"**, Bayeux (France): Editions Heimdal, 1984

German Remote Controlled Vehicles, Ground Power, ? (Japan): Delta Publishing, 1998

Grove, Eric, **Panzerkampfwagen I und II**, ? (Great Britain): Almark Publishing Co Ltd., 1979

Nehring, Walter, **Die Geschichte der deutschen Panzerwaffe 1916-1945**, ? (Germany): Propyläen Verlag, 1969

Guderian, Heinz, **Erinnerungen eines Soldaten**, Stuttgart: Motorbuch-Verlag, 1994

Haupt, Werner, **Endkampf im Westen 1945**, Wölfersheim-Berstadt (Germany): Podzun-Verlag 1979

——-: Krim, **Stalingrad, Kaukasus — Die Heeresgruppe Süd 1941-1945**, Wölfersheim-Berstadt (Germany): Podzun-Verlag, 1977

——-, **Leningrad, Wolchow, Kurland 1941-1945**, Wölfersheim-Berstadt (Germany): Podzun-Verlag, 1976

——-, **Die deutschen Luftwaffen-Felddivisionen 1941-1945**, Wölfersheim-Berstadt (Germany): Podzun-Verlag,1993

——-, **Königsberg, Breslau, Wien, Berlin 1945**, Wölfersheim-Berstadt (Germany): Podzun-Verlag, ?

Healy, Mark, **Kursk 1943, Campaign Series No. 16**, ? (Great Britain): Osprey Military, 1992

Hinze, Rolf, **19.Infanterie- und Panzer-Division — Divisionsgeschichte,** ? (Germany), Verlag Dr. R.Hinze, 1988

Jaugitz, Markus, **Die deutsche Fernlenktruppe 1940-1943,** Wölfersheim-Berstadt (Germany): Podzun-Verlag, 1994

——-, **Die deutsche Fernlenktruppe 1943-1945,** Wölfersheim-Berstadt (Germany): Podzun-Verlag, 1995

Jentz, Thomas L., **Panzertruppen 1933-1942,** Atglen (Pennsylvania, USA): Schiffer Military History, 1996

——-, **Panzertruppen 1943-1945,** Atglen (Pennsylvania, USA): Schiffer Military History, 1998

Jentz / Doyle, **Germany's Tiger Tanks - VK 45.02 to Tiger II,** Atglen (Pennsylvania, USA): Schiffer Military History, 1997

Jentz / Doyle / Sarson, **Tiger I, New Vanguard No. 1,** ? (Great Britain): Osprey Military, 1993

——-, **Tiger II, New Vanguard No. 5,** ? (Great Britain): Osprey Military, ?

Kleine / Kühn, **"Tiger" - Die Geschichte einer legendären Waffe 1942-1945,** Stuttgart: Motorbuch-Verlag, 1976

Kramp, Hans, **Ruhrfront 1944/45,** ? (Germany): SelbstVerlag, 1992

von Krannhals, Hanns, **Der Warschauer Aufstand 1944,** ? (Germany): Bernard & Graefe Verlag, 1962

Kubisch / Jansen: **Borgward - Ein Blick zurück,** ? (Germany): Elefanten-Press, 1984

Kurowski, Franz/ Tornau, Gottfried, Sturmartillerie 1939-1945 , Stuttgart: Motorbuch-Verlag, 1978

Lefevre, Eric, **"Les Panzers" Normandie 1944,** Bayeux (France): Edition Heimdal, 1978

Leiwig, Heinz, **Finale 1945 Rhein-Main,** ? (Germany): Droste Verlag, 1985

Le Tissier, Tony, **Durchbruch an der Oder,** ? (Germany) Ullstein-Verlag, 1995

Le Tissier, Tony, **Der Kampf um Berlin 1945,** ? (Germany): Ullstein-Verlag, 1994

von Manteuffel, Hasso, **Die 7. Panzer-Division,** Wölfersheim-Berstadt (Germany): Podzun-Verlag, 1965

Mc Nair, Ronald, **Repli Sur La Seine,** Bayeux (France): Editions Heimdal, 1995

Niehaus, Werner, **Entscheidungsschlacht an Rhein und Ruhr 1945,** Stuttgart: Motorbuch-Verlag, 1995

Meyer, Hubert, **Kriegsgeschichte der 12. SS Panzer-Division "Hitlerjugend",** Osnabrück (Germany): Munin-Verlag, 1982

Milsom, John, **German Halftracked Vehicles of World War II,** ?: Arms and Armour Press, 1975

Münch, Karlheinz, **Combat History of the schwere Panzerjäger-Abteilung 653,** Winnipeg (Manitoba, Canada): J.J. Fedorowicz Publishing, Inc., 1997

Müller, F., **Leitfaden der Fernlenkung,** ? (Germany): Deutsche RADAR Verlagsgesellschaft m.b.H., 1955

Niehorster, Leo W.G., **German World War II Organizational Series — Volume 3/II,** Hanover (Germany): Eigenverlag, 1992

Nosbüsch, Johannes, **Damit es nicht vergessen wird...,** ? (Germany): Pfälzische Verlagsanstalt GmbH, Landau1986

Otte, Alfred, **"Die weißen Spiegel" Hermann Göring,** Wölfersheim-Berstadt (Germany): Podzun-Verlag, ?

Oswald, Werner, **Kraftfahrzeuge und Panzer der Reichswehr, Wehrmacht und Bundeswehr,** Stuttgart: Motorbuch-Verlag, 1982

Pawlas, Karl R., **Waffen - Revue - Verschiedene Beträge** from issues 1 to 105, ? (Germany): Verlag Journal-Schwend GmbH, various

——-, **Liste der Fertigungskennzeichen für Waffen, Munition und Gerät, Originalgetreuer Nachdruck,** ? (Germany): Publizistisches Archiv für Militär- und Waffenwesen, 1977

Pallud, J.P., **Battle of the Bulge — Then and Now,** London: After the Battle, 1984

——-, **Blitzkrieg in the West — Then and Now,** London: After the Battle, 1991

Piekalkiewicz, Janusz, **Unternehmen Zitadelle,** ? (Germany): Lübbe-Verlag, 1983

Quarrie, Bruce, **Fallschirmpanzer-Division "Hermann Göring",** Vanguard No. 4, ? (Great Britain): Osprey 1978

Ritgen, Helmut, **Die Geschichte der Panzer-Lehr-Division im Westen 1944-1945,** Stuttgart: Motorbuch-Verlag, 1979

Sawodny, Michael, **Deutsche Spezialpanzer,** Wölfersheim-Berstadt (Germany): Podzun-Pallas-Verlag, 1981

Scheibert, Horst, **Die "Tiger"-Familie,** Wölfersheim-Berstadt (Germany): Podzun-Pallas-Verlag, 1979

——-, **Panzer-Grenadier-Division "Großdeutschiand",** Squadron-Signal-Publications, 1977

Schmitz / Thies, **Die Truppenkennzeichen 1939-1945 (Band 1 - 3),** Osnabrück (Germany): Biblio Verlag, 1988 - 1991

Schneider, Wolfgang, **Tigers in Combat I**, Winnipeg (Manitoba, Canada): J.J. Fedorowicz Publishing, Inc., 1994

——-, **Tigers in Combat II**, Winnipeg (Manitoba, Canada): J.J. Fedorowicz Publishing, Inc., 1998

Späeter, Helmuth, **Panzerkorps "Großdeutschland"**, Wölfersheim-Berstadt (Germany): Podzun-Pallas-Verlag, 1984

Spielberger, Walter J., **Der Panzerkampfwagen I und II und ihre Abarten**, Stuttgart: Motorbuch-Verlag, 1974

——-, **Der Panzerkampfwagen III und seine Abarten**, Stuttgart: Motorbuch-Verlag, 1974

——-, **Der Panzerkampfwagen Panther und seine Abarten**, Stuttgart: Motorbuch-Verlag, 1978

——-, **Der Panzerkampfwagen Tiger und seine Abarten**, Stuttgart: Motorbuch-Verlag, 1977

——-, **Spezial - Panzerfahrzeuge**, Stuttgart: Motorbuch-Verlag, 1987

——-, **Sturmgeschütze - Entwicklung und Fertigung der sPaK**, Stuttgart: Motorbuch-Verlag, 199

Spielberger / Doyle /Jentz. **Leichte Jagdpanzer, Entwicklung - Fertigung - Einsatz**, Stuttgart: Motorbuch-Verlag, 1992

——-, **Schwere Jagdpanzer, Entwicklung - Fertigung - Einsatz**, Stuttgart: Motorbuch-Verlag,1993

——-, **Begleitwagen Panzerkampfwagen IV**, Stuttgart: Motorbuch-Verlag, 1998

Spielberger / Feist, **Militärfahrzeuge**, ?: Aero Publishers Inc., 1970

Spielberger / Feist, **Sonderpanzer**, ?: Aero Publishers lne.,1968

Stoves, Rolf, **Die gepanzerten und motorisierten deutschen Großverbände 1935-1945**, Wölfersheim-Berstadt (Germany): Podzun-Pallas- Verlag, 1986

Tessin, Georg, **Verbände und Truppen der Deutschen Wehrmacht und Waffen SS 1939-1945 (Band 1-14)**, Osnabrück (Germany): Biblio Verlag, 1979-1988

——-, **Kriegsstärke-Nachweisungen-Taktische Zeichen-Traditionspflege**, Osnabrück (Germany): Biblio Verlag, 1988

Tieke, Wilhelm, **Das Ende zwischen Oder und Elbe — Der Kampf um Berlin 1945**, Stuttgart: Motorbuch-Verlag, 1992

Touzin, Pierre, **Les Vehicules Blindes Francais 1900-1945**, ? (France): Edition EPA, 1979

Trees, Wolfgang, **Schlachtfeld Rheinland**, Aachen (Germany): Zeitungsverlag Aachen GmbH, 1977

Trenkle, Fritz, **Die deutschen Funklenkverfahren bis 1945**, ? (Germany): Hüthig-Verlag, 1982

United States War Department Technical Manual 15 March 1945 Handbook on German Military Forces

Wendt, Kurt, **"Windhunde" — Ein Bildband der 116.Panzer-Division**, ? (Germany): Verlag Ahrweiler & Goftschalk, 1976

Winkler, Walter, **Der Kampf um** Sewastopol, Starnberger See (Germany): Vowinckel Verlag, 1984

Zaloga, Steven, **Bagration 1944, The Destruction of the Army Group Center**, ? (Great Britain): Osprey Military, 1996

Ziemke, Earl F., **Battle of Berlin**, New York: Ballentine Books Inc., 1969

Unpublished Sources

From the Archives of the Bundesarchiv/Militärarchiv Freiburg: RH 8 / 1 ff; RH 10 / 1 ff; RH 12-2 / 46 ff; RH 20-4 / 159 ff; RH 20-9 / 21 ff; RH 20-14 / 137 ff; RH 20-18 / 77 ff; RH 21-2 / 927 ff; RH 21-3 / 43 ff; RH 21-4 / 14 ff; RH 24-8 / 43 ff; RH 24-30 / 67 ff; RH 24-39 / 261 ff; RH 24-48 / 115 ff; RH 24-54 / 40 ff; RH 24 / 57; RH 27-7 / 53 ff; RH 27-12 / 28 ff; RH 27-20 / 2 ff; RH 27-301 / 7 ff; and, RHD 13

Writings of Ludwig Schreiner; Günther Hopf; Eberhard Held; and, Jus Fischer (†)

Diary entries of Heinz Prenzlin (†); Helmut Greiner; and, Gottfried Beierlein (†)

Tele-Controlled Land Vehicles, Capt. D.I. Symons, WAR OFFICE GS (W) 1 (C), STD

Schuscherowo
Komarowka
Otrada
Sarja Shisni
Swch. 2 Pjatiletka
Viehzuchtstation
Roshdestweno
Sadowaja
Krassn. Poljana
Glasunowka PC
Selenaja Roschtscha
Glebowskij
Lowtschikowo
M.T.St.
Pochwalnaja
Gremjatschewo CC
Glebowo
Kriwyje Werchi
Kunatsch
Technikum Paraschino
Trubizyno
Kamenka
Rshawez
Kudejaroff
Martynoff
Schtscherbatowo
Plotawez
Mal. Bobriki
(Trussowo)
Borsenki
(Wysselki)
Alexandrowka
(Borsenki)
Kriwzowo
Ladyrewo
Wassiljewka
(Schinschino)
Prosorowskij
Ssadowyj
Krassn. Gorka
Wassiljewskij
(Gratschewyj)
Chitrowa
Glasunowa
Trossna
Ssenkowo
Maslowo
Ssoglassnyj
Archangelskoje
Nikolskaja
(Scherbatowo)
Protassowo
Now. Chutor
Wesselyj Bereshok
Bf. Maloarchangelsk
Kleninskij
Pawlowka
Sidorowka
Ssemenowka
Busuluk
Ssokolniki
Otschki
Ssorewnowanije
Petrowka
Judinka
Podlesnaja
Schirokoje Boloto
Stepnaja
Butyrki
Alexandrowka
Fedorowka
Dobrowody
Ponyri 1
Swch. Ponyrowskij
Drushowezkij
Rshawez
Ponyri 1
Drushkowskij
Maja 1 (1 Mai)
Ssnowa
Hanfverarb. Fbr.
Ponyri PC
Prilepy
Beresowyj Log
Goreloje
Prilepy
Bitjug
Beresowskije Wysselki